Institute of Applied Plant Nutrition

Göttingen

04/2021

Water use efficiency of arable and grassland crops in legume-based intercropping systems

Dissertation
zur Erlangung des Doktorgrades
der Fakultät für Agrarwissenschaften
der Georg-August-Universität Göttingen

vorgelegt von

Annika Kathrin Meißner, geb. Lingner

geboren in Göttingen

Göttingen, Juni 2021

Bibliografische Information der Deutschen Nationalbibliothek
Die Deutsche Nationalbibliothek verzeichnet diese Publikation in der Deutschen Nationalbibliografie; detaillierte bibliographische Daten sind im Internet über http://dnb.d-nb.de abrufbar.
1. Aufl. - Göttingen: Cuvillier, 2021
Zugl.: Göttingen, Univ., Diss., 2021

D7

1. Referent:	Prof. Dr. Klaus Dittert (Abt. Pflanzenernährung und Ertragsphysiologie)
2. Korreferent:	Prof. Dr. Johannes Isselstein (Abt. Graslandwissenschaften)

Tag der mündlichen Prüfung: 10.07.2018

Nonnenstieg 8, 37075 Göttingen
Telefon: 0551-54724-0
Telefax: 0551-54724-21
www.cuvillier.de

1. Auflage, 2021
Gedruckt auf umweltfreundlichem, säurefreiem Papier aus nachhaltiger Forstwirtschaft.

ISBN 978-3-7369-7447-0
eISBN 978-3-7369-6447-1

Contents

List of manuscripts

Subsequent manuscripts of the present cumulative doctoral thesis are already published, accepted for publication, prepared for submission or drafted:

Meißner A, Granzow S, Wemheuer F, Pfeiffer B (2021) The cropping system matters – Contrasting responses of winter faba bean (*Vicia faba* L.) genotypes to drought stress. J. Plant Physiol. (accepted for publication)

Granzow S, **Meißner A**, Pfeiffer B, Daniel R, Vidal S, Wemheuer F. Crop genotype and plant compartment determine the response of the active bacterial community towards water deficit

Granzow S, **Meißner A**, Pfeiffer B, Daniel R, Vidal S, Wemheuer F. Response of the active bacterial and fungal communities in the rhizosphere differ towards water deficit

Granzow S, Jayanti H, **Meißner A**, Pfeiffer B, Daniel R, Vidal S, Wemheuer F. Crop species and cropping system alter the effect of *Metarhizium brunneum* seed application on plant-associated bacterial and fungal communities

Meißner A, Tränkner M, Dittert K. Intercropping with winter faba bean in arable land and white clover in grassland improves the water use efficiency of the cropping system

Senbayram M, Wenthe C, **Lingner A**, Isselstein J, Steinmann H, Kaya C, Köbke S (2016) Legume-based mixed intercropping systems may lower agricultural born N_2O emissions. Energy Sustain. Soc. 6. https://doi.org/10.1186/s13705-015-0067-3

Meißner A, Streit J, Koops D, Isselstein J, Dittert K. Remotely assessed vegetation development and water use of arable and grassland intercropping with legumes

List of relevant abbreviations

$\delta^{13}C$	Normalized ratio of the stable carbon isotopes ^{13}C to ^{12}C
asl	Above sea level
DAO	Days after onset of the experiment
DM	Dry matter
dT	Temperature difference between canopy surface and air
DWD	Deutscher Wetterdienst
ET	Evapotranspiration
FW	Fresh weight
IMPAC³	*Novel genotypes for mixed cropping allow for improved sustainable land use across arable land, grassland and woodland*
LA	Leaf area
NDVI	Normalized difference vegetation index
NEE	Net ecosystem exchange of CO_2
NIR	Near infrared
R^2	Coefficient of determination
RWC	Relative water content of leaves
UAV	Unmanned aerial vehicle
VPD	Vapor pressure deficit
WUE	Water use efficiency

Chapter 1: Prologue

Agricultural farming practice in Germany has undergone changes towards less diverse systems with short rotations and few main crops during the last decades (Stein and Steinmann, 2018). Ensuring sustainability with simultaneously increasing yield potential is therefore one of the main challenges for future agriculture (Tilman et al., 2002). Nevertheless, yield progress has declined for several crops in the last years (Fischer and Edmeades, 2010). To further enhance the productivity of agricultural systems, the utilization of genetic resources and biodiversity services has to be improved.

1.1 Legumes in cropping systems

In this context, cropping of legumes can widen existing crop rotations and thus enhance the biodiversity of agricultural landscape. Additionally, fertilizer supported production increases and improvements in independence on fertilizer application is needed (Cazenave, 2018). Legumes with their capability of nitrogen (N) fixation are widely known to be more efficient in the use of soil N and the amount of additional fertilizer can therefore be reduced (Rubiales and Mikic, 2015). As a consequence, seed yield of various crops is increased when the preceding crop was a legume (Tanaka et al., 2007).

Another advantage of legumes is their potential in ecological services (Altieri, 1999; Jensen et al., 2010; Köpke and Nemecek, 2010). The integration of legumes in crop rotations leads to ameliorated soil structure (Rochester et al., 2001), increased microbial activity (Biederbeck et al., 2005) and provision of habitats for beneficial insects (Crist et al., 2006).

Cropping of legumes like pea and faba bean nowadays receives more interest in Europe due to its potential to replace soy bean as protein source for feed and food. To date, Europe is highly dependent on soy bean imports (de Visser et al., 2014; Jensen et al., 2010). Nonetheless, in Germany, legumes such as faba bean (*Vicia faba* L. var. minor) are rarely cultivated as crop rotations are based on cereals. Legumes generate higher complexity for farming management due to soil-borne diseases and possible nitrogen losses via leaching or emissions (Jensen et al., 2010; Reckling et al., 2016).

However, high yielding and stress tolerant legume cultivars are scarce due to low investment in legume breeding (Reckling et al., 2016). There are ongoing studies by several research institutes to translate available genetic resources into novel winter-hardy varieties of faba bean (Link et al., 2010). Those varieties are suitable for a broad range of cropping areas with the ability to make use of a longer vegetation period (Link and Arbaoui, 2005). Frost tolerance plays a major role in breeding of these cultivars, while other factors influencing yield stability receive increasing importance. These are aspects regarding growth conditions, i.e. drought tolerance, as well as pest resistance (Khan et al., 2010; Torres et al., 2006).

1.2 Intercropping systems

At the aim of increasing biodiversity of cropping systems, intercropping of different species provides an opportunity to diversify crop rotations and agricultural landscape. Intercropping systems are defined as growing different crops simultaneously in the same field in variable densities such as substitutive in alternating rows (Andrews and Kassam, 1976). Most often, mixtures of different species result in overyielding of the whole crop stand compared to pure stands such as due to increased interception of light under low interspecific competition (Bilalis et al., 2010; Jolliffe and Wanjau, 1999). Resilience of the system, weed and disease suppression and other factors are also enhanced by the complementarity of legumes and cereals (Carton et al., 2018; Hauggaard-Nielsen et al., 2008, 2001). This complementarity, particularly in differences in plant habitus and flowering characteristics, leads to enhanced biodiversity of associated flora and fauna (Altieri, 1999; Weißhuhn et al., 2017).

Therefore, mixtures of arable crops are to a small share present in existing agricultural systems, especially in organic farms. Where reduced tillage and cropping of legumes is already applied, intercropping systems bear the potential for reduced pesticide inputs (Lemken et al., 2017; Theunissen, 1997). Species mixtures are very common in grassland systems because of yield stability under spatial and seasonal variations (Fowler, 1982). They improve forage quality and higher productivity at late stages of the season (Sleugh et al., 2000). In those systems, nitrogen is used more effectively due to direct transfer from legume to non-legume as well as long term availability by mineralization of crop residues (Olesen et al., 2002; Xiao et al., 2004). That way, nitrogen losses are mitigated and the need for additional fertilizer input is reduced. This nitrogen is transformed into higher protein contents of the harvested products (Bedoussac and Justes, 2010). Also mobilization of soil phosphorus can be promoted in legume-based intercropping systems by associations of mycorrhiza with legume roots or by root-induced pH changes (Betencourt et al., 2012; Ren et al., 2013).

In Germany, selection and breeding for legumes, e.g. faba bean and white clover, is usually performed in pure stands and not in mixtures. Comparing these two crop stands, trait expressions can vary and the best observed performance in pure stands can be different from the best performance in mixtures. This is due to the fact that beneficial effects of intercropping systems interfere with competition for growth resources such as light and water (Gao et al., 2009; Lithourgidis et al., 2011; Malézieux et al., 2009; Mushagalusa et al., 2008). The performance of the system thus depends on environmental conditions and genotypic characteristics (Passioura, 2006). The suitability of a specific genotype is governed by complex interactions. Traits as winter hardiness, competition abilities in comparison to the non-legume, i.e. wheat and ryegrass, and in case of grassland the persistence of the sward therefore need to be tested in mixtures. Consequently, knowledge driven improvement of the performance of such intercropping systems is needed.

1.2.1 Intercropping with faba bean

The cultivation of faba bean is a promising option to replace the high input demanding soybean as locally produced protein source for food and feed (Köpke and Nemecek, 2010). The quality and protein content of cereals can be significantly enhanced when faba bean was intercropped with wheat or durum wheat (Ghanbari-Bonjar and Lee, 2002; Tosti and Guiducci, 2010). Moreover, the yield stability of intercropped faba bean is greater than that of sole cropped faba bean (Hauggaard-Nielsen et al., 2008).

Intercropping of winter types of faba bean and winter wheat have the potential of higher yields due to a longer growth period (López-Bellido et al., 2005) and a better weed suppression than spring types as early canopy closure to compete out emerging weeds (Haymes and Lee, 1999). Yet, the potential of intercropping autumn-sown faba bean and wheat is to date not fully explored as there are only very few winter-hardy varieties on the market, which are able to benefit from rainfall events during winter. On the German market there is only one winter hardy faba bean cultivar available (cv. Hiverna) (Bundessortenamt, 2017). In order to grow faba bean in autumn-sown mixtures with winter wheat, breeding for more adapted cultivars is necessary.

Especially drought adaptation is an issue of recent research as faba bean is a very sensitive crop to water limitations (Belachew et al., 2018; Khan et al., 2010, 2007). Consequently, genotype-specific properties of different faba been genotypes to water deficit can vary in their stress adaptation and drought tolerance (Mwanamwenge et al., 1998). This again would affect the performance of faba bean in the mixture as well as the performance of the intercropping system as a whole.

Structural complexity of faba bean and wheat intercropping systems leads to weed suppression (Ghanbari-Bonjar and Lee, 2003). Nevertheless, weed suppression is less important for intercropping of winter varieties compared to spring varieties, due to its advantage of an early soil cover before weed emergence (Haymes and Lee, 1999). To avoid interspecific competition between faba bean and wheat it is necessary to choose tall growing wheat varieties for better light interception and less shading (Haymes and Lee, 1999). Otherwise, facilitated faba bean growth in intercropping suppresses the accompanying wheat (Lithourgidis and Dordas, 2010). If this is considered, positive effects like an increased proportion of nitrogen derived from N_2 fixation in intercropping of faba bean and wheat compared to pure stands of faba bean are expected (Fan et al., 2006).

1.2.2 Intercropping with white clover and chicory

In grassland, higher plant species richness as present in intercropping systems, leads to an increased biomass production and less investment in root systems, resulting in overyielding of mixtures compared to monocultures (Bessler et al., 2009). Intercropping is therefore a common practice worldwide for pastures, forage and biomass production. Environmental benefits include an increase in biodiversity and a reduced soil erosion by a permanent ground cover (Halty et al., 2017) as well as a decreased risk for annual weeds due to dense vegetation structure and repeated cutting (Weißhuhn et al., 2017). Thereby, high proportions of grasses enhance carbon sequestration by belowground biomass, while high proportions of legumes accumulate more nitrogen (McElroy et al., 2016). This nitrogen can be transferred to non-legumes, especially grasses with fibrous roots (Pirhofer-Walzl et al., 2012).

White clover (*Trifolium repens* L.) as a fast growing species is a strong competitor in grassland systems and therefore avoids weed invasion, leading to increased yield stability (Frankow-Lindberg et al., 2009). Over several years it was shown by Høgh-Jensen and Schjoerring (1997) that after establishment of intercropping, white clover substantially contributes to nitrogen supply for perennial ryegrass in an increasing manner. Additionally, interspecific competition with the accompanying species and seasonal water deficit can lead to declined proportions of white clover in intercropping systems (Hutchinson et al., 1995).

In this context, including species with deep root systems leads to enhanced water availability and in the long term to soil aeration and drainage (Weißhuhn et al., 2017). Deep rooting species such as chicory (*Cichorium intybus* L.) with its tap root system have the capability to take up water from deep soil layers and transfer it to the shallow rooting species (Skinner, 2008). Chicory is used as vegetable but can also be integrated as forage crop into grassland mixtures as chicory has a high yield potential (Hume et al., 1995).

1.3 Drought and water use

The efficient use of water resources is of particular importance under current shifts in precipitation patterns towards extreme weather events (Brouder and Volenec, 2008). This requires sustainable agricultural systems with improved stress tolerance to excess water as well as water scarcity. The latter will be the focus of the present research. In European agricultural systems, the water deficit mostly occurs in a range where plants are still able to maintain leaf water potential and turgor pressure due to stomatal adjustment, root development and reduction in leaf growth (Tardieu, 1996). Thereby transpiration is reduced and water acquisition increased. Yet, these processes lead to reduced yield production (Fita et al., 2015).

For cropping systems with faba bean as well as white clover, water is often the limiting resource (Hutchinson et al., 1995; Jensen et al., 2010). However, legume-based intercropping systems may increase the efficiency of light and water use due to differing stand architecture and alternate rooting (Morris and Garrity, 1993). This way, they contribute to a better realization of yield potential of the crops even under drought conditions.

The yield potential under unfavorable growth conditions can generally be achieved by the adaptation derived from genetic variance transferred into productivity (Boyer 1982). Lobell et al. (2014), however, reported that an increment in yields by breeding for ideal conditions is accompanied by increased sensitivity to drought stress under high vapor pressure deficit. As a consequence, cropping systems nowadays need to gain more stability towards varying climatic conditions.

Breeding strategies for e.g. white clover aim at realizing genetic potential by understanding physiological processes in terms of adaptation to water deficit (Jahufer et al., 2002). Consequently, in order to ensure stable crop production under water limitations, breeding of crops with higher drought tolerance and improved water use efficiency is necessary (Chaerle et al., 2005).

1.3.1 Water use efficiency on the physiological level

Water use efficiency (WUE) can be defined on various scales, depending on the time frame and plant component referred to. It is determined on the leaf level, on the whole plant level and on the crop stand level as general ratio of carbon gain per water loss. On the levels of whole plants as well as crop stands, most definitions of water use efficiency are either related to biomass production and water consumption in various time intervals or to gas exchange of CO_2 and water vapor.

Instantaneous WUE is related to photosynthetic carbon assimilation and water loss via transpiration, which are conversely regulated by stomatal adjustment (Gong et al., 2011). Converted to instantaneous measurements of crop stands in the field, this ratio is expressed as net ecosystem exchange of CO_2 (NEE) per evapotranspiration (ET). A time-integrated approach related to carbon assimilation and transpiration is the stable isotope discrimination against ^{13}C, an indirect parameter to estimate instantaneous WUE (Tambussi et al., 2007).

These measurements of WUE are only reflecting short time intervals and therefore don't include processes of unproductive water loss e.g. during the night. Consequently, WUE evaluated within a short time frame does not necessarily reflect time-integrated WUE, calculated as ratios of biomass per water consumption (Jákli et al., 2016; Tränkner et al., 2016). Here, also parameters such as leaf anatomy, biochemical processes in carbon fixation and night-time respiration contribute to the characteristics of the WUE (Jákli et al., 2017).

1.3.2 Water use efficiency of the crop stand via remote sensing

Information about the production of the cropping system and its water use can be provided by non-invasive techniques such as remote sensing. These techniques are applied in fast screenings of crop stands (Candiago et al., 2015) as it is needed for example as phenotyping tool within the breeding process. The generated information is then used as selection criterion to differentiate genotypes under stress conditions or in various environments (Fiorani and Schurr, 2013).

Other areas of application are the use of near-infrared spectroscopy to predict growth conditions as the carbon and nitrogen content in the soil (Zhang et al., 2017). Furthermore, plant traits for the productivity of the crops in environments with limited resources can be detected (Fiorani and Schurr, 2013). Sensors mounted on tractors then determine the spatial and temporal optimized donation of fertilizer or pesticides to optimize plant growth. The implementation of indices as the Normalized Difference Vegetation Index (NDVI) in this context support the early detection of stress symptoms caused by diseases or other stressors (Behmann et al., 2015).

The NDVI is calculated from light reflectance in the red and the near-infrared range. In the range of the visible light, reflectance is dependent on the absorption of the pigments, foremost chlorophyll, while reflectance in the near-infrared is determined by leaf structure (Carter and Knapp, 2001).

A more distant approach of remote sensing techniques is the use of unmanned aerial vehicles (UAV) in order to estimate yield, species composition and nutrient status of arable and grassland crops (Bendig et al., 2014; Möckel et al., 2014; Möckel et al., 2016). Here, thermal cameras can be used in addition to the abovementioned spectral sensors. These thermal cameras measure the emittance of infrared radiation, detecting the surface temperature and thus the transpiration of crops (Chaerle and Van Der Straeten, 2000).

The canopy surface temperature in the field is highly depending on weather conditions (Mahlein, 2016): wind removes humidity gradient around stomata and increases transpiration, whereas precipitation leads to elevated air humidity and wet surfaces, thus decreasing vapor pressure deficit and reducing transpiration. Elevated air temperature in combination with relative humidity on the other hand increases transpiration demand of the crops.

Consequently, phenotyping of crop stands for productivity as estimated by NDVI and water use as estimated by thermography bear the potential to estimate their water use efficiency and to characterize the system's sustainability. Evaluated under different growth conditions, e.g. several sites and years, these image-based parameters can generate a better understanding of intercropped species and their sustainable production systems.

1.4 Objectives and structure of the thesis

The breeding of new cultivars of crops is traditionally performed in pure stands. This way, the performance of novel genotypes in intercropped stands and responsible traits are not considered. Consequently, growers of arable crops lack information for best management practice of these mixtures. In this framework, the project IMPAC[3] (*Novel genotypes for mixed cropping allow for improved land use across arable land, grassland and woodland*) aims at improving knowledge about species mixtures for future breeding of cultivars suitable for intercropping systems.

For the aspect of genotype-specific performance under water limitations, the effects of water partitioning between component crops were evaluated. In this regard, investigations on the efficiency in water use of intercropped vs. pure stands of species mixtures were conducted in arable land and grassland. Due to the complexity of intercropping systems, it is necessary to assess parameters for water use and drought stress tolerance on different scales. Biochemical and physiological mechanisms of plants grown under controlled conditions in the greenhouse provide precise information about responses of the crops to water scarcity. In contrast, data assessment of field experiments considers more complex growth conditions such as weather events, variations in soil structure and occurrence of pests. These findings of both approaches need to be related to each other to obtain detailed understanding of the dynamics in intercropping systems.

Due to its compatible use of resources, intercropping of multiple species is likely to enhance the WUE. This especially accounts for seasons with less water availability. Although there are several studies investigating water use of single cropped faba bean in arable systems (e.g. Abid et al., 2017; French, 2010; Siddiqui et al., 2015), little is known about species interactions in multi species crop stands. This thesis therefore aims at improving the current knowledge on interaction effects of cropping system, species and genotypes with respect to their water use.

In a set of greenhouse experiments, reported in the second chapter of this thesis, contrasting genotypes of winter faba bean were tested under various conditions. The aim was to evaluate their genotype-specific physiological properties and their suitability for intercropping with winter wheat. Growth conditions varied in terms of water availability and fungi inoculation in order to identify stress adaptation with regard to the drought tolerance and the plant microbiome.

In field experiments, various genotypes of winter faba bean as well as white clover in pure stands and intercropped stands with non-legumes (i.e. winter wheat; ryegrass and chicory) were evaluated. In the third chapter, different approaches were used in order to estimate whether the water use and the water use efficiency (WUE) are improved in intercropping systems. These evaluations were additionally compared to non-legumes where nitrogen fertilizer was applied. In this context, it was tested whether intercropping systems have the potential to reduce nitrous oxide emissions in contrast to conventional cropping systems (chapter four). In the same field experiments, drone-based remote sensing approaches of spectral and thermal imaging were applied to observe differences in growth development and water use of the crop stands (chapter five).

Overall, it was to test whether intercropping of the leguminous and non-leguminous species has advantages in comparison to pure stands. Therefore, the cropping systems in general as well as the various genotypes in particular were observed under contrasting growth conditions in order to reveal differences in the performance.

References

Abid, G., Hessini, K., Aouida, M., Aroua, I., Baudoin, J.-P., Muhovski, Y., Mergeai, G., Sassi, K., Machraoui, M., Souissi, F., others, 2017. Agro-physiological and biochemical responses of faba bean (Vicia faba L. var.'minor') genotypes to water deficit stress. Biotechnol. Agron. Société Environ. Biotechnol. Agron. Soc. Environ. 21.

Altieri, M.A., 1999. The ecological role of biodiversity in agroecosystems, in: Paoletti, M.G. (Ed.), Invertebrate Biodiversity as Bioindicators of Sustainable Landscapes. Elsevier, Amsterdam, pp. 19–31. https://doi.org/10.1016/B978-0-444-50019-9.50005-4

Andrews, D., Kassam, A., 1976. The importance of multiple cropping in increasing world food supplies. American Society of Agronomy, Crop Science Society of America, and Soil Science Society of America.

Bedoussac, L., Justes, E., 2010. The efficiency of a durum wheat-winter pea intercrop to improve yield and wheat grain protein concentration depends on N availability during early growth. Plant Soil 330, 19–35. https://doi.org/10.1007/s11104-009-0082-2

Behmann, J., Mahlein, A.-K., Rumpf, T., Römer, C., Plümer, L., 2015. A review of advanced machine learning methods for the detection of biotic stress in precision crop protection. Precis. Agric. 16, 239–260. https://doi.org/10.1007/s11119-014-9372-7

Belachew, K.Y., Nagel, K.A., Fiorani, F., Stoddard, F.L., 2018. Diversity in root growth responses to moisture deficit in young faba bean (Vicia faba L.) plants. PeerJ 6, e4401. https://doi.org/10.7717/peerj.4401

Bendig, J., Bolten, A., Bennertz, S., Broscheit, J., Eichfuss, S., Bareth, G., 2014. Estimating Biomass of Barley Using Crop Surface Models (CSMs) Derived from UAV-Based RGB Imaging. Remote Sens. 6, 10395–10412. https://doi.org/10.3390/rs61110395

Bessler, H., Temperton, V.M., Roscher, C., Buchmann, N., Schmid, B., Schulze, E.-D., Weisser, W.W., Engels, C., 2009. Aboveground overyielding in grassland mixtures is associated with reduced biomass partitioning to belowground organs. Ecology 90, 1520–1530. https://doi.org/10.1890/08-0867.1

Betencourt, E., Duputel, M., Colomb, B., Desclaux, D., Hinsinger, P., 2012. Intercropping promotes the ability of durum wheat and chickpea to increase rhizosphere phosphorus availability in a low P soil. Soil Biol. Biochem. 46, 181–190. https://doi.org/10.1016/j.soilbio.2011.11.015

Biederbeck, V.O., Zentner, R.P., Campbell, C.A., 2005. Soil microbial populations and activities as influenced by legume green fallow in a semiarid climate. Soil Biol. Biochem. 37, 1775–1784. https://doi.org/10.1016/j.soilbio.2005.02.011

Bilalis, D., Papastylianou, P., Konstantas, A., Patsiali, S., Karkanis, A., Efthimiadou, A., 2010. Weed-suppressive effects of maize–legume intercropping in organic farming. Int. J. Pest Manag. 56, 173–181. https://doi.org/10.1080/09670870903304471

Brouder, S.M., Volenec, J.J., 2008. Impact of climate change on crop nutrient and water use efficiencies. Physiol. Plant. 133, 705–724. https://doi.org/10.1111/j.1399-3054.2008.01136.x

Bundessortenamt, 2017. Beschreibende Sortenliste - Getreide, Mais, Öl- und Faserpflanzen, Leguminosen, Rüben, Zwischenfrüchte.

Candiago, S., Remondino, F., De Giglio, M., Dubbini, M., Gattelli, M., 2015. Evaluating Multispectral Images and Vegetation Indices for Precision Farming Applications from UAV Images. Remote Sens. 7, 4026–4047. https://doi.org/10.3390/rs70404026

Carter, G.A., Knapp, A.K., 2001. Leaf Optical Properties in Higher Plants: Linking Spectral Characteristics to Stress and Chlorophyll Concentration. Am. J. Bot. 88, 677. https://doi.org/10.2307/2657068

Carton, N., Naudin, C., Piva, G., Baccar, R., Corre-Hellou, G., 2018. Differences for traits associated with early N acquisition in a grain legume and early complementarity in grain legume–triticale mixtures. AoB PLANTS 10. https://doi.org/10.1093/aobpla/ply001

Cazenave, A.-B., 2018. Facing changes today for an agriculture tomorrow. Adv Agr Environ Sci. 1, 00004. https://doi.org/10.13140/RG.2.2.15062.96322

Chaerle, L., Saibo, N., Van Der Straeten, D., 2005. Tuning the pores: towards engineering plants for improved water use efficiency. Trends Biotechnol. 23, 308–315. https://doi.org/10.1016/j.tibtech.2005.04.005

Chaerle, L., Van Der Straeten, D., 2000. Imaging techniques and the early detection of plant stress. Trends Plant Sci. 5, 495–501. https://doi.org/10.1016/S1360-1385(00)01781-7

Crist, T.O., Pradhan-Devare, S.V., Summerville, K.S., 2006. Spatial variation in insect community and species responses to habitat loss and plant community composition. Oecologia 147, 510–521. https://doi.org/10.1007/s00442-005-0275-1

de Visser, C.L.M., Schreuder, R., Stoddard, F., 2014. The EU's dependency on soya bean import for the animal feed industry and potential for EU produced alternatives. OCL 21, D407. https://doi.org/10.1051/ocl/2014021

Fan, F., Zhang, F., Song, Y., Sun, J., Bao, X., Guo, T., Li, L., 2006. Nitrogen Fixation of Faba Bean (Vicia faba L.) Interacting with a Non-legume in Two Contrasting Intercropping Systems. Plant Soil 283, 275–286. https://doi.org/10.1007/s11104-006-0019-y

Fiorani, F., Schurr, U., 2013. Future Scenarios for Plant Phenotyping. Annu. Rev. Plant Biol. 64, 267–291. https://doi.org/10.1146/annurev-arplant-050312-120137

Fischer, R.A. (Tony), Edmeades, G.O., 2010. Breeding and Cereal Yield Progress. Crop Sci. 50, S-85. https://doi.org/10.2135/cropsci2009.10.0564

Fita, A., Rodríguez-Burruezo, A., Boscaiu, M., Prohens, J., Vicente, O., 2015. Breeding and Domesticating Crops Adapted to Drought and Salinity: A New Paradigm for Increasing Food Production. Crop Sci. Hortic. 978. https://doi.org/10.3389/fpls.2015.00978

Fowler, N., 1982. Competition and Coexistence in a North Carolina Grassland: III. Mixtures of Component Species. J. Ecol. 70, 77. https://doi.org/10.2307/2259865

Frankow-Lindberg, B.E., Halling, M., Höglind, M., Forkman, J., 2009. Yield and stability of yield of single- and multi-clover grass-clover swards in two contrasting temperate environments. Grass Forage Sci. 64, 236–245. https://doi.org/10.1111/j.1365-2494.2009.00689.x

French, R.J., 2010. The risk of vegetative water deficit in early-sown faba bean (Vicia faba L.) and its implications for crop productivity in a Mediterranean-type environment. Crop Pasture Sci. 61, 566. https://doi.org/10.1071/CP09372

Gao, Y., Duan, A., Sun, J., Li, F., Liu, Zugui, Liu, H., Liu, Zhandong, 2009. Crop coefficient and water-use efficiency of winter wheat/spring maize strip intercropping. Field Crops Res. 111, 65–73. https://doi.org/10.1016/j.fcr.2008.10.007

Ghanbari-Bonjar, A., Lee, H.C., 2003. Intercropped wheat (Triticum aestivum L.) and bean (Vicia faba L.) as a whole-crop forage: effect of harvest time on forage yield and quality. Grass Forage Sci. 58, 28–36. https://doi.org/10.1046/j.1365-2494.2003.00348.x

Ghanbari-Bonjar, A., Lee, H.C., 2002. Intercropped field beans (Vicia faba) and wheat (Triticum aestivum) for whole crop forage: effect of nitrogen on forage yield and quality. J. Agric. Sci. 138. https://doi.org/10.1017/S0021859602002149

Gong, X.Y., Chen, Q., Lin, S., Brueck, H., Dittert, K., Taube, F., Schnyder, H., 2011. Tradeoffs between nitrogen- and water-use efficiency in dominant species of the semiarid steppe of Inner Mongolia. Plant Soil 340, 227–238. https://doi.org/10.1007/s11104-010-0525-9

Halty, V., Valdés, M., Tejera, M., Picasso, V., Fort, H., 2017. Modeling plant interspecific interactions from experiments with perennial crop mixtures to predict optimal combinations. Ecol. Appl. 27, 2277–2289. https://doi.org/10.1002/eap.1605

Hauggaard-Nielsen, H., Ambus, P., Jensen, E.S., 2001. Interspecific competition, N use and interference with weeds in pea–barley intercropping. Field Crops Res. 70, 101–109. https://doi.org/10.1016/S0378-4290(01)00126-5

Hauggaard-Nielsen, H., Jørnsgaard, B., Julia Kinane, Jensen, E.S., 2008. Grain legume–cereal intercropping: The practical application of diversity, competition and facilitation in arable and organic cropping systems. Renew. Agric. Food Syst. 23, 3–12. https://doi.org/10.1017/S1742170507002025

Haymes, R., Lee, H.., 1999. Competition between autumn and spring planted grain intercrops of wheat (Triticum aestivum) and field bean (Vicia faba). Field Crops Res. 62, 167–176. https://doi.org/10.1016/S0378-4290(99)00016-7

Høgh-Jensen, H., Schjoerring, J.K., 1997. Interactions between white clover and ryegrass under contrasting nitrogen availability: N2 fixation, N fertilizer recovery, N transfer and water use efficiency. Plant and Soil 187–199.

Hume, D.E., Lyons, T.B., Hay, R.J.M., 1995. Evaluation of 'Grasslands Puna' chicory (Cichorium intybus L.) in various grass mixtures under sheep grazing. N. Z. J. Agric. Res. 38, 317–328. https://doi.org/10.1080/00288233.1995.9513133

Hutchinson, K., King, K., Wilkinson, D., 1995. Effects of rainfall, moisture stress, and stocking rate on the persistence of white clover over 30 years. Aust. J. Exp. Agric. 35, 1039. https://doi.org/10.1071/EA9951039

Jahufer, M.Z.Z., Cooper, M., Ayres, J.F., Bray, R.A., 2002. Identification of research to improve the efficiency of breeding strategies for white clover in Australia - a review. Aust. J. Agric. Res. 53, 239–257. https://doi.org/10.1071/ar01110

Jákli, B., Tavakol, E., Tränkner, M., Senbayram, M., Dittert, K., 2017. Quantitative limitations to photosynthesis in K deficient sunflower and their implications on water-use efficiency. J. Plant Physiol. 209, 20–30. https://doi.org/10.1016/j.jplph.2016.11.010

Jákli, B., Tränkner, M., Senbayram, M., Dittert, K., 2016. Adequate supply of potassium improves plant water-use efficiency but not leaf water-use efficiency of spring wheat. J. Plant Nutr. Soil Sci. 179, 733–745. https://doi.org/10.1002/jpln.201600340

Jensen, E.S., Peoples, M.B., Hauggaard-Nielsen, H., 2010. Faba bean in cropping systems. Field Crops Res., Faba Beans in Sustainable Agriculture 115, 203–216. https://doi.org/10.1016/j.fcr.2009.10.008

Jolliffe, P.A., Wanjau, F.M., 1999. Competition and productivity in crop mixtures: some properties of productive intercrops. J. Agric. Sci. 132, 425–435.

Khan, H.R., Link, W., Hocking, T.J., Stoddard, F.L., 2007. Evaluation of physiological traits for improving drought tolerance in faba bean (Vicia faba L.). Plant Soil 292, 205–217. https://doi.org/10.1007/s11104-007-9217-5

Khan, H.R., Paull, J.G., Siddique, K.H.M., Stoddard, F.L., 2010. Faba bean breeding for drought-affected environments: A physiological and agronomic perspective. Field Crops Res. 115, 279–286. https://doi.org/10.1016/j.fcr.2009.09.003

Köpke, U., Nemecek, T., 2010. Ecological services of faba bean. Field Crops Res. 115, 217–233. https://doi.org/10.1016/j.fcr.2009.10.012

Lemken, D., Spiller, A., von Meyer-Höfer, M., 2017. The Case of Legume-Cereal Crop Mixtures in Modern Agriculture and the Transtheoretical Model of Gradual Adoption. Ecol. Econ. 137, 20–28. https://doi.org/10.1016/j.ecolecon.2017.02.021

Link, W., Arbaoui, M., 2005. NEUES von der Göttinger Winter-Ackerbohne. Presented at the 56. Tagung 2005 der Vereinigung der Pflanzenzüchter und Saatgutkaufleute Österreichs, Raumberg - Gumpenstein, pp. 1–8.

Link, W., Balko, C., Stoddard, F.L., 2010. Winter hardiness in faba bean: Physiology and breeding. Field Crops Res. 115, 287–296. https://doi.org/10.1016/j.fcr.2008.08.004

Lithourgidis, A.S., Dordas, C.A., 2010. Forage Yield, Growth Rate, and Nitrogen Uptake of Faba Bean Intercrops with Wheat, Barley, and Rye in Three Seeding Ratios. Crop Sci. 50, 2148. https://doi.org/10.2135/cropsci2009.12.0735

Lithourgidis, A.S., Vlachostergios, D.N., Dordas, C.A., Damalas, C.A., 2011. Dry matter yield, nitrogen content, and competition in pea–cereal intercropping systems. Eur. J. Agron. 34, 287–294. https://doi.org/10.1016/j.eja.2011.02.007

Lobell, D.B., Roberts, M.J., Schlenker, W., Braun, N., Little, B.B., Rejesus, R.M., Hammer, G.L., 2014. Greater Sensitivity to Drought Accompanies Maize Yield Increase in the U.S. Midwest. Science 344, 516–519. https://doi.org/10.1126/science.1251423

López-Bellido, F.J., López-Bellido, L., López-Bellido, R.J., 2005. Competition, growth and yield of faba bean (Vicia faba L.). Eur. J. Agron. 23, 359–378. https://doi.org/10.1016/j.eja.2005.02.002

Mahlein, A.-K., 2016. Plant Disease Detection by Imaging Sensors – Parallels and Specific Demands for Precision Agriculture and Plant Phenotyping. Plant Dis. 100, 241–251. https://doi.org/10.1094/PDIS-03-15-0340-FE

Malézieux, E., Crozat, Y., Dupraz, C., Laurans, M., Makowski, D., Ozier-Lafontaine, H., Rapidel, B., Tourdonnet, S., Valantin-Morison, M., 2009. Mixing plant species in cropping systems: concepts, tools and models. A review. Agron. Sustain. Dev. 29, 43–62. https://doi.org/10.1051/agro:2007057

McElroy, M., Papadopoulos, Y.A., Glover, K.E., Dong, Z., Fillmore, S.A.E., Johnston, M.O., 2016. Interactions between Cultivars of Legumes Species (Trifolium pratense L., Medicago sativa L.) and Grasses (Phleum pratense L., Lolium perenne L.) Under Different Nitrogen Levels. Can. J. Plant Sci. https://doi.org/10.1139/CJPS-2016-0130

Möckel, T., Dalmayne, J., Prentice, H., Eklundh, L., Purschke, O., Schmidtlein, S., Hall, K., 2014. Classification of Grassland Successional Stages Using Airborne Hyperspectral Imagery. Remote Sens. 6, 7732–7761. https://doi.org/10.3390/rs6087732

Möckel, T., Dalmayne, J., Schmid, B., Prentice, H., Hall, K., 2016. Airborne Hyperspectral Data Predict Fine-Scale Plant Species Diversity in Grazed Dry Grasslands. Remote Sens. 8, 133. https://doi.org/10.3390/rs8020133

Morris, R.A., Garrity, D.P., 1993. Resource capture and utilization in intercropping: water. Field Crops Res. 34, 303–317. https://doi.org/10.1016/0378-4290(93)90119-8

Mushagalusa, G.N., Ledent, J.-F., Draye, X., 2008. Shoot and root competition in potato/maize intercropping: Effects on growth and yield. Environ. Exp. Bot. 64, 180–188. https://doi.org/10.1016/j.envexpbot.2008.05.008

Mwanamwenge, J., Loss, S.P., Siddique, K.H.M., Cocks, P.S., 1998. Growth, seed yield and water use of faba bean (Vicia faba L.) in a short-season Mediterranean-type environment. Aust. J. Exp. Agric. 38, 171. https://doi.org/10.1071/EA97098

Olesen, J.E., Rasmussen, I.A., Askegaard, M., Kristensen, K., 2002. Whole-rotation dry matter and nitrogen grain yields from the first course of an organic farming crop rotation experiment. J. Agric. Sci. 139, 361–370. https://doi.org/10.1017/S002185960200268X

Passioura, J., 2006. Increasing crop productivity when water is scarce—from breeding to field management. Agric. Water Manag., Special Issue on Water Scarcity: Challenges and Opportunities for Crop Science 80, 176–196. https://doi.org/10.1016/j.agwat.2005.07.012

Pirhofer-Walzl, K., Rasmussen, Jim, Høgh-Jensen, H., Eriksen, J., Søegaard, K., Rasmussen, Jesper, 2012. Nitrogen transfer from forage legumes to nine neighbouring plants in a multi-species grassland. Plant Soil 350, 71–84. https://doi.org/10.1007/s11104-011-0882-z

Reckling, M., Bergkvist, G., Watson, C.A., Stoddard, F.L., Zander, P.M., Walker, R.L., Pristeri, A., Toncea, I., Bachinger, J., 2016. Trade-Offs between Economic and Environmental Impacts of Introducing Legumes into Cropping Systems. Front. Plant Sci. 7. https://doi.org/10.3389/fpls.2016.00669

Ren, L., Lou, Y., Zhang, N., Zhu, X., Hao, W., Sun, S., Shen, Q., Xu, G., 2013. Role of arbuscular mycorrhizal network in carbon and phosphorus transfer between plants. Biol. Fertil. Soils 49, 3–11. https://doi.org/10.1007/s00374-012-0689-y

Rochester, I.., Peoples, M.., Hulugalle, N.., Gault, R.., Constable, G.., 2001. Using legumes to enhance nitrogen fertility and improve soil condition in cotton cropping systems. Field Crops Res. 70, 27–41. https://doi.org/10.1016/S0378-4290(00)00151-9

Rubiales, D., Mikic, A., 2015. Introduction: Legumes in Sustainable Agriculture. Crit. Rev. Plant Sci. 34, 2–3. https://doi.org/10.1080/07352689.2014.897896

Siddiqui, M., Al-Khaishany, M., Al-Qutami, M., Al-Whaibi, M., Grover, A., Ali, H., Al-Wahibi, M., Bukhari, N., 2015. Response of Different Genotypes of Faba Bean Plant to Drought Stress. Int. J. Mol. Sci. 16, 10214–10227. https://doi.org/10.3390/ijms160510214

Skinner, R.H., 2008. Yield, Root Growth, and Soil Water Content in Drought-Stressed Pasture Mixtures Containing Chicory. Crop Sci. 48, 380. https://doi.org/10.2135/cropsci2007.04.0201

Sleugh, B., Moore, K.J., George, J.R., Brummer, E.C., 2000. Binary Legume–Grass Mixtures Improve Forage Yield, Quality, and Seasonal Distribution. Agron. J. 92, 24. https://doi.org/10.2134/agronj2000.92124x

Stein, S., Steinmann, H.-H., 2018. Identifying crop rotation practice by the typification of crop sequence patterns for arable farming systems – A case study from Central Europe. Eur. J. Agron. 92, 30–40. https://doi.org/10.1016/j.eja.2017.09.010

Tambussi, E. a., Bort, J., Araus, J. l., 2007. Water use efficiency in C3 cereals under Mediterranean conditions: a review of physiological aspects. Ann. Appl. Biol. 150, 307–321. https://doi.org/10.1111/j.1744-7348.2007.00143.x

Tanaka, D.L., Krupinsky, J.M., Merrill, S.D., Liebig, M.A., Hanson, J.D., 2007. Dynamic Cropping Systems for Sustainable Crop Production in the Northern Great Plains. Agron. J. 99, 904–911. https://doi.org/10.2134/agronj2006.0132s

Tardieu, F., 1996. Drought perception by plants Do cells of droughted plants experience water stress? Plant Growth Regul. 20, 93–104. https://doi.org/10.1007/BF00024005

Theunissen, J., 1997. Application of Intercropping in Organic Agriculture. Biol. Agric. Hortic. 15, 250–259. https://doi.org/10.1080/01448765.1997.9755200

Tilman, D., Cassman, K.G., Matson, P.A., Naylor, R., Polasky, S., 2002. Agricultural sustainability and intensive production practices. Nature 418, 671–677. https://doi.org/10.1038/nature01014

Torres, A.M., Román, B., Avila, C.M., Satovic, Z., Rubiales, D., Sillero, J.C., Cubero, J.I., Moreno, M.T., 2006. Faba bean breeding for resistance against biotic stresses: Towards application of marker technology. Euphytica 147, 67–80. https://doi.org/10.1007/s10681-006-4057-6

Tosti, G., Guiducci, M., 2010. Durum wheat–faba bean temporary intercropping: Effects on nitrogen supply and wheat quality. Eur. J. Agron. 33, 157–165. https://doi.org/10.1016/j.eja.2010.05.001

Tränkner, M., Jákli, B., Tavakol, E., Geilfus, C.-M., Cakmak, I., Dittert, K., Senbayram, M., 2016. Magnesium deficiency decreases biomass water-use efficiency and increases leaf water-use efficiency and oxidative stress in barley plants. Plant Soil. https://doi.org/10.1007/s11104-016-2886-1

Weißhuhn, P., Reckling, M., Stachow, U., Wiggering, H., 2017. Supporting Agricultural Ecosystem Services through the Integration of Perennial Polycultures into Crop Rotations. Sustainability 9, 2267. https://doi.org/10.3390/su9122267

Xiao, Y., Li, L., Zhang, F., 2004. Effect of root contact on interspecific competition and N transfer between wheat and fababean using direct and indirect 15N techniques. Plant Soil 262, 45–54. https://doi.org/10.1023/B:PLSO.0000037019.34719.0d

Zhang, L., Yang, X.M., Drury, C.F., Chantigny, M., Gregorich, E., Miller, J., Bittman, S., Reynolds, W., Yang, J., 2017. Infrared spectroscopy prediction of organic carbon and total nitrogen in soil and particulate organic matter from diverse Canadian agricultural regions. Can. J. Soil Sci. https://doi.org/10.1139/CJSS-2017-0070

Chapter 2

Knowledge about the performance of legume-based intercropping systems under water deficit conditions is to date limited as the underlying processes of water partitioning in species mixtures are complex. Detailed analyses were therefore conducted in order to deepen our understanding on the facilitation processes in the use of water resources within intercropped stands in comparison to pure stands. Investigations on the effects on the plant microbiome were included to get insights into the functioning of intercropping systems under water deficit.

In a first experiment, it was tested whether intercropping of arable crops, i.e. winter faba bean and winter wheat, has a beneficial effect on the system's water use efficiency and drought tolerance (chapter 2.1). Moreover, the performances of contrasting winter faba bean genotypes grown in either pure or intercropped stands were evaluated. Several biochemical, physiological and growth parameters were analyzed to improve our knowledge on water partitioning and drought stress responses of the crop stands.

In addition to the abovementioned methods, bacterial and fungal communities in the rhizosphere as well as bacterial communities in the leaf endosphere were examined (chapter 2.2 and 2.3) as those microorganisms are closely related to ecosystem functioning. These methods were applied to evaluate alterations in the active microbial community towards water deficit in intercropped stands with regard to species- and genotype-specific responses. The aim was to achieve a better understanding of how intercropping systems and water deficit affect the plant microbiome and subsequently plant growth and biomass production.

Following that, obtained results were further evaluated in another experiment, where seeds of winter faba bean and winter wheat in both crop stands were inoculated with *Metarhizium brunneum* (chapter 2.4). This entomopathogenic fungus is a biocontrol agent that can be used as alternative pesticide. It may therefore promote the diversity in microbial communities, whereas information about its effect in combination with intercropping of species are scarce. In this context, it was evaluated whether the inoculation leads to corresponding shifts in the bacterial and fungal communities of the rhizosphere soil as well as the root and leaf endosphere of winter faba bean and winter wheat.

Chapter 2.1:
The cropping system matters – Contrasting responses of winter faba bean (*Vicia faba* L.) genotypes to drought stress

Annika Meißner, Sandra Granzow, Franziska Wemheuer, Birgit Pfeiffer

Accepted for publication in
Journal of Plant Physiology

The cropping system matters – Contrasting responses of winter faba bean (*Vicia faba* L.) genotypes to drought stress

Meißner, Annika[a,b,*,1], Granzow, Sandra[b,c,2], Wemheuer, Franziska[c], Pfeiffer, Birgit[d,3]

[a]Institute of Applied Plant Nutrition, University of Goettingen, Carl-Sprengel-Weg 1, D-37075 Goettingen, Germany

[b]Center of Biodiversity and Sustainable Land Use, University of Goettingen, Buesgenweg 1, D-37077 Goettingen, Germany

[c]Agricultural Entomology, Department of Crop Sciences, University of Goettingen, Grisebachstrasse 6, D-37077 Goettingen, Germany

[d]Division of Plant Nutrition and Crop Physiology, Department of Crop Sciences, University of Goettingen, Carl-Sprengel-Weg 1, D-37075 Goettingen, Germany

*** Correspondence:**
Annika Meißner, M.Sc.
annika.meissner@mein.gmx

Present Addresses:

[1]KWS SAAT SE & Co. KGaA, Grimsehlstrasse 31, D-37574 Einbeck, Germany

[2]Faculty of agriculture, University of Goettingen, Buesgenweg 5, D-37077 Goettingen, Germany

[3]Institute of Microbiology and Genetics, Department of Genomic and Applied Microbiology, University of Goettingen, Grisebachstr. 8, D-37077 Goettingen, Germany

Highlights

- Drought stress was analyzed by numerous physiological and biochemical plant traits
- Intercropping increased biomass production even under water deficit
- Synergistic effects of species mixtures were dependent on the faba bean genotype
- Effects of genotypes were most clearly in comparative analyses of various traits
- Faba bean genotypes and consolidated traits should be considered in future studies

Summary

Intercropping of legumes and cereals provides many ecological advantages and contributes to a sustainable agriculture. These agricultural systems face ongoing shifts in precipitation patterns and seasonal drought. Although the effect of drought stress on legumes has been frequently studied, knowledge about water deficits influencing legumes under different cropping systems is still limited. Therefore, we investigated the impact of water deficit and re-irrigation on two winter faba bean genotypes (S_004 and S_062) and winter wheat (var. Genius) in pure and intercropped stands under greenhouse conditions. Various physiological and biochemical parameters, such as canopy surface temperature, leaf relative water content and proline content, were collected at three time points (beginning of water deficit, end of water deficit, after re-irrigation). In addition, water use efficiency (WUE) was analyzed at the end of the experiment. The overall drought stress tolerance was determined as conceptual analysis of all measured parameters. Water deficit significantly affected WUE, surface temperature and proline content of both winter faba bean genotypes. Interestingly, intercropping with wheat resulted in an overall high drought tolerance of genotype S_004, while genotype S_062 had a high drought tolerance in pure stands. Under water deficit, pure stands of S_062 substantially increased WUE by 30.5 %. Intercropping of genotype S_004 increased the dry matter per plant by 31.7 % compared to pure stands under water deficit. Contrary, intercropping of genotype S_062 did not improve the dry matter production. Our findings indicate that genotype S_004 benefits from resource complementarity in intercropping systems with wheat, whereas S_062 is better suitable for pure stands due to competitive effects. Thus, our study highlights that the drought tolerance of winter faba bean genotypes depends on the cropping system, leading to a demand for drought-adapted cultivars specifically selected for intercropping.

Keywords: intercropping, faba bean genotype, *Vicia faba*, drought stress, water deficit, re-irrigation.

1 Introduction

Drastic changes in regional precipitation patterns are predicted to occur with an increased frequency of extreme weather events, due to ongoing climate change (Brouder and Volenec, 2008; Spinoni et al., 2015). These extreme weather events account for half of the yield fluctuations worldwide (Fahad et al., 2017; Zampieri et al., 2017). As a consequence, there is a growing demand for sustainable and productive agricultural systems. Intercropping systems, defined as growing two or more species simultaneously on the same field (Vandermeer, 1989), are well known to meet both demands (reviewed in Malézieux et al., 2009). Despite difficulties in plant protection and technical barriers, especially farmers practicing reduced tillage and growing legumes are most willing to take advantage of these growing systems (Lemken et al., 2017). Legume-cereal mixtures are most widespread as they can enhance yield stability and the exploitation of available resources such as water (e.g. Hauggaard-Nielsen et al., 2008; Lithourgidis et al., 2011; Reynolds et al., 1994).

Previous studies investigating intercropping systems and plant responses to deficits in water availability found contrasting results in dependent of crop specie characteristics. For instance, the water use efficiency (WUE) of intercropped plant species with different root and shoot architecture such as maize (*Zea mays* L.) and pea (*Pisum sativum* L.) or wheat (*Triticum aestivum* L.) and faba bean (*Vicia faba* L.) increased compared to sole cropped legumes or cereals due to synergistic effects (Chai et al., 2014; Chapagain and Riseman, 2015; Morris and Garrity, 1993). Contrary to this, studies on cowpea (*Vigna unguiculata* L.)/pearl millet (*Pennisetum glaucum* L.), potato (*Solanum tuberosum* L.)/maize and pea/maize observed interspecific competition for water resources, leading to dominance in water uptake of one species over the other or in few cases even to yield reduction (Mao et al., 2012; Mushagalusa et al., 2008; Zegada-Lizarazu et al., 2006).

To date, there are only few studies on water allocation in faba bean/wheat intercropping. Grown in pure stands faba bean is very sensitive to water availability limitations (Amede and Schubert, 2003). This leads to high yield variability of faba bean throughout the years (Khan et al., 2007; Rubiales and Mikic, 2015). Nevertheless, growing faba bean is highly valuable as it increases the soil microbial activity and carbon mineralization (Aschi et al., 2017). The biological nitrogen fixation of faba bean reduces external fertilizer needs and CO_2 emissions while the root system improves physical soil properties such as porosity (Karkanis et al., 2018). Drought tolerant and high water use efficient cultivars are therefore essential to assure high productivity and to maintain yield stability in dry seasons and dry areas.

In periods of early plant growth, autumn sown faba bean plants have advantages compared to spring sown varieties due to their capability to use water resources early after the frost period in winter. Those winter-hardy and frost tolerant faba bean genotypes increase yield and yield stability in North Europe and North America compared to spring sown genotypes (Maalouf et al., 2019) as the vegetation period of winter crops is longer than that of spring crops. Furthermore, winter forms have the advantage that they can accompany winter wheat in intercropping systems, in which wheat growth and its productivity is stimulated (Xiao et al., 2018). However, winter-hardy faba bean varieties are rare and more extensive breeding is needed (Sallam et al., 2017).

Within the breeding and selection process for winter-hardy faba bean genotypes, maintained photosynthesis and biomass formation as well as reduced water use are important indicators for plant drought tolerance (Link et al., 1999). In pure stands, drought stress significantly influences parameters such as leaf temperature, grain yield and WUE of faba bean genotypes, which differ in their sensitivity to drought stress (Alghamdi et al., 2015).

Although several drought tolerance-related traits of faba bean have been extensively studied in field and greenhouse experiments, our knowledge about the drought tolerance of faba bean genotypes in intercropped systems is still limited as most previous research has focused on pure stands (e.g. Ali et al., 2016; Belachew et al., 2017; Khazaei et al., 2013). In studies on faba bean/wheat intercropping the main focus is most often on overyielding, nitrogen use, root growth and microbial aspects. Drought and its impacts on faba bean/wheat intercropping are less studied, be it in the field or in greenhouses. The latter topics, however, are crucial to understand as the mixing partners in faba bean/wheat intercropping improve or deteriorate their use of water resources when comparing pure to intercropped stands (Chapagain and Riseman, 2015).

Hence, we investigated the impact of water deficit on of two winter faba bean genotypes (S_004 and S_062) and winter wheat (var. Genius) in pure and intercropped stands. To control the amounts of water applied during the water deficit phase, we focused on a greenhouse set-up for an unadulterated evaluation of several physiological and biochemical parameters. We hypothesized that (1) winter faba bean genotypes with contrasting habitus and maturity differ in their response towards water deficit and thus (2) differentiate the tolerance to water deficit in intercropping with winter wheat under controlled greenhouse conditions. Obtained results will further deepen our understanding of how the drought tolerance of winter faba bean is determined by interacting effects of genotype and cropping system.

2 Material and methods

2.1 Plant material

Higher growth rates and accordingly bigger leaf area are indicators for suitability in intercropping (Semere and Froud-Williams, 2001). As a consequence, we chose the two winter faba bean genotypes S_004 and S_062, which differ in their growth parameters. Both are experimental lines from the Division of Plant Breeding at the University of Goettingen (Siebrecht-Schöll, 2019). The genotype S_004 has a high yield production, medium height, low tillering and late maturity and is therefore expected to be most suitable for intercropping, whereas S_062 is characterized with a short height, small leaves, high tillering and early maturity, suggesting that this genotype might be more suitable for pure stands.

The two genotypes of winter faba bean were selected from a set of field trial-tested inbred lines used within the IMPAC³ project (*Novel genotypes for mixed cropping allow for improved sustainable land use across arable land, grassland and woodland*). They were provided by the Institute of Plant breeding at the University of Goettingen. Both winter faba bean genotypes were grown in pure stands or intercropped with winter wheat (var. Genius) under two different water supply conditions, i.e., under water deficit as well as under sufficiently irrigated control conditions. The wheat genotype Genius was provided by Norddeutsche Pflanzenzucht Hans-Georg Lembke KG.

Seeds were surface-sterilized by serial washing according to Andreote et al. (2010). In brief, seeds were immersed in 70 % ethanol for 2 minutes, in 2 % sodium hypochlorite for 3 minutes and in 70 % ethanol for 30 seconds. Finally, the seeds were rinsed four times in sterile distilled water. After disinfection, seeds were placed on wetted sterile tissues and germinated at 7 °C in the dark until the seedlings developed roots of approximately 4 cm in length. The pre-germination allowed for the identification of dead seed material and ensured the same plant numbers in each pot. The inserting of the seeds in the soil is defined as day zero.

2.2 Soil material and experimental design

Plants were grown in polypropylene pots (Sunware; 45.5 x 36 x 24 cm) for a period of six weeks. Each pot contained field soil from the experimental study site Reinshof, Germany (51.48° N, 9.92° E and 157 m asl.). The soil was classified as Gleyic Fluvisol according to the FAO classification system and contained 21 % clay, 68 % silt and 11 % sand with pH 7.3 and 2.8 % humus. The soil was air-dried and sieved (< 10 mm) prior experimental start to avoid plant residues and bigger soil particles. The soil volume of each pot accounted for approximately 20 L with a dry weight of 18 kg. To prevent soil compaction, the filling of the pots was performed in layers by adding distilled water to each layer. After emergence of the seedlings, the soil was covered by gravel to minimize water losses through evaporation. Phosphorus (50 mg P/kg dry soil) and potassium concentrations (140 mg K/kg dry soil) were in an optimal range according to the German nutrient-availability class system (Kuchenbuch and Buczko, 2011) and were measured according to VDLUFA (2009) by ICP-OES (Vista-RL ICP-OES, Varian, Palo Alto, USA). Sufficient availability of nitrogen was regularly surveyed through evaluating chlorophyll concentration by SPAD readings (SPAD-502Plus, Konica Minolta, Japan) on the youngest fully expanded leaves (data not shown).

Five different crop stands were established: intercropping of winter faba bean S_004 with winter wheat, intercropping of winter faba bean S_062 with winter wheat, pure stand of S_004, pure stand of S_062 and pure stand of winter wheat. Pure stands of each winter faba bean genotype consisted of six rows with 5 seeds each (in total 30 seeds/pot; Fig. 1). Pure stands of winter wheat consisted of six rows with 12 seeds each (in total 72 seeds/pot). In intercropping systems, 15 faba bean and 36 wheat seedlings per pot were sown in distinct rows in a substitutive design (Vandermeer, 1989). Half of the pots of each crop stand was treated with optimal irrigation (control treatments) or with a period of reduced irrigation (water deficit treatments).

2.3 Growth conditions and water management

In the greenhouse, photosynthetic photon flux density was 400 µmol m^{-2} s^{-1} at plant level with a 10/14 h day/night photoperiod. The day length represents the early growth phase of winter faba bean in the field before the frost period. The CO_2 concentration reached around 450 ppm, the average air temperature was 23 °C and there was an average relative humidity of 50 % (Supplementary Figure S.1). To avoid placement effects and heat accumulation beneath the lamps, the air was constantly circulated within the greenhouse by ventilators installed 2 m above the plants.

During the experiment, water loss by transpiration was documented by placing the pots permanently on scales (TQ30, ATP Messtechnik, Germany). The weight reduction was measured every 30 minutes in order to constantly determine water consumption. Additionally, volumetric soil water content was monitored in all treatments using time-domain reflectometry (TDR) probes (EC-5 Moisture Sensor, Decagon Device, USA; Fig. 2). These systems avoid hidden drought due to elevated transpiration of increased biomass (Senbayram et al., 2015) and water that is taken up and accumulated in the plants instead of being transpired.

The plants of all treatments were sufficiently irrigated with distilled water to 90 % field capacity depending on plant growth and water consumption for a growing period of 24 days when faba bean plants reached the four leaf-stage (BBCH 14/34; Lancashire et al., 1991) (Fig. 3). In water deficit treatments, reduced irrigation was applied over a period of ten days. First, we reduced the amount of irrigated water to 75 % compared to those of the control treatments. At day 28, we reduced the water amount to 25 %. At day 34, water deficit pots were re-irrigated for seven days with an adequate amount of water depending on plant growth and water consumption. Control pots were sufficiently irrigated over the entire experimental period (Supplementary Figure S.2).

The total duration of the experiment was six weeks until the developmental stage of seven leaves of the winter faba bean (BBCH 17/37). This duration was chosen not to risk uneven flower development as vernalization of the seeds might not have happened completely which would then cause bias in the results. Moreover, it is known that faba bean investigated in greenhouses grow faster and produce less biomass compared to faba bean grown in the field (Amalfitano et al., 2018). We therefore grew the plants until the beginning of flowering to reflect plant development in the field as much as possible. Further evaluations of relevant developmental stages such as full flowering and pod filling need to be conducted in follow-up studies under field conditions.

2.4 Harvests and determination of drought stress-related parameters

Three partial harvests were conducted during the experiment to analyze the three stages of water deficit (Fig. 3). Various plant parameters such as leaf relative water content, proline content, canopy surface temperature and gas exchange of CO_2 and H_2O were monitored throughout the three stages. The first partial harvest (beginning of water deficit) was performed on day 29 when the soil was slightly dried due to ongoing transpiration by the plants and first wilting symptoms on the leaves occurred. The second partial harvest (end of water deficit) was performed on day 34 when the water deficit intensified and considerable wilting symptoms became visible. The third partial harvest (re-irrigation) was conducted on day 38. Plants were randomly selected from each pot and crop species. Individual leaf samples of each plant were collected for the analysis of relative water content and proline content. Subsequently, remaining aboveground material of these plants was removed.

At the end of the experiment on day 41, six representative plants per pot and crop species were harvested. Leaf area of the harvested plants was determined with a LiCor 3100 leaf area meter (Licor, NE, USA). Dry matter (DM) of these plants was determined after drying the plant biomass at 105 °C until weights were constant. Water use efficiency (WUE) in g DM L^{-1} was calculated based on the total aboveground biomass per pot as well as total water consumption, in which the water consumption was the amount of water used for irrigation throughout the experiment:

$$WUE\ [g\ DM\ L^{-1}] = \frac{total\ dry\ matter\ per\ pot}{total\ water\ consumption}$$

2.4.1 Gas exchange and thermal images

Gas exchange of CO_2 and H_2O in terms of net ecosystem exchange (NEE) and evapotranspiration (ET) of the crop stands was measured between 9 am and 6 pm under light conditions. The NEE and ET were determined according to Lindner et al. (2015) in a closed system by covering all plants including the pot with a transparent chamber (base area 0.36 m^2). Changes in CO_2 and H_2O concentrations compared to the initial ambient air within the empty chamber were measured by using a GFS 3000 (Heinz Walz GmbH, Germany) and calculated from the slopes of these curves according to actual temperature and air volume in the chamber.

The canopy surface temperature was determined as an estimator for water loss via the stomata, which is accompanied by the cooling of the leaves (Khan et al., 2007). Thermal images were taken with a T640 infrared camera (FLIR Systems, OR, USA). The surface temperature of the canopies was evaluated by analyzing the images with the software FLIR ResearchIR version 3.3.12277.1002 (FLIR Systems, OR, USA). Both methods were applied four times during the experimental phase, including an initial measurement before reducing the irrigation in water deficit treatments (Fig. 3).

2.4.2 Relative water content and proline content of leaves

For the determination of turgidity, the relative water content (RWC) of the leaves was investigated according to Barrs and Weatherley (1962). The RWC estimates the cellular hydration and indirectly describes the osmotic adjustment of plants and their ability to absorb soil water (Siddiqui et al., 2015). The second fully expanded leaf was sampled and the fresh weight (FW) was recorded around solar noon. The leaf samples were incubated in closed boxes with distilled water at 23 °C for three hours. Afterwards, the turgid weight (TW) was determined and the leaf samples were dried at 60 °C for 24 hours to examine the dry weight (DW). Finally, the RWC was calculated as follows:

$$RWC\ [\%] = \frac{FW - DW}{TW - DW} * 100$$

Accumulation of the amino acid proline is considered a common physiological response to water scarcity (Verbruggen and Hermans, 2008). The proline content in the leaves was measured according to a modified protocol of Bates, Waldren, and Teare (1973). In brief, the third fully expanded leaf was sampled, immediately frozen in liquid nitrogen and then freeze-dried. Ground samples were dissolved in an aqueous solution of 3 % sulfosalicylic acid. After centrifuging (10.000 rpm, 20 min), aliquots of the extracts were added to a solution of 2.5 % acid-ninhydrin and glacial acetic acid and then incubated in a 100 °C water bath for 1 hour to develop a color reaction. This color reaction was terminated by placing the samples on ice. The proline color complex was extracted from the solution by adding toluene. This was then measured with a spectrophotometer (V-650, Jasco Corporation, Japan) at a wavelength of 520 nm. L-proline was used in different concentrations for standard curve settings.

2.5 Statistical analyses

Statistical analyses were performed using R version 3.4.1 (R Core Team, 2017) and the R package *agricolae* version 1.2-8 (Mendiburu, 2015). For repeated measurements, data were tested separately for each measurement day. Within these measurement days, data were tested for normal distribution with Shapiro-Wilk-Test (Shapiro and Wilk, 1965) and for homogeneity of variance with Levene-Test (Levene, 1960). A compliance of the requirements was given. Three-way analysis of Variance (ANOVA) was performed to determine differences between all treatments, followed by Duncan's post-hoc test (Duncan, 1955) with a significance level of $\alpha = 0.05$. The ten treatments (see chapter 2.2) were replicated four times in a fully randomized design. Winter faba bean and winter wheat were tested separately.

3 Results

3.1 Biomass production and water use efficiency

Plant productivity was evaluated as DM, leaf area and WUE, which were analyzed and calculated at the final harvest. The DM of faba bean ranged from 1.27 ± 0.08 to 1.89 ± 0.2 g per plant, with no significant influence of the water deficit treatments (Fig. 4; Supplementary Table S.1). Among the crop stands, genotype S_004 had the greatest dry matter per plant in intercropping. Under water deficit, there was a significant decrease by 31.7 % comparing the intercropped stands with pure stands of S_004. Wheat was neither significantly affected by the water supply nor by genotype of intercropped faba bean.

Leaf area of faba bean and wheat in general showed a similar pattern as the DM, with genotype S_004 having the biggest leaf area of about 340 cm^2 per plant in control treatments of both crop stands (Fig. 4; Supplementary Table S.2). Additional to the trends in DM, significant decreases in the leaf area were observed comparing control and water deficit treatments of pure and intercropped stands of S_004 and S_062, respectively. For wheat under water deficit, intercropping significantly increased the leaf area by 45 % (with S_062) and 65 % (with S_004) compared to pure stands. Those trends in DM and LA were already observed in a pre-experiment with the same experimental set-up (Supplementary Figure S.3).

This pre-experiment served to test and establish the planned irrigation scheme, as well as to analyze the feasibility of sampling and its influence on the experiment.

At the end of the experiment, WUE was calculated on the crop stand level. In line with the results described above, water deficit led to a significant increase in WUE in intercropped stands of faba bean S_004 (45 %) and in pure stands of faba bean S_062 (38 %) (Fig. 5). We observed no differences between control and water deficit treatment in pure stands of genotype S_004 (average 3.23 g DM/L). In addition, WUE of both pure stands of S_004 was significantly higher compared to the intercropped control plants. Pure stands of wheat had a generally lower WUE with an average of 1.85 g DM/L and were not affected by water supply (Supplementary Table S.3).

3.2 Leaf gas exchange of CO_2 and H_2O

Net ecosystem exchange of CO_2 (NEE) and evapotranspiration (ET) were measured four times throughout the experiment as shown in Fig. 3. The largest NEE for all crops and all treatments was measured before the initiation of the water deficit treatment (Fig.6A; Supplementary Table S.4 - S.5). During water deficit and re-irrigation, values of NEE ranged from -0.6 to 6.6 µmol CO_2 m^{-2} s^{-1}, while ET varied between 0.4 and 2.5 mmol H_2O m^{-2} s^{-1}. Implementing a water deficit significantly reduced NEE in intercropped and pure stands of both faba bean compared to those of the control plants during the entire period of the deficit. Comparing crop stands, pure stands of genotype S_062 had significantly larger NEE than the intercropped plants under both water supply conditions. Towards the end of water deficit, intercropping of S_004 had significantly larger NEE of on average 3.3 µmol CO_2 m^{-2} s^{-1} than intercropping of S_062. The NEE values in pure stands of S_062 was larger than in those of S_004. After re-irrigation, all crop stands including those of the faba bean reached the same NEE level of around 2 µmol CO_2 m^{-2} s^{-1}.

Similar to NEE, the crop stands in all treatments had the highest levels of ET before initiation of water deficit (Fig.6B; Supplementary Table S.6 - S.7). During the drought period, water deficit significantly reduced ET compared to the respective controls except for the pure stands of S_004. At the beginning of water deficit, those differences in intercropped treatments were smaller for genotype S_004 (0.5 mmol H_2O m^{-2} s^{-1}) than for S_062 (0.8 mmol H_2O m^{-2} s^{-1}). In general, the ET values increased towards the end of the water deficit. Pure faba bean S_062 then had the highest ET, which was significantly more than pure stands of S_004. After re-irrigation, all intercropped and pure stands of faba bean congregated at a level of about 1 mmol H_2O m^{-2} s^{-1}, only intercropping genotype S_062 had a significantly lower ET of 0.7 mmol H_2O m^{-2} s^{-1}.

Evapotranspiration and NEE of wheat in pure stands ranged from 0.4 to 1 mmol H_2O m^{-2} s^{-1} and from 0.2 to 2.4 µmol CO_2 m^{-2} s^{-1}, respectively, after the initiation of water deficit. Both parameters were lower under water deficit, being significant only at the end of this phase.

3.3 Thermal images

In correspondence to a reduced ET, surface temperatures of the water deficit treated plants during the drought period were elevated compared to the initial values (23 °C ± 0.9), and significantly increased compared to control treatments (Fig. 6C; Supplementary Table S.8 - S.9). Pure stands of faba bean displayed a genotype-dependent response at the beginning of water deficit: S_062 did not differ among water supply conditions (average 26.2 °C), while S_004 showed a 2.9 °C significantly lower surface temperature under control conditions. Similar to the results for NEE and ET, water deficit treatments of both cropping systems reached surface temperatures of 24.7 °C after re-irrigation and were thus equal to the respective controls. In pure stands of wheat, the water deficit treatment was constant throughout the measurement period (23.5 to 24.7 °C) and displayed no significant difference compared to the control.

3.4 Leaf relative water content

In general, genotype S_004 showed comparably high and stable RWC under both water supply conditions and in both cropping systems (Fig. 7; Supplementary Table S.10 - S.11). Contrastingly, water deficit reduced the RWC of genotype S_062. This effect was most pronounced in intercropping, where the reduction of 11 % and 24.3 % was significant at beginning and end of water deficit, respectively. Pure stands of S_062 showed a significant decrease of 11 % at the end of water deficit only. After re-irrigation, all faba bean in pure stands reach the same level of RWC (average of 91 %). However, intercropped S_004 plants had a significantly higher RWC (average 93.8 %) than intercropped S_062 plants (average 89 %) at that phase.

In general, the RWC of wheat reached high values (above 90 %). Wheat intercropped with faba bean S_062 showed values of 108.6 % RWC under controlled conditions, which was significantly higher than wheat intercropped with S_004.

3.5 Proline content

Proline content of all treatments varied between 300 and 600 µg g^{-1} fresh weight (FW) without effects of the water supply at the beginning of water deficit. (Fig. 8; Supplementary Table S.12 - S.13). At the end of water deficit, the proline content of S_004 in pure stands as well as that of S_062 in intercropped stands significantly increased in water deficit treatments (1047 and 1690 µg g^{-1} FW, respectively) compared to controls (489 and 329 µg g^{-1} FW, respectively). After re-irrigation, the proline content of all faba bean treatments ranged from 400 to 800 µg g^{-1} FW. Intercropping significantly increased the proline content of genotype S_062 but not of genotype S_004, while the opposite was observed in pure stands.

Proline content of wheat was differently affected by water deficit. Pure stands of wheat under control conditions had significantly higher proline content compared to the water deficit treatment (end of water deficit). After re-irrigation, wheat intercropped with faba bean S_062 had a higher proline content (726 µg g^{-1} FW) than wheat intercropped with faba bean S_004 (586 µg g^{-1} FW).

3.6 Conceptional analysis of diverse results

As the aforementioned plant traits showed several similar trends but also some discrepancies, the results of all parameters measured at the end of water deficit were summarized (Table 1). Significant responses to water deficit in comparison to control treatments were classified to get a comprehensive overview on the trends among the applied methodologies. The classification of the results into 'drought tolerant' (positive response) or 'drought susceptible' (negative response) followed the statement of Link et al. (1999), i.e. drought tolerant responses were defined as maintained production parameters under reduced water use.

We observed that intercropping of S_004 as well as pure stands of S_062 predominantly responded positively in terms of drought tolerance. Intercropping of S_004 showed positive responses to water deficit for seven out of eight parameters while pure stands of S_062 responded positively for six parameters. Contrastingly, intercropping of S_062 and pure stands of S_004 predominantly displayed negative characteristics under water deficit.

4 Discussion

The identification and selection of drought tolerant winter faba bean cultivars suitable for intercropping are of crucial interest to assure yield stability under changing climatic conditions with increasing risk for drought events. The main aim of the present study was the evaluation of two contrasting genotypes and their responses under water deficit in pure and intercropping systems.

Increased proline content in leaves as stress signal

Crops exposed to water deficit depend on acclimation mechanisms to maintain productivity. Changes in the proline content are often mentioned as one of the first responses of plants under drought, although the function of the amino acid in this context is not clear. In our experiment, the proline content of plant leaves drastically increased for faba bean genotype S_062 in intercropping and S_004 in pure stands. As proline plays an important role as a signaling molecule and in recovery processes (Kavi Kishor, 2015; Szabados and Savouré, 2010), we suggest that the observed proline accumulation is a signal for drought induced stress and therefore requires acclimation mechanisms.

The proline accumulation might be emphasized by the observed trend in biomass reduction of faba bean. This probably caused a concentration effect of proline and other important osmotically active substances as shown for *Salvadora persica* L. by Parida et al. (2016). If biomass production was sustained, concurrently low levels of proline content in pure stands of S_062 and in intercropping of faba bean S_004 were observed. Here, we suggest that proline turnover instead of accumulation is indirectly linked to drought tolerance as discussed in Bhaskara et al. (2015). The low levels of proline content thus indicate low levels of plant stress and vice versa.

Faba bean genotypes determine RWC in cropping systems

The functioning of osmoregulation and cellular hydration in the leaves was quantified through looking at RWC. The RWC of faba bean S_004 in both crop stands was not affected by water deficit at any of the harvest dates. This observation is surprising as faba bean usually shows wilting symptoms at early stages of drought stress (McDonald and Paulsen, 1997). This loss of turgor has been shown in a similar experiment to be indicative of decreased RWC of faba bean leaves (Siddiqui et al., 2015). The observed maintenance of RWC and turgidity under water deficit in faba bean S_004 indicates good stress adaption of this genotype in terms of intact physiological processes, as demonstrated for other faba bean genotypes in pure stands (Khazaei et al., 2013).

Contrastingly, the RWC of faba bean S_062 was decreased by water deficit to different extents, pointing to a comparably bad stress adaption. This effect in RWC was observed as an indicator for a reduced ability in osmotic adjustment in soybean and tobacco (*Nicotiana tabacum* L.) (Flexas et al., 2006; Meyer and Boyer, 1981), and decreased water potential and loss of turgor in faba bean (Mwanamwenge et al., 1999). Although an actively induced decrease in plant water potential might be advantageous for water uptake, adjustments in leaf water potential cannot be used as indicators for drought tolerance (Martínez-Vilalta and Garcia-Forner, 2017). Therefore, we considered a reduced RWC under water deficit as disadvantageous when plants grow in a water scarce environment.

In line with these contrasting findings for S_062 and S_004, Abid et al. (2017) found different responses in RWC as dependent on the faba bean genotype. The authors explained this by different sensitivities in physiological processes such as stomatal adjustment and photosynthesis which was also observed in our study.

Equivalent to the sensitivity of intercropped S_062 in terms of stress-induced proline accumulation, the decline in RWC of S_062 was further intensified when intercropped with wheat. These observations point to negative effects of intercropping on the susceptible osmoregulation of genotype S_062. We suggest that faba bean S_062 does not benefit from the mixture with wheat but rather suffers from competition. Dominance effects were also shown for switchgrass (*Panicum virgatum* L.) and milkvetch (*Astragalus adsurgens* Pall.) as well as for maize and pea (Mao et al., 2012; Xu et al., 2008), where milkvetch and maize where the species with greater productivity. More pronounced changes in RWC under increasing stress intensity were also observed for faba bean in a greenhouse experiment (Abid et al., 2017), although this study was conducted in pure stands. We conclude that the combination of water deficit and competition observed in our study led to an inferior performance of genotype S_062 in intercropped stands.

Intercropping promotes gas exchange of faba bean S_004

The high RWC of faba bean S_004 suggests active physiological processes. One of the first physiological responses of grain legumes and other crops to drought is stomatal closure in order to avoid dehydration and water loss and to maintain the water status (Flexas et al., 2006; Stoddard et al., 2006). A strong interrelation therefore exists between NEE, ET and canopy surface temperature. Correlations between ET and canopy surface temperature are displayed in Supplementary Figure S.4.

In our study, canopy surface temperatures of faba bean increased in both stands in response to water deficit, which is an indication of decreased transpiration rates and stomatal closure (Chaves et al., 2002; Farooq et al., 2010; Khan et al., 2007). Overall, stomatal closure was also clearly indicated by NEE and ET. This proves physiological acclimation for all plants to water deficit in our experiment. However, faba bean genotypes S_062 and S_004 responded to a different extent and were also affected by the cropping system.

Generally, physiological acclimation in terms of maintenance of CO_2 assimilation supporting high biomass production has been described as desirable for faba bean under low water supply (Link et al., 1999). This principle, however, was developed for plants grown in pure stands. In our study, intercropping improved the physiological activity and drought acclimation of faba bean S_004. The influence of different crop architecture of faba bean and wheat on the amount of unproductive water loss, i.e. evaporation, as described in a study with several cereals (Tambussi et al., 2007) can be neglected as the soil was covered with gravel. However, the ET is dependent on the microclimate within the crop stand (Jákli et al., 2016). In our study, this effect is crucial in intercropped stands where the species complement one another. This especially accounts for intercropped stands with S_004, where bigger leaflets lead to a denser canopy. Through having constant air circulation above the plants, the high sowing density further contributed to this effect.

Water use efficiency and growth are determined by genotype and cropping system

In correspondence to previously discussed results, we also observed differences between the genotypes regarding their WUE in our experiment. More precisely, we found again combined effects of genotype and cropping system. The observation of an increased WUE in pure stands of S_062 and intercropped stands of S_004 is in line with stable proline contents, leaf area and partly stable RWC as well as reduced transpiration. The results therefore clearly point to a specific cropping system-dependent drought stress response of the different genotypes.

Reduced gas exchange and elevated canopy temperature under water deficit were indications of stomatal closure, which led to increases in WUE by 30.5 %. This response to water deficit is in accordance with a pot experiment on pure faba bean, where WUE was enhanced by about 50 % under drastic reductions in water supply (Zabawi and Dennet, 2010). The stomatal closure did surprisingly not lead to growth inhibition due to impaired CO_2 assimilation as shown in field-grown sugar beet (*Beta vulgaris* L.) (Jákli et al., 2017). This might be due to different characteristics of the plant species or the time span of the growth period, which was shorter in our experiment than the vegetation period in the field. However, the observed results could be due to high levels of previous stomata conductance, which may have been reduced to a certain extent without dramatic negative effects on CO_2 assimilation and thus physiological activity and biomass production.

The observed variation in RWC, however, did not translate into differing biomass production levels, which was found for common bean (*Phaseolus vulgaris* L.) harvested after eight days of water deficit in a pot experiment by Ramos et al. (2003). This might be attributable to a concentration effect mentioned by Teulat et al. (1997). The authors stated there was an accumulation of osmotically active compounds due to impaired leaf expansion under water deficit. In fact, leaf area was smaller in drought susceptible treatments, i.e. S_004 in pure stands and S_062 in intercropped stands. As the plants did not lose leaves, this effect can be attributed to reduced leaf expansion under water deficit compared to the controls. According to Mwanamwenge et al. (1999), the reduction of leaf expansion is a common drought stress response of faba bean. It maintains turgor pressure and allows for an improved ability to recover after re-irrigation. In our experiment, however, the effect of reduced leaf expansion seems to be rather disadvantageous. The ability to recover was not affected as our plants recovered in terms of all parameters to the same level compared to the respective control treatments after re-irrigation. This sensitivity and concurrent ability to recover of faba bean was also observed by Mwanamwenge et al. (1999) at the same developmental stage (six leaf-stage).

Other reasons for the absence of significant water deficit effects on DM production might be the relatively short period of water deficit and the young age of the plants. At this age, only the tall growing genotype S_004 benefited from water use in intercropping when water was scarce. These contrasting responses of the genotypes suggest that intercropping does not per se improve the drought tolerance of plants but that the level of inter-specific competition with wheat can differ between faba bean genotypes. A dependence of plant characteristics on competition and facilitation has been already observed in field experiments on different grain legume-cereal intercrop systems (Hauggaard-Nielsen et al., 2008). In the intercropping system of S_004, the interaction between the two crop species avoided a competitive scenario and resulted in a positive mixture effect, even at a key plant development stage (six leaf stage). This was probably due to niche complementarity in rooting systems and water uptake that occurs in pot experiments with faba bean-wheat intercropping (Bargaz et al., 2016). The opposing results of the genotypes observed in our study suggest different faba bean genotype suitability when intercropped with wheat. Caviglia et al. (2004) argues that other selection criteria for cultivars are needed for intercropping than for pure stands. Although our results were generated under controlled conditions, we support this statement and propose that a cropping system-targeted breeding is necessary for the successful integration of winter faba bean into intercropping systems.

Wheat benefits from intercropping with faba bean S_004

In order to investigate whether differences in intercropping are derived from the performance of the faba bean genotype only or further affected by the responses of wheat, we evaluated competitive effects on the non-legume partner wheat. When intercropped with S_004, wheat showed greater DM per plant than in pure stands. Although this effect could partly be attributed to a lower amount of total irrigation water, it was similar to the positive responses of faba bean. This indicates a beneficial mixture effect and thus superior performance of intercropping S_004 with wheat with regard to biomass production. Moreover, the previously discussed overall suitability of faba bean S_004 for intercropping positively affected the proline content of wheat. In the recovery phase, the aforementioned stress signal of proline accumulation in wheat was lower in the mixture with S_004 than in the mixture with S_062. We thus conclude that both species can increase their productivity in the mixture if suitable genotypes are chosen, as shown in several studies (e.g. Bargaz et al., 2016; Reynolds et al., 1994; Yang et al., 2011).

In pure wheat stands, the surface temperature, RWC and WUE as well as the DM were not markedly affected by water deficit. This suggests wheat yields are highly stable in intercrop and pure stands. Similar results were found in other studies on wheat, barley (*Hordeum vulgare* L.) and rice (Kumar et al., 2004; Schonfeld et al., 1988; Teulat et al., 1997). In these studies, high yield stability under drought conditions was derived from a maintained water balance, which was reflected by high RWC of leaves. According to Vassileva et al. (2011), drought tolerant wheat genotypes are those with maintained physiological mechanisms, reduced water loss and fast recovery rates, which is comparable to the definition of drought tolerant faba bean genotypes (Link et al., 1999). Other drought tolerant wheat genotypes should be further investigated as part of intercropping systems with faba bean as the combination of faba bean and wheat genotypes is the main driver of successful intercropping systems (Brooker et al., 2015).

Study limitations

The irrigation aimed at having the same amount of water available for all treatments, which was at ≥ 30 % soil volumetric water content (VWC) in the control treatments and eventually reduced to ≤ 20 % VWC in the water deficit treatments (Fig.2). This caused clear wilting in the water deficit treatments while the controls did not show any drought stress symptoms. However, this was difficult to achieve for all crop species and crop systems as the treatments differed in their amount of water used between the irrigation intervals. The fluctuations in VWC where so high, that the amount of irrigated water needed to be increased for those treatments with highest fluctuations (i.e. pure stands of faba bean S_062) in order to avoid repeated drought scenarios shortly before each irrigation. Additionally, the control treatment of intercropping with faba bean S_004 dropped to low VWC between day 34 and 38 and the drought treatment of intercropping with faba bean S_062 also had generally low VWC. From these circumstances, different intensities of drought stress might have caused variations in water extraction, finally leading to some of the observed differences between pure and intercropped treatments.

However, the levels of VWC suddenly dropped or raised in different phases between measurements without changes in the irrigation scheme. This results most likely from plugging in and out of the sensors from the data logger to conduct measurements. Afterwards, the accuracy of ± 3 % of the sensors (Kanso et al., 2020) probably caused shifting of the data. The lower level of VWC in intercropping with faba bean S_062 might also be derived from the accuracy of the sensors as both faba bean in intercropping overall received similar amounts of irrigated water (Supplementary figure S.2). In faba bean S_062, these amounts of irrigation did not translate into biomass production (DM and LA) comparable to faba bean S_004. From transpiration measurements and RWC we can exclude excessive water loss by intercropped S_062 which would explain the lower VWC. Thus, we conclude that intercropped S_062 experienced water deficit on a similar level as intercropped S_004 and the observed genotype and cropping system effects are not derived from differences in drought severity.

Another key limitation of this study is that it was conducted under greenhouse conditions. Drought in the field is often unpredictable in duration and severity. We focused on a greenhouse set-up with distinguishable plant responses that were not masked by environmental conditions. Muktadir et al. (2020) state in their review on faba bean breeding for drought adaptation that selection under controlled conditions provides reliable and repeatable results of traits such as stomatal conductance, RWC and isotopic carbon discrimination. However, in many cases plant growth in greenhouses significantly differs from plant growth in the field (Poorter et al., 2016). Future studies need to address this research topic under natural (field) conditions to verify our results and to further prove these plant responses in other specific growth stages such as flowering and ripening.

5 Conclusion

Intercropping of winter faba bean and winter wheat has the potential to increase the sustainability and productivity of agricultural systems. In such intercropping systems, winter types of faba bean have a higher yield potential and a more optimized use of water resources compared to spring types. However, the impact of different genotypic characteristics on drought stress tolerance within these complex cropping systems is poorly understood.

The general idea of enhanced drought tolerance due to positive synergistic effects in intercropping was only partly supported by our study, as we observed beneficial effects for intercropping faba bean S_004 with wheat, while intercropped genotype S_062 did not respond in the expected manner. In contrast, genotype S_062 revealed a better ability to adapt to water scarce environments in pure stands. The drought tolerance of the crop stands was highly dependent on the combined effect of the cropping system and winter faba bean genotype.

As a consequence, we conclude that for the drought tolerance of intercropping systems, selection of the best winter faba bean genotype regarding complementary stand architecture and rooting patterns matters. The genotype-dependent interactions in intercropping further point to a demand for selection specific to intercropping in order to develop suitable cultivars. Here, our study provides a better understanding of the complex physiological plant responses to water deficits and we recommend that future studies consider several traits together. More detailed research on replicability in the field with root development and yield production under natural conditions is of crucial importance for the development of drought-tolerant genotypes and improvement of intercropping systems.

6 Abbreviations

DAO: Days after onset of the experiment

DM: Dry matter

DW: Dry weight

ET: Evapotranspiration

FW: Fresh weight

LA: Leaf area

NEE: Net ecosystem exchange of CO_2

RWC: Relative water content

TW: Turgid weight

WUE: Water use efficiency

7 Acknowledgments

We would like to acknowledge the Division of Plant breeding at the University of Goettingen for providing the seed material. Furthermore, we thank Isa Bulut for his practical assistance especially in the methodology of proline content. Our thanks also go to William Nelson for English proof reading of the manuscript. The presented data was first published as part of a PhD thesis (see Granzow 2019).

Funding: This work was funded by the Federal Ministry of Education and Research (FKZ 031A351A). IMPAC[3] is a project of the Center of Biodiversity and sustainable Land Use at the University of Goettingen. The funding source had no particular involvement in this study.

8 Author Contributions

Annika Meißner: Conceptualization, Methodology, Investigation, Formal analysis, Visualization, Writing - original draft, Writing - review & editing. **Sandra Granzow:** Conceptualization, Writing - review & editing. **Franziska Wemheuer:** Conceptualization, Writing - review & editing. **Birgit Pfeiffer:** Conceptualization, Methodology, Supervision, Writing - review & editing.

Tables

Table 1. Summary of all measured parameters at the end of water deficit of the two winter faba bean genotypes in intercropped and pure stands. Arrows: significant differences at $p < 0.05$, minus: no differences between control and water deficit treatment. Green: considered as positive response, black: considered as negative response.

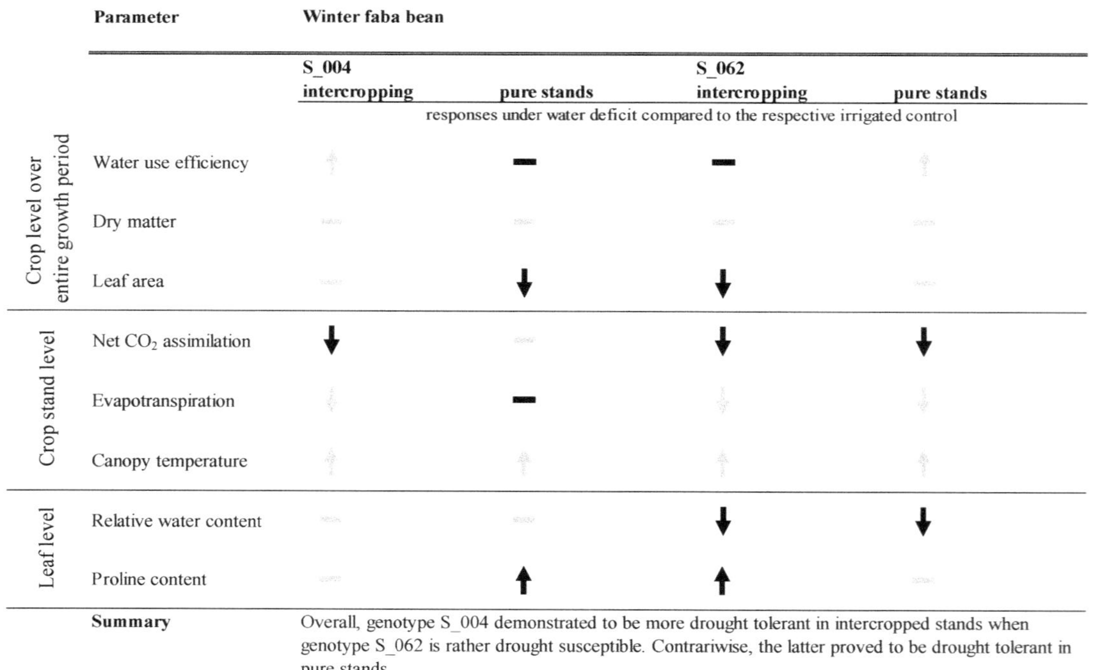

	Parameter	Winter faba bean			
		S_004 intercropping	S_004 pure stands	S_062 intercropping	S_062 pure stands
		responses under water deficit compared to the respective irrigated control			
Crop level over entire growth period	Water use efficiency	↑ (green)	–	–	↑ (green)
	Dry matter	– (green)	– (green)	– (green)	– (green)
	Leaf area	– (green)	↓	↓	– (green)
Crop stand level	Net CO_2 assimilation	↓	– (green)	↓	↓
	Evapotranspiration	↓ (green)	–	↓ (green)	↓ (green)
	Canopy temperature	↑ (green)	↑ (green)	↑ (green)	↑ (green)
Leaf level	Relative water content	– (green)	– (green)	↓	↓
	Proline content	– (green)	↑	↑	– (green)
	Summary	Overall, genotype S_004 demonstrated to be more drought tolerant in intercropped stands when genotype S_062 is rather drought susceptible. Contrariwise, the latter proved to be drought tolerant in pure stands.			

Figures

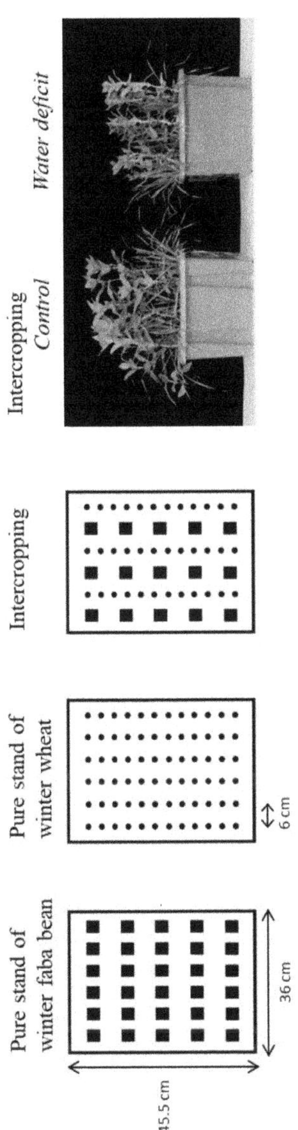

Figure 1. Seeding scheme of winter faba bean and winter wheat in pure stands and intercropping with pot dimensions and row distance. Squares: winter faba bean, circles: winter wheat. The picture (right) shows intercropped stands from control and water deficit treatments at the end of the experiment.

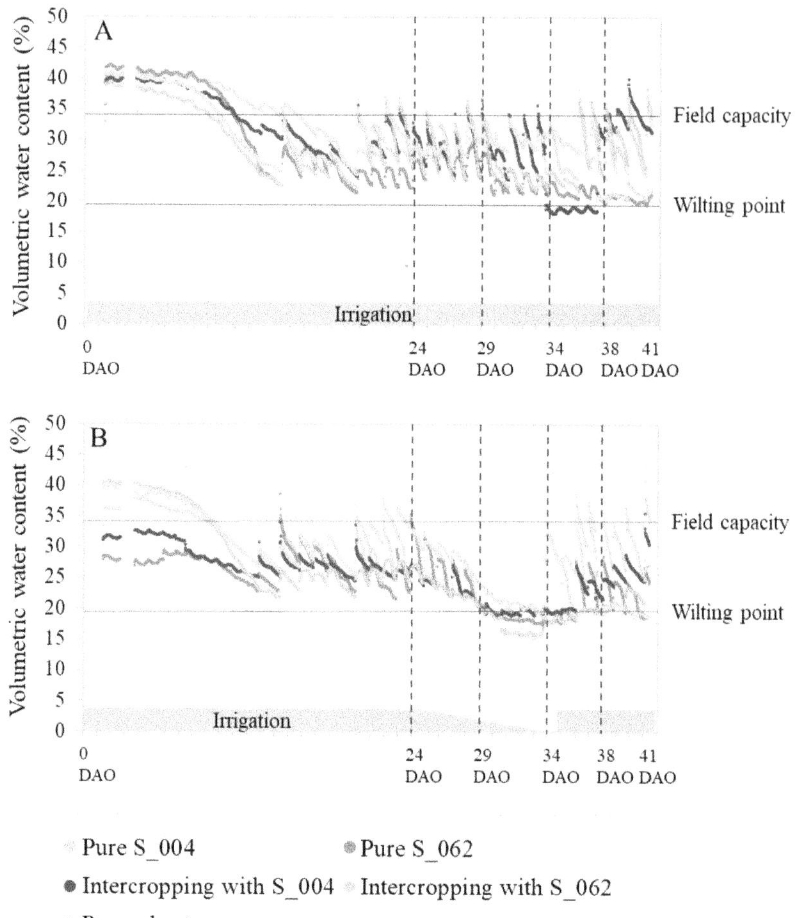

Figure 2. Soil volumetric water content of winter faba bean genotypes S_004 and S_062 as well as winter wheat in intercropped and pure stands during the course of the experiment. A: control, B: water deficit. DAO: days after onset of the experiment. Dashed lines indicate days of measurements: 24 DAO "before water deficit", 29 DAO "beginning of water deficit", 34 DAO "end of water deficit", 38 DAO "re-irrigation".

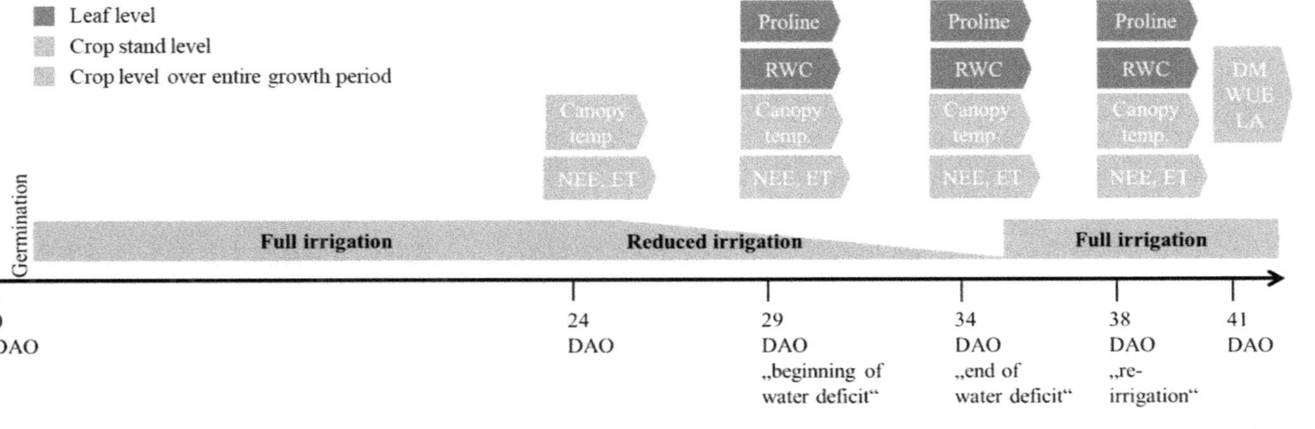

Figure 3. Timeline of the experiment indicating harvests and measurements of canopy surface temperature and net gas exchange of CO_2 (NEE) and H_2O (ET) as well as leaf samplings for relative water content (RWC) and proline content; including the final harvest for dry matter (DM), leaf area (LA) and biomass water use efficiency (WUE). Brown: instantaneous measurements on the leaf level; green: instantaneous measurements on the crop stand level; orange: measurements on the crop stand level reflecting the whole growth period.

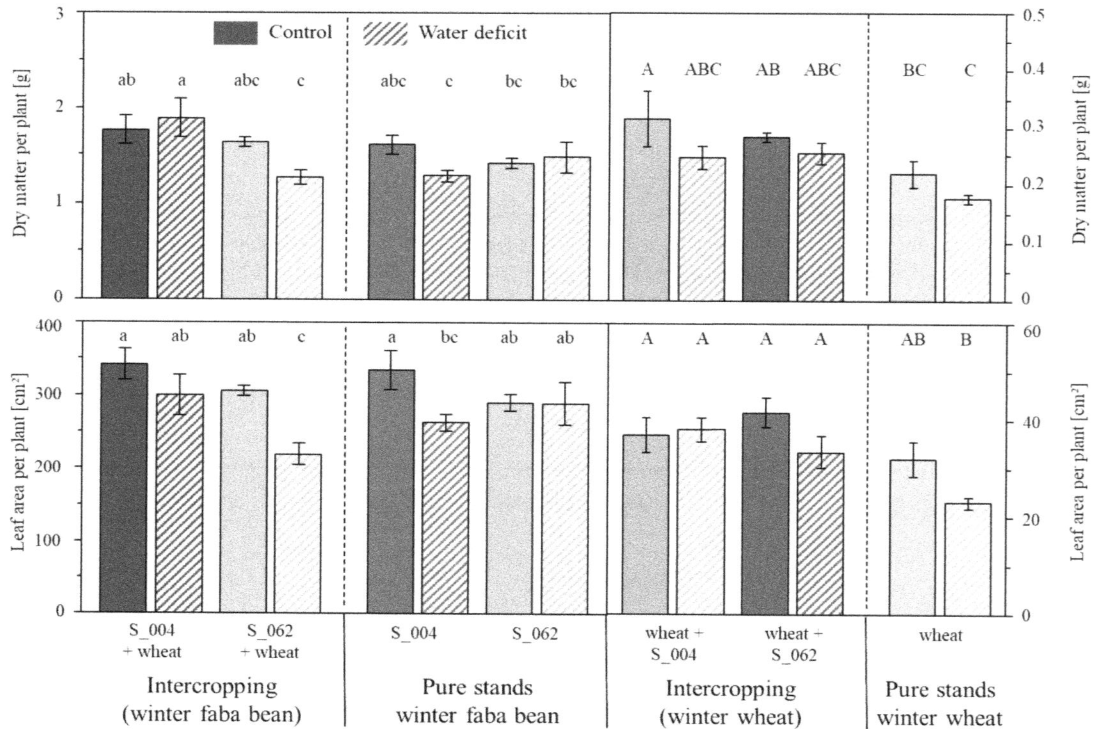

Figure 4. Dry matter and leaf area of winter faba bean genotypes S_004 and S_062 as well as winter wheat in intercropped and pure stands at the end of the experiment. Error bars display the standard error. Different letters indicate significant differences as calculated by the Duncan-test ($p < 0.05$, n = 4).

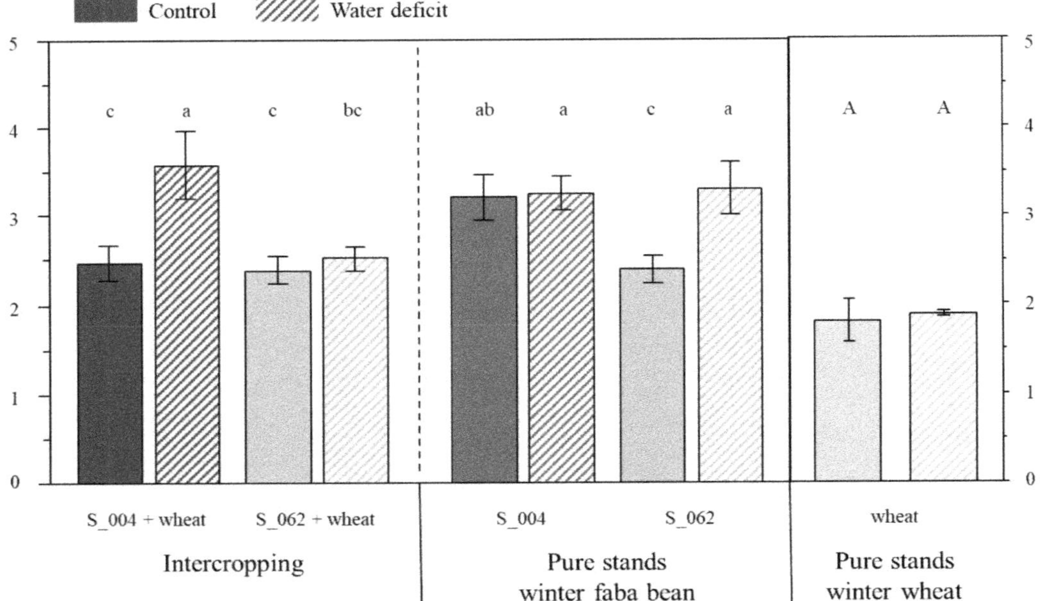

Figure 5. Water use efficiency of winter faba bean genotypes S_004 and S_062 as well as winter wheat in intercropped and pure stands at the end of the experiment (day 41). Intercropping considers the canopy of both species together. Error bars display the standard error. Different letters indicate significant differences as calculated by the Duncan-test ($p < 0.05$, n = 4).

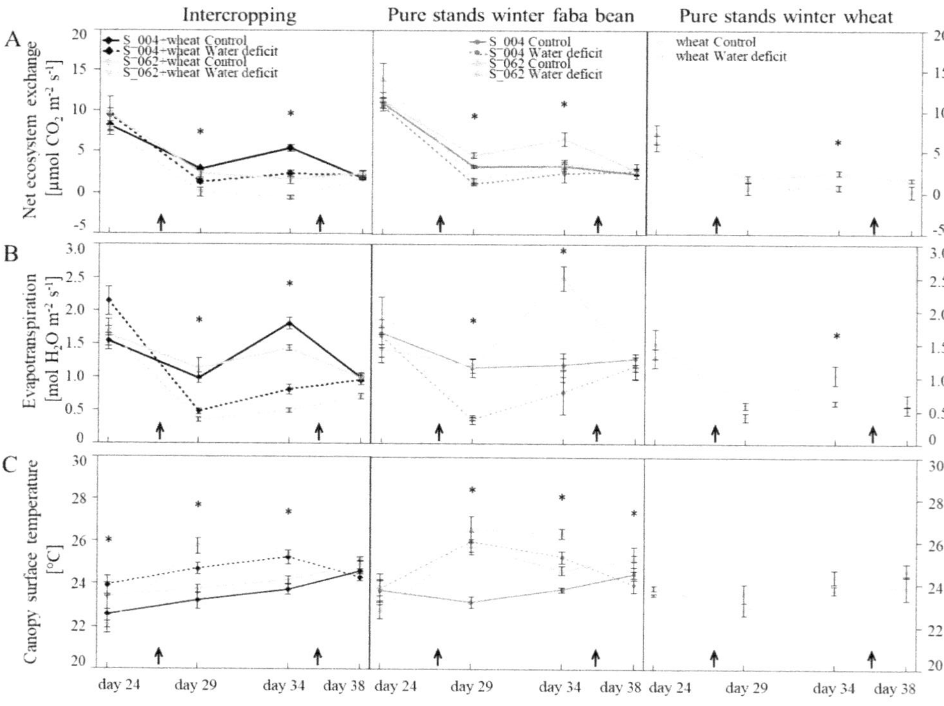

Figure 6. Net ecosystem exchange of CO_2 (A), evapotranspiration (B) and canopy surface temperature (C) of intercropping and pure stands of winter faba bean and winter wheat. Intercropping considers the canopy of both species together. Arrows mark the beginning and end of the water deficit period respectively. Error bars display the standard error (n = 4). Asterisks indicate significant differences among treatments within one day; see also Supplementary Tables S.5, S.7 and S.9.

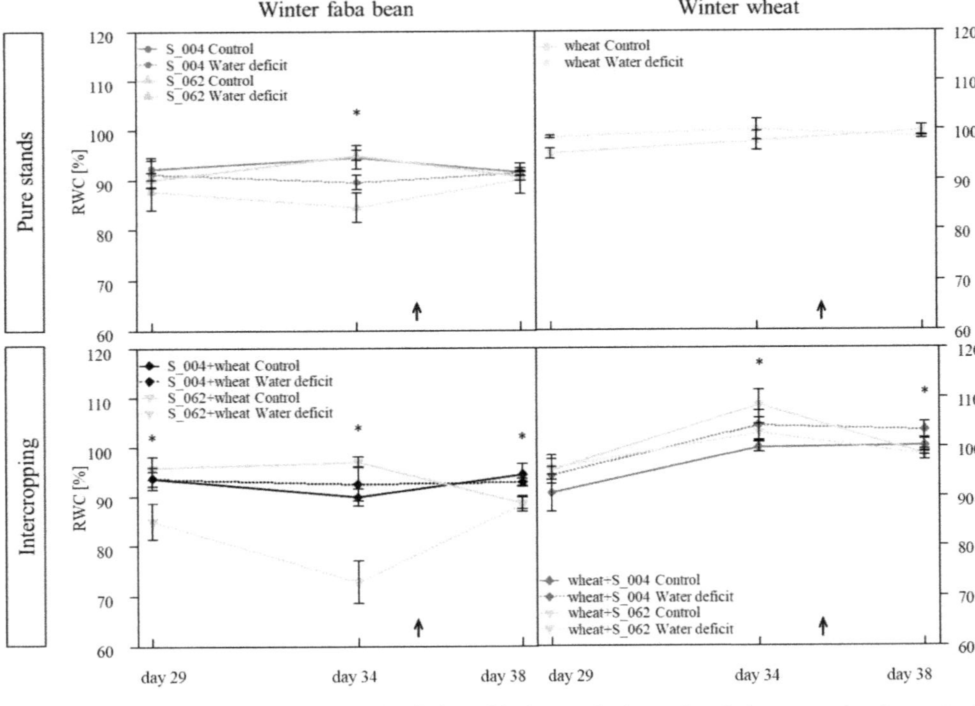

Figure 7. Leaf relative water content (RWC) of winter faba bean and winter wheat in intercropped and pure stands. The arrow marks the end of the water deficit period. Error bars display the standard error (n = 4). Asterisks indicate significant differences among treatments within one day; see also Supplementary Table S.11.

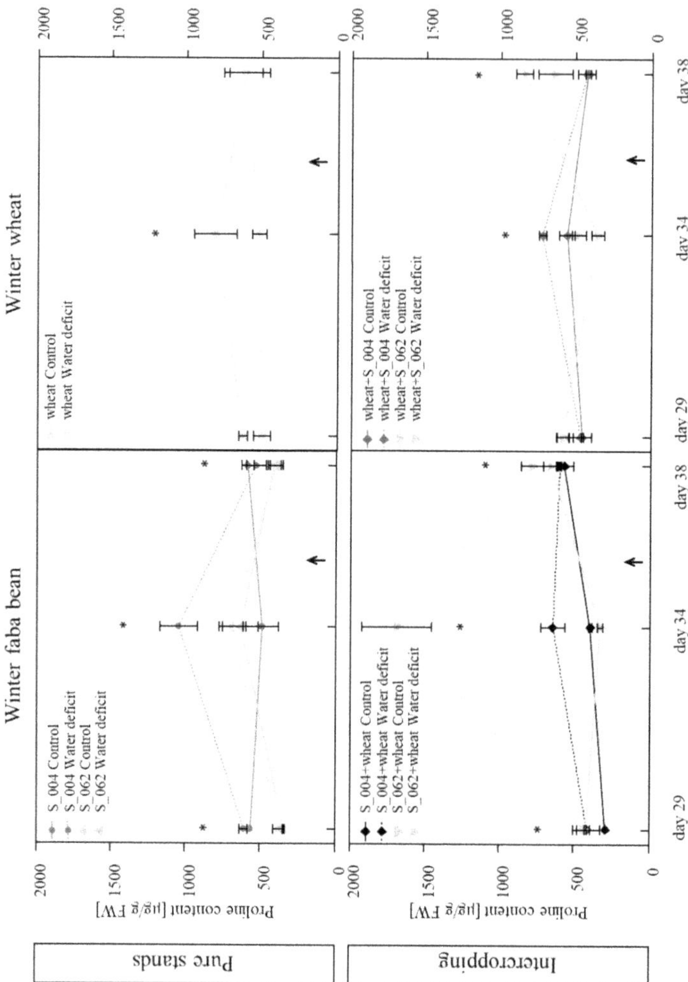

Figure 8. Proline content of winter faba bean and winter wheat in intercropped and pure stands. The arrow marks the end of the water deficit period. Error bars display the standard error (n = 4). Asterisks indicate significant differences among treatments within one day; see also Supplementary Table S.13.

9 References

Abid, G., Hessini, K., Aouida, M., Aroua, I., Baudoin, J.-P., Muhovski, Y., Mergeai, G., Sassi, K., Machraoui, M., Souissi, F., others, 2017. Agro-physiological and biochemical responses of faba bean (Vicia faba L. var.'minor') genotypes to water deficit stress. Biotechnol. Agron. Soc. Environ. 21.

Alghamdi, S.S., Al-Shameri, A.M., Migdadi, H.M., Ammar, M.H., El-Harty, E.H., Khan, M.A., Farooq, M., 2015. Physiological and Molecular Characterization of Faba bean (Vicia faba L.) Genotypes for Adaptation to Drought Stress. J. Agron. Crop Sci. 201, 401–409. https://doi.org/10.1111/jac.12110

Ali, M.B.M., Welna, G.C., Sallam, A., Martsch, R., Balko, C., Gebser, B., Sass, O., Link, W., 2016. Association Analyses to Genetically Improve Drought and Freezing Tolerance of Faba Bean (Vicia faba L.). Crop Sci. 56, 1036–1048. https://doi.org/10.2135/cropsci2015.08.0503

Amalfitano, C., Agrelli, D., Borrelli, C., Cuciniello, A., Morano, G., Caruso, G., 2018. Production system effects on growth, pod yield and seed quality of organic faba bean in southern Italy. Folia Hortic. 30, 375–385. https://doi.org/10.2478/fhort-2018-0033

Amede, T., Schubert, S., 2003. Mechanisms of drought resistance in grain legumes I: Osmotic adjustment. Ethiop. J. Sci. 26, 37–46.

Andreote, F.D., Rocha, U.N. da, Araújo, W.L., Azevedo, J.L., Overbeek, L.S. van, 2010. Effect of bacterial inoculation, plant genotype and developmental stage on root-associated and endophytic bacterial communities in potato (Solanum tuberosum). Antonie Van Leeuwenhoek 97, 389–399. https://doi.org/10.1007/s10482-010-9421-9

Aschi, A., Aubert, M., Riah-Anglet, W., Nélieu, S., Dubois, C., Akpa-Vinceslas, M., Trinsoutrot-Gattin, I., 2017. Introduction of Faba bean in crop rotation: Impacts on soil chemical and biological characteristics. Appl. Soil Ecol. 120, 219–228. https://doi.org/10.1016/j.apsoil.2017.08.003

Bargaz, A., Isaac, M.E., Jensen, E.S., Carlsson, G., 2016. Nodulation and root growth increase in lower soil layers of water-limited faba bean intercropped with wheat. J. Plant Nutr. Soil Sci. 179, 537–546. https://doi.org/10.1002/jpln.201500533

Barrs, H.D., Weatherley, P.E., 1962. A re-examination of the relative turgidity technique for estimating water deficits in leaves. Aust. J. Biol. Sci. 15, 413–428.

Bates, L.S., Waldren, R.P., Teare, I.D., 1973. Rapid determination of free proline for water-stress studies. Plant Soil 39, 205–207.

Belachew, K.Y., Nagel, K.A., Stoddard, F.L., 2017. Root and Shoot Traits for Drought Tolerance in Faba Bean (Vicia faba L.). Presented at the Plant Organ Growth Symposium, Elche, Spain.

Bhaskara, G.B., Yang, T.-H., Verslues, P.E., 2015. Dynamic proline metabolism: importance and regulation in water limited environments. Front. Plant Sci. 6. https://doi.org/10.3389/fpls.2015.00484

Brooker, R.W., Bennett, A.E., Cong, W.-F., Daniell, T.J., George, T.S., Hallett, P.D., Hawes, C., Iannetta, P.P.M., Jones, H.G., Karley, A.J., Li, L., McKenzie, B.M., Pakeman, R.J., Paterson, E., Schöb, C., Shen, J., Squire, G., Watson, C.A., Zhang, C., Zhang, F., Zhang, J., White, P.J., 2015. Improving intercropping: a synthesis of research in agronomy, plant physiology and ecology. New Phytol. 206, 107–117. https://doi.org/10.1111/nph.13132

Brouder, S.M., Volenec, J.J., 2008. Impact of climate change on crop nutrient and water use efficiencies. Physiol. Plant. 133, 705–724. https://doi.org/10.1111/j.1399-3054.2008.01136.x

Caviglia, O.P., Sadras, V.O., Andrade, F.H., 2004. Intensification of agriculture in the south-eastern Pampas. Field Crops Res. 87, 117–129. https://doi.org/10.1016/j.fcr.2003.10.002

Chai, Q., Qin, A., Gan, Y., Yu, A., 2014. Higher yield and lower carbon emission by intercropping maize with rape, pea, and wheat in arid irrigation areas. Agron. Sustain. Dev. 34, 535–543. https://doi.org/10.1007/s13593-013-0161-x

Chapagain, T., Riseman, A., 2015. Nitrogen and carbon transformations, water use efficiency and ecosystem productivity in monocultures and wheat-bean intercropping systems. Nutr. Cycl. Agroecosystems 101, 107–121. https://doi.org/10.1007/s10705-014-9647-4

Chaves, M.M., Pereira, J.S., Maroco, J., Rodrigues, M.L., Ricardo, C.P.P., Osório, M.L., Carvalho, I., Faria, T., Pinheiro, C., 2002. How Plants Cope with Water Stress in the Field. Photosynthesis and Growth. Annals of Botany 89, 907–916. https://link.springer.com/article/10.1007/s00442-005-0256-4

Duncan, D.B., 1955. Multiple Range and Multiple F Tests. Biometrics 11, 1–42. https://doi.org/10.2307/3001478

Fahad, S., Bajwa, A.A., Nazir, U., Anjum, S.A., Farooq, A., Zohaib, A., Sadia, S., Nasim, W., Adkins, S., Saud, S., Ihsan, M.Z., Alharby, H., Wu, C., Wang, D., Huang, J., 2017. Crop Production under Drought and Heat Stress: Plant Responses and Management Options. Front. Plant Sci. 8. https://doi.org/10.3389/fpls.2017.01147

Farooq, M., Kobayashi, N., Ito, O., Wahid, A., Serraj, R., 2010. Broader leaves result in better performance of indica rice under drought stress. J. Plant Physiol. 167, 1066–1075. https://doi.org/10.1016/j.jplph.2010.03.003

Flexas, J., Ribas-Carbó, M., Bota, J., Galmés, J., Henkle, M., Martínez-Cañellas, S., Medrano, H., 2006. Decreased Rubisco activity during water stress is not induced by decreased relative water content but related to conditions of low stomatal conductance and chloroplast CO2 concentration. New Phytol. 172, 73–82. https://doi.org/10.1111/j.1469-8137.2006.01794.x

Granzow, S., 2019. Effects of cropping systems on plant-associated microbial communities of faba bean and wheat. Georg-August-Universität Göttingen, Göttingen.

Hauggaard-Nielsen, H., Jørnsgaard, B., Julia Kinane, Jensen, E.S., 2008. Grain legume–cereal intercropping: The practical application of diversity, competition and facilitation in arable and organic cropping systems. Renew. Agric. Food Syst. 23, 3–12. https://doi.org/10.1017/S1742170507002025

Jákli, B., Hauer-Jákli, M., Böttcher, F., Meyer zur Müdehorst, J., Senbayram, M., Dittert, K., 2017. Leaf, canopy and agronomic water-use efficiency of field-grown sugar beet in response to potassium fertilization. J. Agron. Crop Sci. https://doi.org/10.1111/jac.12239

Jákli, B., Tränkner, M., Senbayram, M., Dittert, K., 2016. Adequate supply of potassium improves plant water-use efficiency but not leaf water-use efficiency of spring wheat. J. Plant Nutr. Soil Sci. 179, 733–745. https://doi.org/10.1002/jpln.201600340

Kanso, T., Gromaire, M.-C., Ramier, D., Dubois, P., Chebbo, G., 2020. An Investigation of the Accuracy of EC5 and 5TE Capacitance Sensors for Soil Moisture Monitoring in Urban Soils-Laboratory and Field Calibration. Sensors 20, 6510. https://doi.org/10.3390/s20226510

Karkanis, A., Ntatsi, G., Lepse, L., Fernández, J.A., Vågen, I.M., Rewald, B., Alsiņa, I., Kronberga, A., Balliu, A., Olle, M., Bodner, G., Dubova, L., Rosa, E., Savvas, D., 2018. Faba Bean Cultivation – Revealing Novel Managing Practices for More Sustainable and Competitive European Cropping Systems. Front. Plant Sci. 9. https://doi.org/10.3389/fpls.2018.01115

Kavi Kishor, P.B., 2015. Role of proline in cell wall synthesis and plant development and its implications in plant ontogeny. Front. Plant Sci. 6. https://doi.org/10.3389/fpls.2015.00544

Khan, H.R., Link, W., Hocking, T.J., Stoddard, F.L., 2007. Evaluation of physiological traits for improving drought tolerance in faba bean (Vicia faba L.). Plant Soil 292, 205–217. https://doi.org/10.1007/s11104-007-9217-5

Khazaei, H., Street, K., Santanen, A., Bari, A., Stoddard, F.L., 2013. Do faba bean (Vicia faba L.) accessions from environments with contrasting seasonal moisture availabilities differ in stomatal characteristics and related traits? Genet. Resour. Crop Evol. 60, 2343–2357. https://doi.org/10.1007/s10722-013-0002-4

Kuchenbuch, R.O., Buczko, U., 2011. Re-visiting potassium- and phosphate-fertilizer responses in field experiments and soil-test interpretations by means of data mining. J. Plant Nutr. Soil Sci. 174, 171–185. https://doi.org/10.1002/jpln.200900162

Kumar, R., Malaiya, S., Srivastava, M.N., 2004. Evaluation of morphophysiological traits associated with drought tolerance in rice. Indian J. Plant Physiol. 9, 305–307.

Lancashire, P.D., Bleiholder, H., Boom, T.V.D., Langelüddeke, P., Stauss, R., Weber, E., Witzenberger, A., 1991. A uniform decimal code for growth stages of crops and weeds. Ann. Appl. Biol. 119, 561–601. https://doi.org/10.1111/j.1744-7348.1991.tb04895.x

Lemken, D., Spiller, A., von Meyer-Höfer, M., 2017. The Case of Legume-Cereal Crop Mixtures in Modern Agriculture and the Transtheoretical Model of Gradual Adoption. Ecol. Econ. 137, 20–28. https://doi.org/10.1016/j.ecolecon.2017.02.021

Levene, H., 1960. Robust tests for equality of variances. Contrib. Probab. Stat. 1, 278–292.

Lindner, S., Otieno, D., Lee, B., Xue, W., Arnhold, S., Kwon, H., Huwe, B., Tenhunen, J., 2015. Carbon dioxide exchange and its regulation in the main agro-ecosystems of Haean catchment in South Korea. Agric. Ecosyst. Environ. 199, 132–145. https://doi.org/10.1016/j.agee.2014.09.005

Link, W., Abdelmula, A.A., Kittlitz, E. von, Bruns, S., Riemer, H., Stelling, D., 1999. Genotypic variation for drought tolerance in Vicia faba. Plant Breed. 118, 477–484.

Lithourgidis, A.S., Dordas, C.A., Damalas, C.A., Vlachostergios, D.N., 2011. Annual Intercrops: An Alternative Pathway for Sustainable Agriculture. Aust. J. Crop Sci. 5, 396–410.

Maalouf, F., Hu, J., O'Sullivan, D.M., Zong, X., Hamwieh, A., Kumar, S., Baum, M., 2019. Breeding and genomics status in faba bean (Vicia faba). Plant Breed. 138, 465–473. https://doi.org/10.1111/pbr.12644

Malézieux, E., Crozat, Y., Dupraz, C., Laurans, M., Makowski, D., Ozier-Lafontaine, H., Rapidel, B., Tourdonnet, S., Valantin-Morison, M., 2009. Mixing plant species in cropping systems: concepts, tools and models. A review. Agron. Sustain. Dev. 29, 43–62. https://doi.org/10.1051/agro:2007057

Mao, L., Zhang, L., Li, W., Werf, W. van der, Sun, J., Spiertz, H., Li, L., 2012. Yield advantage and water saving in maize/pea intercrop. Field Crops Res. 138, 11–20. https://doi.org/10.1016/j.fcr.2012.09.019

Martínez-Vilalta, J., Garcia-Forner, N., 2017. Water potential regulation, stomatal behaviour and hydraulic transport under drought: deconstructing the iso/anisohydric concept: Deconstructing the iso/anisohydric concept. Plant Cell Environ. 40, 962–976. https://doi.org/10.1111/pce.12846

McDonald, G.K., Paulsen, G.M., 1997. High temperature effects on photosynthesis and water relations of grain legumes. Plant Soil 196, 47–58.

Mendiburu, F.D., 2015. Agricolae: statistical procedures for agricultural research. R Package Version 1, 2–3. http://CRAN.R-project.org/package=agricolae

Meyer, R.F., Boyer, J.S., 1981. Osmoregulation, solute distribution, and growth in soybean seedlings having low water potentials. Planta 151, 482–489. https://doi.org/10.1007/BF00386543

Morris, R.A., Garrity, D.P., 1993. Resource capture and utilization in intercropping: water. Field Crops Res. 34, 303–317. https://doi.org/10.1016/0378-4290(93)90119-8

Muktadir, M.A., Adhikari, K.N., Merchant, A., Belachew, K.Y., Vandenberg, A., Stoddard, F.L., Khazaei, H., 2020. Physiological and Biochemical Basis of Faba Bean Breeding for Drought Adaptation—A Review. Agronomy 10, 1345. https://doi.org/10.3390/agronomy10091345

Mushagalusa, G.N., Ledent, J.-F., Draye, X., 2008. Shoot and root competition in potato/maize intercropping: Effects on growth and yield. Environ. Exp. Bot. 64, 180–188. https://doi.org/10.1016/j.envexpbot.2008.05.008

Mwanamwenge, J., Loss, S.P., Siddique, K.H.M., Cocks, P.S., 1999. Effect of water stress during floral initiation, flowering and podding on the growth and yield of faba bean (Vicia faba L.). Eur. J. Agron. 11, 1–11. https://doi.org/10.1016/S1161-0301(99)00003-9

Parida, A.K., Veerabathini, S.K., Kumari, A., Agarwal, P.K., 2016. Physiological, Anatomical and Metabolic Implications of Salt Tolerance in the Halophyte Salvadora persica under Hydroponic Culture Condition. Front. Plant Sci. 7. https://doi.org/10.3389/fpls.2016.00351

Poorter, H., Fiorani, F., Pieruschka, R., Wojciechowski, T., van der Putten, W.H., Kleyer, M., Schurr, U., Postma, J., 2016. Pampered inside, pestered outside? Differences and similarities between plants growing in controlled conditions and in the field. New Phytol. 212, 838–855. https://doi.org/10.1111/nph.14243

R Core Team, 2017. R: A language and environment for statistical computing. R Foundation for Statistical Computing, Vienna, Austria.

Ramos, M.L.G., Parsons, R., Sprent, J.I., James, E.K., 2003. Effect of water stress on nitrogen fixation and nodule structure of common bean. Pesqui. Agropecu. Bras. 38, 339–347.

Reynolds, M.P., Sayre, K.D., Vivar, H.E., 1994. Intercropping wheat and barley with N-fixing legume species: a method for improving ground cover, N-use efficiency and productivity in low input systems. J. Agric. Sci. 123, 175. https://doi.org/10.1017/S002185960006843X

Rubiales, D., Mikic, A., 2015. Introduction: Legumes in Sustainable Agriculture. Crit. Rev. Plant Sci. 34, 2–3. https://doi.org/10.1080/07352689.2014.897896

Sallam, A., Ghanbari, M., Martsch, R., 2017. Genetic analysis of winter hardiness and effect of sowing date on yield traits in winter faba bean. Sci. Hortic. 224, 296–301. https://doi.org/10.1016/j.scienta.2017.04.015

Schonfeld, M.A., Johnson, R.C., Carver, B.F., Mornhinweg, D.W., 1988. Water Relations in Winter Wheat as Drought Resistance Indicators. Crop Sci. 28, 526. https://doi.org/10.2135/cropsci1988.0011183X002800030021x

Semere, T., Froud-Williams, R.J., 2001. The effect of pea cultivar and water stress on root and shoot competition between vegetative plants of maize and pea: Root and shoot competition. J. Appl. Ecol. 38, 137–145. https://doi.org/10.1046/j.1365-2664.2001.00570.x

Senbayram, M., Tränkner, M., Dittert, K., Brück, H., 2015. Daytime leaf water use efficiency does not explain the relationship between plant N status and biomass water-use efficiency of tobacco under non-limiting water supply. J. Plant Nutr. Soil Sci. 178, 682–692. https://doi.org/10.1002/jpln.201400608

Shapiro, S.S., Wilk, M.B., 1965. An Analysis of Variance Test for Normality (Complete Samples). Biometrika 52, 591–611. https://doi.org/10.2307/2333709

Siddiqui, M.H., Al-Khaishany, M.Y., Al-Qutami, M.A., Al-Whaibi, M.H., Grover, A., Ali, H.M., Al-Wahibi, M.S., Bukhari, N.A., 2015. Response of Different Genotypes of Faba Bean Plant to Drought Stress. Int. J. Mol. Sci. 16, 10214–10227. https://doi.org/10.3390/ijms160510214

Siebrecht-Schöll, D.J., 2019. Breeding analysis of eight winter faba bean genotypes for mixed cropping with winter wheat. Georg-August-Universität Göttingen, Göttingen.

Spinoni, J., Naumann, G., Vogt, J., 2015. Spatial patterns of European droughts under a moderate emission scenario. Adv. Sci. Res. 12, 179–186. https://doi.org/10.5194/asr-12-179-2015

Stoddard, F.L., Balko, C., Erskine, W., Khan, H.R., Link, W., Sarker, A., 2006. Screening techniques and sources of resistance to abiotic stresses in cool-season food legumes. Euphytica 147, 167–186. https://doi.org/10.1007/s10681-006-4723-8

Szabados, L., Savouré, A., 2010. Proline: a multifunctional amino acid. Trends Plant Sci. 15, 89–97. https://doi.org/10.1016/j.tplants.2009.11.009

Tambussi, E.A., Bort, J., Araus, J.L., 2007. Water use efficiency in C3 cereals under Mediterranean conditions: a review of physiological aspects. Ann. Appl. Biol. 150, 307–321. https://doi.org/10.1111/j.1744-7348.2007.00143.x

Teulat, B., Monneveux, P., Wery, J., Borries, C., Souyris, I., Charrier, A., This, D., 1997. Relationships between relative water content and growth parameters under water stress in barley: a QTL study. New Phytol. 137, 99–107.

Vandermeer, J., 1989. The ecology of intercropping. Cambridge University Press, Cambridge. https://doi.org/10.1017/CBO9780511623523.

Vassileva, V., Signarbieux, C., Anders, I., Feller, U., 2011. Genotypic variation in drought stress response and subsequent recovery of wheat (Triticum aestivum L.). J. Plant Res. 124, 147–154. https://doi.org/10.1007/s10265-010-0340-7

Verbruggen, N., Hermans, C., 2008. Proline accumulation in plants: a review. Amino Acids 35, 753–759. https://doi.org/10.1007/s00726-008-0061-6

Xiao, J., Yin, X., Ren, J., Zhang, M., Tang, L., Zheng, Y., 2018. Complementation drives higher growth rate and yield of wheat and saves nitrogen fertilizer in wheat and faba bean intercropping. Field Crops Res. 221, 119–129. https://doi.org/10.1016/j.fcr.2017.12.009

Xu, B.C., Li, F.M., Shan, L., 2008. Switchgrass and milkvetch intercropping under 2:1 row-replacement in semiarid region, northwest China: Aboveground biomass and water use efficiency. Eur. J. Agron. 28, 485–492. https://doi.org/10.1016/j.eja.2007.11.011

Yang, C., Huang, G., Chai, Q., Luo, Z., 2011. Water use and yield of wheat/maize intercropping under alternate irrigation in the oasis field of northwest China. Field Crops Res. 124, 426–432. https://doi.org/10.1016/j.fcr.2011.07.013

Zabawi, A.G.M., Dennet, M.D.D., 2010. Responses of faba bean (Vicia faba L. cv Maris Bead) to different levels of plant available water. II. Yield, water use and water use efficiency. J. Trop. Agric. and Fd. Sc. 38, 145–152.

Zampieri, M., Ceglar, A., Dentener, F., Toreti, A., 2017. Wheat yield loss attributable to heat waves, drought and water excess at the global, national and subnational scales. Environ. Res. Lett. 12, 064008. https://doi.org/10.1088/1748-9326/aa723b

Zegada-Lizarazu, W., Izumi, Y., Iijima, M., 2006. Water Competition of Intercropped Pearl Millet with Cowpea under Drought and Soil Compaction Stresses. Plant Prod. Sci. 9, 123–132.

Chapter 2.2:
Crop genotype and plant compartment determine the response of the active bacterial community towards water deficit

Sandra Granzow, Annika Meißner, Birgit Pfeiffer, Rolf Daniel, Stefan Vidal, Franziska Wemheuer

Manuscript draft

Crop genotype and plant compartment determine the response of the active bacterial community towards water deficit

Sandra Granzow [1,2*], Annika Meißner[2,3], Birgit Pfeiffer[4], Rolf Daniel[4], Stefan Vidal[1] and Franziska Wemheuer[1]

[1]Division of Agricultural Entomology, Department of Crop Sciences, University of Göttingen, Göttingen, Germany

[2]Center of Biodiversity and Sustainable Land Use, University of Göttingen, Göttingen, Germany

[3] Institute of Applied Plant Nutrition, University of Göttingen, Göttingen, Germany

[4]Genomic and Applied Microbiology and Göttingen Genomics Laboratory, Institute of Microbiology and Genetics, University of Göttingen, Göttingen, Germany

*** Correspondence:**
Dr. Sandra Granzow
sandra.granzow@agr.uni-goettingen.de

Keywords: active bacterial community, intercropping, drought, endophytes, functional profiles

Abstract

Drought is one of the most important environmental stresses and causes severe decline in crop yields. Sustainable agricultural practices such as new crop genotypes or intercropping might have the potential to alleviate prognosticated changes in precipitation and temperatures. As plant-associated microbial communities are important for plant growth and health, it is of crucial interest to understand how changing environmental conditions such as drought will affect the plant microbiome. However, our understanding about the effects of cropping system and drought on plant microbiome is still scarce. In the present study, we investigated how water deficit changes the active bacterial community in the rhizosphere soil and leaf endosphere of winter wheat (genotype: Genius) and two winter faba bean genotypes (S_004; S_062) under different cropping systems. We showed that crop species, genotype and plant compartment significantly influenced the active bacterial community in their composition and diversity. Plant related traits strongly shaped responses of bacteria towards cropping system and water deficit. For example, endophytic bacterial diversity and richness were significantly reduced by water deficit specific for one faba bean genotype which was mainly related in changes of plant physiological parameters such as chlorophyll content and sugar concentration in leaves. In contrast, in the rhizosphere soil alpha-diversity demonstrated a marked resistance towards water deficit, whereas bacterial community composition was significantly altered dependent on faba bean genotype. Predicted functional profiles obtained from 16S rRNA data revealed that similar to composition, crop species and compartment significantly changed functioning; however, cropping system and water deficit did not influence these profiles. Obtained results highlight that there are complex interactions between plants, associated microorganisms and their environment that might influence agricultural productivity.

1 Introduction

Agricultural droughts can create serious threats to food security by reducing crop yields worldwide (Fahad et al., 2017; Zampieri et al., 2017). It is prognosticated that changes in precipitation and temperatures will increase the duration and frequency of drought periods in Europe in the next 20 years (Christensen and Christensen, 2007; Spinoni et al., 2015). Thus, there is a growing demand for drought tolerant cultivars which improve crop yields even under dry climatic conditions (Nuccio et al., 2018). Plant responses to water deficit include root biomass adjustment, stomatal activity, increased water use efficiency (WUE), or the synthesis of osmolytes such as sugar (Bray, 1997; Osakabe et al., 2014). These responses differ between plant species, genotype, and plant development stage (Mwanamewenge et al., 1999; Abid et al., 2017; Ouyang et al., 2017). In addition, beneficial microorganisms can increase the resistance and resilience to stress conditions such as drought (de Zelicourt et al., 2013). The inoculation with endophytic bacteria including *Burkholderia phytofirmans* or *Bacillus subtilis* conferred drought resistance in different plant species including *Zea mays* (Naveed et al., 2014), *Brachypodium distachyon* (Gagné-Bourque et al., 2015) and *Sorghum bicolor* (Xu et al., 2018). For example, Gagné-Bourque et al. (2015) showed that an inoculation of an endophytic *Bacillus subtilis* strain isolated from switchgrass conferred drought resistance in *Brachypodium distachyon* via upregulation of drought-response genes, modulation of the DNA methylation process, and an increase in the soluble sugars and starch content of leaves. In addition, endophytic bacteria contribute to nutrient acquisition, promote plant growth and some also have the potential as biocontrol agents (Hardoim et al., 2015; Tian et al., 2017). Thus, they are important in a sustainable agriculture (Ryan et al., 2008; Berg et al., 2009).

Previous studies reported significant effects of drought and re-watering on microbial communities (e.g., Xu et al. 2018; Nguyen et al. 2018; Naylor et al., 2017). Recently, Xu et al. (2018) found that drought significantly restructured the bacterial composition in roots of sorghum. In addition, bacteria showed a high degree of resilience after re-watering. Similar effects of drought on the overall bacterial community composition in different rice compartments were observed by Santos-Medellin et al. (2017). The authors concluded that the restructuring of the associated microbiome might contribute to plant survival under extreme environmental conditions. According to Nguyen et al. (2018), agricultural practices such as nitrogen fertilization can restrain the resilience of soil bacterial communities after prolonged drought. Other agricultural practices such as intercropping systems have received more attention in the past decades (Yang et al., 2011; Hu et al., 2017) as intercropping of wheat and maize significantly increased water use and water use efficiency compared to sole cropping (Yang et al., 2011).

So far, most studies investigating the response of microbial communities towards environmental stressors such as drought focused on the entire microbial community (Kaurin et al., 2018; Nguyen et al., 2018; but see Barnard et al., 2013). However, the potentially active microbial community is more sensitive to abiotic stresses and thus is more closely related to ecosystem functionality (Blagodatskaya and Kuzyakov, 2013; Herzog et al., 2015; Taschen et al., 2017). Therefore, it is important to understand how water deficit and different cropping systems alter the active plant-associated bacterial community of important crop species. Hence, the aim of the present study was to investigate the influence of water deficit and re-watering on the metabolically active bacterial communities of winter wheat (*Triticum aestivum* L.; genotype: Genius) and two genotypes of winter faba bean (*Vicia faba* L.; S_004 and S_062) under different cropping systems. We hypothesized that (i) composition, diversity and associated taxa of the active bacterial community are affected by crop species, faba bean genotype and plant compartment. We expected further (ii) that the response of the bacterial community towards water deficit and cropping system is dependent on these factors. Based on previous findings (Wemheuer et al., 2017), we hypothesized that (iii) bacterial functioning is altered in a different manner towards cropping system and water deficit as the bacterial community composition

To corroborate these hypotheses, wheat and two genotypes of faba bean were grown in monoculture or in row intercropping with (water deficit treatments) or without drought stress (control treatments). Plant and soil samples were collected at three time points: beginning of water deficit, during water deficit and after re-watering. Bacterial communities in rhizosphere and leaf endosphere were examined by iTag sequencing of bacterial 16S rRNA genes, amplified by two-step reverse transcriptase PCR. Functional profiles of active community members were predicted using Tax4Fun. To our knowledge, this is the first study investigating the combined and separate effect of intercropping and drought stress on the metabolically active plant-associated bacterial community of two important crop species. Obtained results will further deepen our understanding how sustainable agricultural practices and plant-associated microorganisms might mitigate future drought events.

2 Material and Methods

2.1 Plant material

To examine the combined influence of cropping system and water deficit on the active bacterial community in roots and attached soil (here regarded as rhizosphere soil) and aerial (here regarded as leaf) endosphere, a greenhouse experiment was conducted in autumn 2016. Seeds of the two winter faba bean genotypes (genotypes: S_004; S_062) were provided by the Institute of Plant breeding of the University of Göttingen. The two faba bean genotypes S_004 and

S_062) were selected based on previously field trial-tested inbred lines used within the IMPAC[3] project (*Novel genotypes for mixed cropping allow for improved sustainable land use across arable land, grassland and woodland*). The genotype S_004 is characterized by medium height and leaf size, low tillering, late maturity and high yield. In contrast, genotype S_062 is very short with small leaflets, high tillering and early maturing. Seeds of winter wheat (genotype: Genius) were provided by Norddeutsche Pflanzenzucht Hans-Georg Lembke KG. All seeds were surface-sterilized by serial washing according to Andreote et al. (2010) with one modification. Immersion in sterile distilled water was performed four times for 30 s. Surface sterilized seeds were placed on wet sterile tissues and germinated at 7 °C under dark conditions until seedlings developed roots with a length of approximately 4 cm.

2.2 Experimental design and soil substrate

Pre-germinated seeds of faba bean and wheat were sown in monoculture or as mixture in polypropylene containers (Sunware; 45.5 x 36 x 24 cm) in a randomized block design (day 0, DAO, days after onset of experiment). Twelve treatments were established: faba bean monoculture S_004 with or without water deficit (S4_FBM_D/C), faba bean monoculture S_062 with or without water deficit (S62_FBM_D/C), faba bean S_004 intercropped with wheat with or without water deficit (S4_FBIC_D/C; WIC_D/C), faba bean S_062 intercropped with wheat with or without water deficit (S62_FBIC_D/C; WIC_D/C), and wheat monoculture with or without water deficit (WM_D/C; Table 1). Each treatment was replicated four times, resulting in a total of 40 containers. We defined two different cropping systems (monoculture and intercropping), whereas cropping regimes compromised each treatment, e.g. WM_D and FBM_C.

For monocultures, 30 faba bean or 72 wheat seeds per container were sown in six rows. For intercropping systems, 15 faba bean and 36 wheat seeds were sown in alternate rows (Vandermeer, 1992). Each container was filled with air-dried, sieved (< 10 mm) and layered soil from the experimental study site in Reinshof (51.48 ° N, 9.92 ° E and 157 m asl.), Germany. The soil volume of each pot accounted for approximately 20 L with a dry weight of 18 kg. Filling of the pots was performed in layers adding distilled water to each layer to prevent soil compaction. After emergence of the seedlings, the soil was covered by gravel to minimize water loss by evaporation. The soil was classified as Gleyic Fluvisol according to the FAO classification system and contained 21 % clay, 68 % silt and 11 % sand, with pH 7.3 and 2.8 % humus. Nutrients such as phosphorus (50 mg P/kg dry soil) and potassium (140 mg K/kg dry soil) were in an optimal range according to the German nutrient-availability class system (Kuchenbuch and Buczko 2011).

2.3 Water management and growth conditions

During the experiment, photosynthetic photon flux density was 400 μmol m^{-2} s^{-1} at plant level with a 10/14 h day/night photoperiod. Furthermore, the CO_2 concentration reached around 450 ppm. There was a relative humidity of 50 % and an average air temperature of 23 °C. Water loss by transpiration was documented by placing the pots permanently on balances (TQ30, ATP Messtechnik, Germany). The weight reduction was measured every 30 minutes in order to constantly determine water consumption. This system avoids hidden drought due to higher transpiration of increased biomasses (Senbayram et al., 2015). Plants of all treatments were irrigated with distilled water to 90 % field capacity. After a growing period of 24 days and a BBCH of 14/34 for faba bean and a BBCH of 14/15 of wheat plants (Lancashire et al.,1991), the amount of water in water deficit treatments was reduced to 75 % compared to control treatments. At day 28, the amount of water in these treatments was further reduced to 25 %. At day 34, all water deficit treatments were re-watered with the adequate amount of water according to plant growth and water consumption. All control pots were sufficiently irrigated during the whole experimental duration (6 weeks).

2.4 Sampling

Soil and plant samples were collected from control and water deficit treatments at day 29 (beginning of water deficit), day 34 (during water deficit) and at day 38 (after re-watering of water deficit plants) (Figure 1). For microbial community analysis, one faba bean and two wheat plants per container and harvest were randomly sampled which showed no obvious sign of any disease infection. The roots were gently shaken to remove the non-rhizosphere soil. Rhizosphere soil for pH value and C/N was collected by carefully brushing the roots. Rhizosphere soil and roots of each plant species and each pot were pooled for molecular analysis. All samples for molecular analysis were immediately frozen in liquid nitrogen, transferred to the laboratory and stored at -80 °C. In total, 96 faba bean (48 plants of each genotype) and 144 wheat plants were collected. Rhizosphere and aerial plant parts of each crop species and container were pooled, resulting in a total of 96 faba bean and 72 wheat samples (Table 1).

2.5 Edaphic properties and plant stress-related parameters

For determination of edaphic properties such as total organic carbon and total organic nitrogen subsamples of all rhizosphere samples were dried at 60 °C for two days and subsequently sieved to < 2 mm. Carbon and nitrogen concentrations from dried subsamples were determined using a NA-1500N analyser (Thermo Fisher Scientific, Waltham, USA). Afterwards, the carbon-to-nitrogen (C/N) ratio was calculated. The pH values of all rhizosphere soil samples were measured as follows: 10 g of dried and sieved soil was added in a small beaker with 25 ml 0.01 M calcium chloride. Soil solution was homogenized after 30 min and 60 min, and subsequently soil pH_{CaCl} was measured. The height and aerial fresh biomass of all plants used for microbial community analysis were measured. Estimation of chlorophyll concentration was conducted using SPAD-502 Plus meter (Konica Minolta, Japan) on the youngest fully expanded leaf to survey the availability of plant nitrogen. Three faba bean and one wheat plant per container were measured. For determination of soluble sugar content, approximately 50 mg of plant material from the youngest fully expanded leaf was homogenized with 1.8 ml ddH_2O in a thermoshaker at 60 °C and subsequently centrifuged at 13,000 rpm for 45 min. Samples were vortexed every 12-15 min. Afterwards, samples were centrifuged at 14,000 rpm for 20 min. Extracted supernatant was stored at -20 °C until measurement. Soluble sugar content (mg/g, dry weight) was determined using the Sucrose/D-Glucose/D-Fructose kit as recommended by the manufacturer (R-Biopharm, Mannheim) with modifications; all volumes were reduced to ¼. Moreover, glucose was added to sucrose, and an additional cuvette was used only with glucose and extinction one was measured after 5 minutes. The soluble sugar content was measured spectrophotometrically (V-650, Jasco Corporation, Japan) at 340 nm using glucose or glucose and sucrose as control. Details on edaphic properties and plant parameters are provided in Table S1.

2.6 Surface sterilization of pant material

Leaves were surface sterilized according to Wemheuer and Wemheuer (2017). The effectiveness of applied sterilization process was controlled as described previously (Wemheuer et al., 2016). In brief, aliquots of the water used in the final wash step were plated on common laboratory media plates, i.e. Luria-Bertani agar and potato dextrose agar. The plates were incubated in the dark at 25 °C for at least one week. No growth of microorganisms was observed. In addition, water from the same aliquots was subjected to PCR targeting the bacterial 16S rRNA gene as described below for microbial community analysis. No PCR products were detected.

2.7 RNA Extraction and Purification

Environmental RNA of the rhizosphere was extracted from 2 g soil per sample employing the RNA PowerSoil total RNA isolation kit as recommended by the manufacturer (MoBio Laboratories, Carlsbad, CA, USA, now Qiagen, Hilden, Germany). RNA was extracted from 100 - 250 mg plant material according to Weinbauer et al. (2002) with slight modifications: 2 ml tubes were used, and all solution volumes were 10-times reduced. In addition, the first vortexing step was performed with a FastPrep® - 24 Classic Instrument (Biomedicals) at 4 m/s for 60 s. Extracted RNA was purified employing the RNeasy Mini Kit as recommended by the manufacturer (Qiagen, Hilden, Germany) with modifications according to Streit and Daniel (2010). Residual DNA was removed with the TURBO DNA-free™ kit (Thermo Scientific) from the extracted RNA according to the manufacturer's protocol. In addition, 1/40 volume Ribolock RNase Inhibitor (40 U/µL) (Thermo Scientific) was added in the first step of the DNA digestion. The absence of DNA was confirmed by PCR using the partial 16S rRNA as target gene for amplification of bacteria. For details of the PCR reaction and cycling conditions as well as the primers see the first PCR according to Wemheuer and Wemheuer (2017). The DNA-free RNA was further purified according to Streit and Daniel (2012). RNA concentrations were determined using a NanoDrop ND-1000 spectrophotometer (NanoDrop Technologies, Wilmington, DE, USA).

2.8 Synthesis of cDNA from total RNA

Purified RNA from 168 plant and 168 rhizosphere samples were converted to cDNA by employing the SuperScript™III reverse transcriptase Kit as recommended by the supplier (Invitrogen, Karlsruhe, Germany) with two modifications. Same reverse primer 1193r (20 µM) was used for the reaction as for the following PCR. After the last step, 0.5 µl RNase H (5 U/µl; Fermentas) was added, and samples were incubated for 15 min at 37 °C and subsequently for 10 min at 65 °C. The cDNA was stored at -20 °C.

2.9 Amplification of 16S rRNA gene

Bacterial communities in leaves and rhizosphere were assessed by PCR approach targeting the V5-V7 region of the 16S rRNA gene. The following primers were used: 799F (Chelius and Triplett, 2001) and 1193R (Bodenhausen et al., 2013; Hartman et al., 2017) containing MiSeq adaptors (underlined) Miseq-799F 5'-TCGTCGGCAGCGTCAGATGTGTATAAGAGACAG AACMGGATTAGATACCCKG-3'; MiSeq- 1193R 5'GTCTCGTGGGCTCGGAGATGTGT ATAAGAGACAGACGTCATCCCCACCTTCC-3'. The PCR mixture (25 µl) contained 5 µl of five-fold Phusion GC buffer, 200 µM of each of the four deoxynucleoside triphosphates, 4 µM of each primer, 0.5 U of Phusion high fidelity DNA polymerase (Thermo Scientific) and approximately 50 ng of cDNA as template. The following thermal cycling scheme was used: initial denaturation at 98 °C for 30 s, 30 cycles of denaturation at 98 °C for 15 s, annealing at 53 °C for 30 s, followed by extension at 72 °C for 30 s. The final extension was carried out at 72 °C for 2 min. Negative controls were performed using the reaction mixture without template. Genomic DNA of *Escherichia coli* strain DH5α was used as template in the positive control. Three independent PCRs were performed per sample. Obtained PCR products per sample were controlled for appropriate size, pooled in equal amounts, and purified using the NucleoMag NGS Clean up (Macherey-Nagel, Düren, Germany).

Quantification of the purified PCR products was performed using the Quant-iT dsDNA HS assay kit and a Qubit fluorometer (Thermo Scientific) as recommended by the manufacturer. Quantified PCR products were barcoded using the Nextera XT-Index kit (Illumina, San Diego, USA) and the Kapa HIFI Hot Start polymerase (Kapa Biosystems, Wilmington, USA). The Göttingen Genomics Laboratory determined the sequences of the partial 16S rRNA genes employing the MiSeq Sequencing platform and the MiSeq Reagent Kit v3 (2 x 300 cycles) as recommended by the manufacturer (Illumina).

2.10 Processing of bacterial community dataset

Generated sequencing data were initially quality filtered with the Trimmomatic tool version 0.36 (Bolger et al., 2014). Low quality reads were truncated if the quality dropped below 15 in a sliding window of 4 bp. Subsequently, all reads shorter than 100 bp and orphan reads were removed. Remaining sequences were merged, quality-filtered and further processed with USEARCH version 10.0.240 (Edgar, 2010). Filtering included the removal of reads shorter than 300 or longer than 500 bp as well as the removal low quality reads (expected error > 1) and reads with more than one ambitious base. Processed sequences of all samples were concatenated to one file and subsequently dereplicated into unique sequences. These sequences were denoised with the unoise3 algorithm implemented in USEARCH (Edgar, 2010). Chimeric sequences were initially removed in denovo mode during denoising. Subsequently, remaining chimeric sequences were removed using UCHIME (Edgar et al., 2011) in reference mode with the SILVA SSU Ref NR 99 132 database (Quast et al., 2013) as reference data set for bacteria. All zOTUs consisting of one single sequence (singletons) were removed. Filtered sequences were mapped on remaining unique sequences to determine the occurrence and abundance of each unique sequence in every sample. To assign taxonomy of bacteria chimera-free sequences were classified by BLAST alignment against the most recent SILVA database (Quast et al., 2013) with an e-value threshold of 1e-20. All non-bacterial zOTUs were removed based on their taxonomic classification in the respective database. Final zOTU table is provided in Table S2. Only zOTUs occurring in more than one sample were considered for further statistical analysis. Samples with less than 145 sequences per sample were removed prior statistical analysis, resulting in 323 samples for bacteria.

2.11 Statistical Analysis

All statistical analyses were performed using R version 3.4.0 (R Core Team, 2016). Differences were considered as statistically significant with $P \leq 0.05$. Differences in alpha- or beta-diversity as well as sequencing depth with regard to cropping system and water treatment (yes/no) were tested by a Kruskal-Wallis test. There were no significant differences of the mean sequencing depths between cropping system and water treatment. In consequence, zOTU tables were not rarefied as recommended by McMurdie and Holmes (2014).

Alpha-diversity indices (Richness, Shannon index of diversity and Michaelis Menten Fit) were calculated in the *vegan* package version 2.4.4 (Oksanen et al., 2016) and the *drc* package version 3.0-1 (Ritz and Streibig, 2016). zOTU table was rarefied using the *rrarefy* function in *vegan* and samples with less than 2,935 (rhizosphere soil) and 1,790 (leaves) were removed prior alpha-diversity analysis. Sample coverage was estimated using the Michaelis-Menten Fit calculated in R. For this purpose, richness and rarefaction curves were calculated using the *picante* package version 1.6-2 (Kembel et al., 2010). Richness and diversity were calculated using the *specnumber* and *diversity* function, respectively. The Michaelis-Menten Fit was subsequently calculated from generated rarefaction curves using the *MM2* model within the *drc* package version 3.0-1 (Ritz and Streibig, 2016). All alpha-diversity indices were calculated ten times. The average from each iteration was used for further statistical analysis. Final table containing bacterial richness, diversity, Michaelis-Menten Fit and coverage is provided in Table S3.

Data were tested for normal distribution with *shapiro* and homogeneity of variance with *leveneTest* function with the package *car* version 2.1-5 (Fox and Weisberg, 2011). For global differences (for all three harvests) between measured edaphic properties and plant parameters were calculated with a linear mixed model with the function *lme* and the R package *nlme* version 3.1-131 (Pinheiro et al., 2017) with pot number as random factor. Data was log-transformed when not normal distributed. F-values were evaluated with ANOVA and type = "marginal". In addition, each harvest was tested separately with a post hoc test using Dunn's test with *p*-value adjustment "BH" and the function *dunnTest* in the R package *FSA* version 0.8.17 (Ogle, 2016). Alpha-diversity was evaluated with Kruskal-Wallis-test or the post hoc test using *dunnTest.*

Differences in community composition as well as function were investigated by permutational multivariate analysis of variance (PERMANOVA; Anderson, 2001) based on Bray-Curtis distance matrices using 999 random permutations. OTU tables were subsampled ten times and all tables were summed up to account for low abundant species. Global differences (all three harvests) in crop species and compartment were tested with Adonis and specified with "strata = pot". A significant *p*-value in PERMANOVA for beta-diversity can be driven by true biological differences, differences within group (variance) or both (Anderson, 2001). In case of significant *p*-values in PERMANOVA, we tested for differences in homogeneity using permutational analysis of multivariate dispersions (PERMDISP, Anderson, 2006) with 999 permuations. NMDS, PERMANOVA and PERMDISP were run using functions; *metaMDS*, *adonis* and *betadisper*, respectively, in the R package *vegan* (Oksanen et al., 2016). Differences in community composition were visualized using the *metaMDS* function within the *vegan* package (Oksanen et al., 2016). To investigate in differences between cropping regimes, pairwise Adonis with *p*-value adjustment "BH" based on Bray-Curtis distances were used (Martinez Arbizu, 2017).

To identify zOTUs highly associated to cropping regime and crop genotype with regard to plant compartment, multipattern analyses were applied. For that purpose, bacteria were investigated using the *multipatt* function from the *IndicSpecies* package version 1.7.6 (DeCáceres and Legendre, 2009). Only bacterial zOTUs found in at least three samples were used. The biserial coefficients (R) with a particular cropping regime or genotype were corrected for unequal sample size using the function *r.g* (Tichy and Chytry, 2006). For visualization, a bipartite network was generated using the treatment as source nodes and the taxa as target nodes. Network generation was performed using the *edge-weighted spring embedded layout* algorithm in Cytoscape version 3.3.0 (Shannon et al., 2003). The results of the multipattern analyses are provided in Table S4. Functional profiles were predicted from obtained 16S rRNA gene data using Tax4Fun2 (Aßhauer et al., 2015). Tax4Fun transforms the SILVA-based zOTU classification into a taxonomic profile of KEGG organisms, which is subsequently normalized by the 16S rRNA copy number (obtained from NCBI genome annotations). Afterwards, KEGG profiles are converted into artificial metagenomes by combining functional profiles calculated for each of the KEGG genomes. Genes involved in nutrient cycling, plant-growth promoting and stress were identified in the resulting profiles and visualized in a heatmap using *heatmap* function.

3 Results and Discussion

3.1.1 Stress-related plant parameters

As abiotic stress has been shown to affect plant growth and productivity by imposing certain physiological and biochemical changes, we investigated several plant-stress related parameters such as soluble sugars or chlorophyll content (Abid et al., 2017). Moreover, intercropping systems have been shown to enhance plant growth compared to monoculture (Song et al. 2007b). Here, we found that biomass of faba bean genotype S_004 and wheat was significantly higher in WIC/FBM compared to WM/FBIC (LMM; df=13, F=6.73, p=0.022; df=21, F=11.15, p=0.0031) (Table 2). Effects of cropping system on height were only observed during harvest 3. Genotype S_004 was significantly taller in monoculture compared to intercropping system (Kruskal-Wallis (KW)-test, x^2=4.21, df=1, p=0.04). Chlorophyll content of faba bean genotype S_062 and wheat were significantly influenced by water treatment (LMM, df= 13, F=9.64, p=0.0084) and cropping system (LMM, df=21, F=12.71, p=0.0018), respectively (Table 3). Highest chlorophyll values in S_062 were observed for water deficit treatment/FBIC compared to control/FBM. Linear mixed effect model showed that sucrose was significantly influenced by cropping system in wheat leaves (LMM, df=21, F=7.61, p=0.0118). Furthermore, we observed significantly lower values of all three soluble sugar concentrations in S_062 under water deficit compared to control for harvest 2 (Table 4; KW-test, glucose, x^2=6.89, df=1, p=0.008; sucrose, x^2=4.86, p=0.027; fructose, x^2=4.42, p=0.035).

3.1.2 Edaphic parameters

We investigated in several edaphic properties including pH value, total organic carbon and nitrogen, as previous studies have been shown that cropping system or drought can change chemical characteristics in the rhizosphere soil (Song et al., 2007b; Preece and Peñuelas, 2016). We found that cropping system was the most influencing factor on edaphic properties compared to water treatment. Results of linear mixed effect model showed that pH value was significantly influenced by cropping system in the rhizosphere of wheat (LMM, df=21, F=5.72, p=0.00262). This was mainly observed for harvest 2, with lowest pH values under WIC (Table 5). In addition, we observed that pH value was significantly lower in FBIC compared to FBM for both genotypes specific for harvest 2 (KW-test, S_004 x^2=6.37, df=1, p=0.0116; S_062: x^2=10.63, df=1, p=0.0011). C:N ratio as well as carbon were significantly affected by cropping system in the wheat rhizosphere (LMM, C:N ratio: df=21, F=5.96, p=0.023; carbon: df= 21, F=4.47, p=0.046). Total nitrogen and carbon had significantly lower values under WIC compared to WM for harvest 3 whereas the opposite was observed for harvest 2. Cropping system also significantly influenced C:N ratio in S_004 (S_004; LMM, df=13, F=50.54, p<0.0001) and highest C:N ratio was found under FBIC compared to FBM (Table 6).

3.2 Overall bacterial community

The response of bacterial communities of faba bean and wheat towards water deficit under different cropping systems was assessed by Illumina (MiSeq) sequencing targeting the bacterial 16S rRNA gene. After removal of low-quality reads, PCR artefacts (chimeras) and non-target contaminations, a total of 3,592,483 high quality reads were obtained for bacteria (Table S3). Sequence numbers per sample varied between 145 and 71,557 (average 11,120) and obtained sequences were grouped into 6,560 bacterial zOTUs (Table S2). Calculated Michaelis-Menten Fit and rarefaction curves confirmed that the majority of the bacterial community was recovered by the surveying effort (Figures S1, S2, Table S3).

Bacteria were dominated by seven phyla (> 0.5% of all sequences across all samples): Proteobacteria (74.71 %), Actinobacteria (9.56 %), Bacteroidetes (9.49 %), Gemmatimonadetes (2.77 %), Acidobacteria (1.16 %), Chloroflexi (0.72 %) and Entotheonellaeota (0.58 %) (Figure 2, 3). The Proteobacteria were dominated by Gammaproteobacteria (60.12 %), followed by Alpha- (8.75 %) and Deltaproteobacteria (5.84 %). The abundant bacterial phyla were present in all samples and accounted for 99.0 %, of all sequences analysed in this study. At family level, *Burkolderiaceae* (49.55 %), *Microscillaceae* (4.19 %) and *Halomonaceae* (3.11 %) dominated the bacterial dataset. The most frequent bacterial genus was *Curvibacter* (29.34 %). Other abundant genera were, for example, *Mitsuaria* (10.72 %), *Halomonas* (1.85 %), *Rhizobacter* (1.64 %), *Allorhizobium-Neorhizobium-Pararhizobium-Rhizobium* (1.59 %) and *Ohtaekwangia* (1.59 %). Results are in line with previous studies investigating plant-associated bacterial communities (Gdanetz et al., 2017; Naylor et al., 2017; Zhou et al., 2017).

3.3 Bacterial community was strongly affected by crop species and compartment

According to our first hypothesis that crop species, genotype and compartment affect bacterial community, we calculated diversity (represented by the Shannon index H') and richness (number of observed unique sequences). Leaf bacterial richness was significantly influenced by crop species. We observed significantly higher bacterial richness in faba bean plants for harvest 1 and 2 (KW-test, H1, x^2=5.99, df=1, *p*=0.014, H2, x^2=6.50, df=1, *p*=0.011). In the rhizosphere, we found that bacterial richness was significantly higher in wheat compared to faba bean (KW-test, H1, x^2=4.08, df=1, *p*=0.043). Genotype did not influence bacterial alpha-diversity.

In line with our hypothesis, plant compartment and crop species significantly influenced microbial community composition (Table 9). Compartment explained 40.7 % of the variance in the bacterial dataset (PERMANOVA, *p*=0.001), whereas crop species explained 3.3 % and 1.9 % of the variance in the leaf endosphere and rhizosphere soil (PERMANOVA, leaves, *p*=0.001; rhizosphere, *p*=0.003). Faba bean genotype only influenced bacterial community composition in the rhizosphere specific for harvest 1 (Table 9). Here, genotype explained 6.1 % of the variance (PERMANOVA, *p*=0.016). Moreover, several bacterial taxa differed in their abundance with regard to crop species and compartment. For example, *Curvibacter* was the predominant bacterial genus found in the leaf endosphere with an average relative abundance of 53.80 %, whereas in the rhizosphere *Curvibacter* was found with 4.72 % abundance (Figure 2, 3). Higher relative abundance of *Halomonas* was recorded in wheat leaves (4.83 %) compared to faba bean genotypes (S_004: 1.58 % and S_062: 1.29 %). In addition, *Allorhizium-Neorhizobium-Pararhizobium-Rhizobium* was more often found in faba bean rhizosphere (4.08 %) compared to wheat (0.06 %).

In line with our results, previous studies showed significant effects of crop species on bacterial composition and diversity in the rhizosphere soil and plant endosphere (Wemheuer et al., 2017; Zhou et al., 2017). Partly in contrast to our observation, previous studies reported that plant genotype significantly affected bacterial community in the rhizosphere (Mahoney, Yin and Hulbert, 2017; Li et al., 2018) and leaf endosphere (Wagner et al., 2016; Montanari-Coelho et al., 2018). Recently, Mahoney, Yin and Hulbert (2017) found that bacterial community composition in the rhizosphere was significantly influenced by nine wheat genotypes. In addition, only 24 (out of the 1305) most abundant OTUs differed significantly in frequency between genotypes, indicating that host genotype played just a minor but significant role in bacterial diversification. Similar, other studies assumed that plant microbiome composition is more environmentally dependent rather than on the plant genotype *per se* and that host-adapted microbes may have been selected as they provide some selective advantage for their host (Bulgarelli et al., 2012; Vandenkoornhuyse et al., 2015).

3.4 Crop genotype and plant compartment influenced response of bacterial community towards water deficit and cropping system

We further hypothesized that crop species, genotype and compartment would alter the response of bacterial communities towards water deficit and cropping system. Cropping system significantly influenced bacterial diversity and richness but only in the leaf endosphere (Table 8). In leaves of S_062, the cropping system FBIC had significantly higher bacterial diversity and richness compared to FBM only in harvest 1 (KW-test, shannon, x^2=8.37, df=1, p=0.0038; richness, x^2=5.36, p=0.02). In addition, bacterial richness was significantly higher under WIC compared to WM only in harvest 3 (KW-test, x^2=5.1, p=0.023).

Evaluation of bacterial community composition with NMDS based on Bray-Curtis analysis showed no evident influence of cropping system in the rhizosphere; however, a smaller separation between S4_FBIC and S4_FBM was observed in the leaf endosphere specific for harvest 1 (Figure 4). PERMANOVA found that cropping system significantly influenced bacterial community composition in faba bean S_004, and explained 16.0 % of the variance (PERMANOVA, p=0.026).

Previous studies reported that intercropping significantly influenced bacterial community composition and increased bacterial diversity in soil (Song et al., 2007b; Yang et al., 2016). Yang et al. (2016) investigated ten different spring crops grown in monoculture and intercropping system and found that intercropping increased bacterial diversity in the rhizosphere, however, responses were also crop species dependent. So far, most research focused on bacterial communities in the soil, but studies which investigated the leaf endosphere under different cropping systems are scarce (Gdanetz et al., 2017; Granzow et al., 2017). For example, Granzow and coworkers (2017) investigated different plant compartments of faba bean and wheat grown in monoculture, row intercropping and mixed intercropping system. They reported that changes in bacterial diversity and richness were mainly recorded between mixed and row intercropping or mixed intercropping and monoculture in soil and root endosphere. Furthermore, leaf endophytes were not affected in their diversity or composition through cropping system (Granzow et al., 2017) which is in contrast to our results. We speculate that different soil type, plant species/genotypes or plant growth stage might contribute to the contradictory results, as these factors have been shown to influence microbial communities (Wagner et al., 2016; Gdanetz et al., 2017).

Furthermore, we observed that water deficit significantly influenced bacterial alpha-diversity but only in the leaf endosphere (Table 8). In faba bean leaves of S_062, bacterial diversity and richness significantly decreased under water deficit compared to well-watered plants specific for harvest 2 (KW-test, shannon, x^2=5.33, df=1, p=0.021; richness, x^2=8.04, p=0.0046). Moreover, the combination of water deficit and cropping system significantly affected leaf bacterial diversity (KW-test, x^2=7.95, df=3, p=0.05) and richness (KW-test, x^2=9.49, df=3, p=0.02) in S_062 for harvest 1. Dunn's test showed that the cropping regime S62_FBIC_D had significantly higher richness compared to S62_FBM_D, whereas diversity was significantly higher in S62_FBIC_D compared to S62_FBM_C.

Bacterial community composition in the rhizosphere of faba bean genotype S_062 showed a distinct clustering between control and water deficit treatments in the NMDS analysis for harvest 1 (Figure 5). PERMANOVA also confirmed that water deficit significantly influenced bacterial community composition in the rhizosphere of genotype S_062 specific or harvest 1, which explained 20.9 % of the variance (PERMANOVA, p=0.001). However, dispersion among water treatments was not homogenous (PERMDISP, F=18.26, p=0.002). Cropping regimes did not affect bacterial community composition.

Partly in accordance to our observation, previous studies demonstrated that water deficit decreased bacterial diversity in the root endosphere of several plant species (Naylor et al., 2017; Fitzpatrick et al., 2018). For example, Naylor and coworkers (2017) observed that water deficit decreased bacterial diversity in different plant compartments of 18 grass species including the root endosphere. However, they also indicated that water deficit effects were dependent on plant growth stage, and strongest response of bacterial diversity was displayed at early flowering compared to late flowering. On the other hand, Fitzpatrick et al. (2018) reported that bacterial diversity was not significantly influenced by water deficit in the root endosphere of 30 different angiosperms; however, a slight decrease was observed in drought treated plants. For the present study, we speculate that drought-induced plant responses including physiological and molecular changes (Abid et al., 2017) were responsible for the genotype specific effect on the leaf microbiome. In line with this assumption, we observed that chlorophyll concentration and all three soluble sugars (glucose, sucrose and fructose) in leaves significantly changed according to water deficit in genotype S_062 specific for harvest 2 similar to alpha-diversity (Table 3, 4).

As previous studies have shown, plant stress through drought can change root exudation pattern dependent on plant species (Henry et al., 2007; Preece and Peñuelas, 2016). Thus, soil microbial communities might be affected indirectly through rhizodeposition (Preece and Peñuelas, 2016). Similar, we recorded changes in bacterial community composition towards water deficit dependent on faba bean genotype. Partly in line with our finding, Santos-Medellin and coworkers (2017) investigated in four different rice cultivars and plant compartments and recorded compartment-specific cultivar effects on the drought response for the bacterial community composition. However, they only found few individual OTUs which showed differential responses to drought based on genotype, indicating that communities assembled in each cultivar responded relatively similar towards drought.

In contrast to our results, studies found that effect of water deficit on bacterial community composition was most pronounced in plant endosphere compared to rhizosphere (Naylor et al., 2017; Santos-Medellin et al., 2017; Fitzpatrick et al., 2018). As each plant compartment represents distinct niches for microbial communities in terms of exposure towards drought (Wallace et al., 2018), we also assume that responses can differ. However, it is a complex interplay of plant specific stress response and plant-microbe associations towards abiotic stress. Thus, bacterial response towards drought might differ dependent on crop species, genotype or plant growth stage as already indicated in previous research (Naylor et al., 2017; Fitzpatrick et al., 2018). Furthermore, contradictory findings between studies might be attributed to experimental settings such as drought intensities or soil characteristics, and precipitation history in soil which has been shown to affect soil microbial communities and their response towards drought (Kaisermann et al., 2015; Kaisermann et al. 2017; Santos-Medellin et al., 2017).

3.5 Associated bacterial taxa are altered by cropping regimes and genotypes

To identify bacterial taxa responsible for the observed differences among water deficit and cropping system or genotypes, we performed a multipattern analysis to investigate which microorganisms are significantly associated with those treatments (Table S4). In general, soil communities harbored more associated zOTUs than endophyte communities which is most probably related with higher sequence numbers in soil compared to endosphere samples. The number of significantly associated taxa in the leaf endosphere decreased in the water deficit treatments for harvest 2 compared to harvest 1 (Figure 6, 7). In the rhizosphere soil, wheat cropping regimes harbored the greatest number of associated bacterial taxa whereas faba bean cropping regimes the least number. In addition, we found that drought cropping regimes especially from faba bean showed more associated bacterial taxa from the phylum Actinobacteria than well-watered plants which was most pronounced for harvest 1 in both compartments. Similar to the community composition, we found significantly associated bacterial taxa dependent on genotype and water treatment (Figure 7). We recorded most associated taxa for genotype S_062 under drought in the leaf endosphere and rhizosphere soil for harvest 1. Here, associated taxa were mainly assigned to Actinobacteria that belonged to the bacterial genus *Mycobacterium* spp. in the leaf endosphere and the bacterial family *Gaiellales* in the rhizosphere soil. For harvest 2, the genus *Blastococcus* spp. (Actinobacteria) was mainly associated with the genotype S_004 under drought.

In line with our observation, previous studies indicated that specific bacterial phyla are enriched under drought conditions especially in the endosphere (Naylor et al., 2017; Santos-Medellin et al., 2017; Fitzpatrick et al., 2018). Naylor et al. (2017) performed quantification of absolute abundance of Actinobacteria in roots under drought and control conditions using qPCR. Results demonstrated that Actinobacteria exhibited a marked increase in absolute abundance in drought-treated roots for different C3 and C4 grass species. In addition, they also confirmed that significant indicator taxa for drought-treated plants belonged mainly to Actinobacteria which was similar to our results (Naylor et al., 2017). A possible explanation for these findings might be that Actinobacteria increase in abundance under drought, whereas sensitive taxa diminish because this bacterial phylum is well-known to be highly tolerant for life under stressful abiotic conditions like drought (Bull and Asenjo, 2013; Kavamura et al., 2013). We further hypothesize that crops under water deficit selected competent microorganisms which provide the crops some degree of tolerance or assist in their development through growth promotion (Goh et al., 2013; Coleman-Derr and Tringe, 2014). In accordance with this assumption, we found the plant-associated bacterium *Mycobacteria* significantly associated with drought plants which has been reported to possess ACC deaminase activity that is responsible in enhancing plant growth (Cardinale et al., 2015). In addition, the bacterial genus *Blastococcus* has been often found in arid microbiome surveys as endophyte or soil inhabitant which has also been described as plant growth promoter (Hamedi and Mohammadipanah, 2015; Tahtamouni et al., 2016). Observed taxa are frequently described in plant microbiome surveys (Gdanetz et al., 2017; Naylor et al., 2017) but their specific roles in association with plants under water deficit remains unclear.

3.6 Functional profiles of bacterial communities are altered by crop species and plant compartment

We further hypothesized that bacterial functioning is altered in a different manner as bacterial community composition towards water deficit and cropping system. To clarify this hypothesis, functional profiles were predicted from obtained 16S rRNA gene data using Tax4Fun2 (Table S5). Functional profiles significantly differed between crop species (PERMANOVA, leaves, $R^2 = 5.5$ %, $p = 0.001$; rhizosphere, $R^2 = 8.7$ %, $p = 0.001$) and compartment (PERMANOVA, $R^2 = 31.9$ %, $p = 0.001$); however, cropping system and water deficit did not alter overall functioning.

To gain deeper insights into bacterial functioning, we focused on predicted abundances of genes involved in nitrogen cycling, i.e. nitrite reductase, plant growth promotion, i.e. amidase or in stress, i.e. catalase (Figure 8, 9). In general, genes involved in dissimilatory nitrate reduction such as nitrate reductase were more abundant in wheat leaves compared to faba bean. In addition, the gene acetolactate decarboxylase [EC: 4.1.1.5] putatively involved in plant growth promotion was more abundant in wheat rhizosphere compared to faba bean. We also found differences between cropping regimes. For example, for harvest 1 we observed in S4_FBM_D rhizosphere higher abundances of predicted genes involved in nitrification such as ammonia monooxygenase [EC: 1.14.99.39] compared to S4_FBM_C. For harvest 2, in average higher predicted abundances of genes involved in stress, plant growth promotion and nitrogen metabolism were found in S4_FBM_D compared to S4_FBM_C. In the leaf endosphere, we found higher predicted abundance of genes involved in dissimilatory nitrate reduction [EC: 1.7.1.15; EC: 1.7.5.1 1.7.99.-] in the cropping regime WIC_C compared to WM_C and WIC_D for harvest 1.

As we already confirmed for the bacterial community composition, crop species was an important factor in changing functional profiles in the leaf endosphere and rhizosphere soil. In line with this, Wemheuer et al. (2017) demonstrated that the functional profiles of the bacterial endophytic community differed significantly between three different grass species. Moreover, they also reported that response of endophyte community composition and diversity in comparison to functioning differed towards agricultural practices, indicating that function and phylogeny of different bacteria are not necessarily related to each other (Wemheuer et al., 2017). This assumption might further explain our observation that water deficit and cropping system changed community composition but not functioning. Moreover, Vandenkkoornhyuse et al. (2015) suggested that functional differences between crop species might be related to an accessory microbiome unique for each plant. An accessory microbiome contains more dispensable functions or microorganisms whose presence is related to interactions with the surrounding environmental conditions (Vandenkoornhuyse et al., 2015).

4 Conclusion

To date, the combined effect of cropping system and water deficit on active bacterial communities in leaf endosphere and rhizosphere soil of two important crop species have not been studied using large-scale metabarcoding. In line with our hypotheses, we demonstrated that crop species, genotype and plant compartment significantly influenced the active bacterial community in their composition, diversity and associated taxa. These plant related traits strongly shaped the response of bacteria towards water deficit and cropping system. In accordance with our third hypothesis, functional profiles were not affected by cropping system and water deficit but crop species and compartment altered functioning. Obtained results highlight that there are complex interactions between plants, associated microorganisms and their environment that might influence agricultural productivity.

5 Acknowledgment

This study is part of the project IMPAC[3] and was funded by the Federal Ministry of Education and Research (FKZ 031A351A). The authors thank Prof. Dr. Wolfgang Link from the Working Group "Breeding Research Faba Bean" (Division of Plant Breeding at the University of Göttingen) for providing the seed material.

Tables

Table 1. Sampling numbers for each container and harvest.

Treatments /Compartments	ID	Rhizosphere	Leaves	Plants/ Treatment
		Harvest 1		
Faba bean monoculture S_004	**S4_FBM**	1 (8, n=8)	1 (8, n=8)	8
Faba bean monoculture S_062	**S62_FBM**	1 (8, n=8)	1 (8, n=8)	8
Faba bean intercropping S_004	**S4_FBIC**	1 (8, n=8)	1 (8, n=8)	8
Faba bean intercropping S_062	**S62_FBIC**	1 (8, n=8)	1 (7, n=8)	8
Wheat monoculture	**WM**	2 (8, n=8)	2 (7, n=8)	16
Wheat intercropped	**WIC**	2 (15, n=16)	2 (16, n=16)	32
		Harvest 2		
Faba bean monoculture S_004	**S4_FBM**	1 (7, n=8)	1 (8, n=8)	8
Faba bean monoculture S_062	**S62_FBM**	1 (7, n=8)	1 (8, n=8)	8
Faba bean intercropping S_004	**S4_FBIC**	1 (8, n=8)	1 (8, n=8)	8
Faba bean intercropping S_062	**S62_FBIC**	1 (8, n=8)	1 (8, n=8)	8
Wheat monoculture	**WM**	2 (8, n=8)	2 (8, n=8)	16
Wheat intercropped	**WIC**	2 (15, n=16)	2 (16, n=16)	32
		Harvest 3		
Faba bean monoculture S_004	**S4_FBM**	1 (8, n=8)	1 (7, n=8)	8
Faba bean monoculture S_062	**S62_FBM**	1 (7, n=8)	1 (7, n=8)	8
Faba bean intercropping S_004	**S4_FBIC**	1 (8, n=8)	1 (7, n=8)	8
Faba bean intercropping S_062	**S62_FBIC**	1 (8, n=8)	1 (8, n=8)	8
Wheat monoculture	**WM**	2 (8, n=8)	2 (8, n=8)	16
Wheat intercropped	**WIC**	2 (15, n=16)	2 (14, n=16)	32
Total (for each harvest)		64	64	32(FB), 48(W)
Total (all)		192	192	240

WM, wheat in monoculture; FBM, faba bean in monoculture, FBIC, faba bean samples in intercropping; WIC, wheat samples in intercropping. Numbers before brackets refer to sampled plants per pot. Numbers in brackets refer to the number of samples left after removal of samples with too low sequencing numbers. Harvest 1 refers to "beginning of water deficit", harvest 2 refers to "during water deficit" and harvest 3 refers to "re-watering".

Table 2. Height [cm] and biomass [g] of faba bean and wheat plants.

	Height			Biomass		
Treatment	Harvest 1	Harvest 2	Harvest 3	Harvest 1	Harvest 2	Harvest 3
Wheat_C	37.8±0.53	37.4±0.93	37.2±0.78	1.2±0.18	1.2±0.09	1.4±0.12
Wheat_D	37.5±0.68	36.7±0.83	37.4±1.02	1.4±0.15	1.0±0.09	1.3±0.11
WM	37.2±0.83	36.4±0.90	36.5±1.06	1.0±0.04	**0.9±0.06a**	**1.1±0.10a**
WIC	37.9±0.49	37.4±0.81	37.7±0.78	1.5± 0.16	**1.2± 0.08b**	**1.5± 0.10b**
S4_C	38.2±2.30	45.8±2.92	46.3±3.27	9.2±1.60	12.8±2.02	15.4±2.88
S4_D	39.5±1.82	44.5±1.54	46.0±2.65	10.1±1.67	12.4±1.49	12.4±2.06
S4_FBM	41.3±1.53	47.0±2.27	**50.5±2.53a**	**12.1±1.55a**	13.9±1.52	17.2±2.71
S4_FBIC	36.4±2.15	43.3±2.23	**41.7±2.40b**	**7.2±1.13b**	11.3±1.87	10.6±1.64
S62_C	**37.3±1.79A**	**44.4±2.41AB**	**47.4±2.62B**	9.3±1.03	**10.8±0.63a**	12.0±1.52
S62_D	36.6±1.68	40.1±2.20	40.6±2.39	8.3±1.19	**8.7±0.74b**	9.8±0.56
S62_FBM	38.7±1.19	43.7±1.70	46.0±1.65	8.3±1.09	9.4±0.88	10.6±0.47
S62_FBIC	35.2±1.95	40.8±2.91	42.0±3.48	9.3±1.14	10.0±0.67	12.1±1.70

Different letters in columns indicate statistically significant differences between treatments (Dunn'sTest or Kruskal-Wallis-test; $p \leq 0.05$, means ± SE). Abbreviations: FBM/WM, faba bean/wheat grown in monoculture; FBIC/WIC, faba bean/wheat intercropped; C, control treatment; D, water deficit treatment. Harvest 1, beginning of water deficit; Harvest 2, during water deficit; Harvest 3, re-watering.

Table 3. Chlorophyll content index measured with a SPAD meter.

Treatment	Harvest 1	Harvest 2	Harvest 3
Wheat_C	37.5±1.48	36.6±1.00	34.4±1.29
Wheat_D	35.4±1.59	37.1±1.03	37.7±1.37
WIC_C	**38.9±1.78a**	38.0±1.15	35.8±1.42
WIC_D	**37.7±1.12ab**	38.6±1.02	39.6±1.45
WM_C	**34.8±2.38ab**	33.9±1.02	31.6±2.20
WM_D	**29.2±2.87b**	34.1±1.51	33.9±1.92
S4_C	37.1±1.33	39.7±1.54	40.4±1.40
S4_D	**37.6±0.99B**	**42.6±1.57AB**	**44.9±1.43A**
S4_FBIC_C	37.6±0.89	40.7±2.61	41.4±2.26
S4_FBIC_D	38.2±1.24	41.8±2.13	45.0±2.44
S4_FBM_C	36.8±2.71	38.9±1.94	39.6±1.87
S4_FBM_D	37.0±1.65	43.6±2.54	44.9±1.89
S62_C	**32.0±0.60A**	**36.3±0.96aB**	**38.2±1.31aB**
S62_D	**33.4±0.75A**	**42.1±0.75bB**	**43.3±1.31bB**
S62_FBIC_C	31.8±1.05	**35.4±1.30a**	**37.4±2.55b**
S62_FBIC_D	33.5±1.14	**43.2±0.33b**	**46.4±0.64a**
S62_FBM_C	32.3±0.73	**37.2±1.45ab**	**39.2±1.02b**
S62_FBM_D	33.5±1.15	**41.0±1.31ab**	**40.3±1.22ab**

Different small and large letters in columns and rows indicate statistically significant differences between treatments (Dunn'sTest or Kruskal-Wallis-test; $p \leq 0.05$, means ± SE). Abbreviations: FBM/WM, faba bean/wheat grown in monoculture; FBIC/WIC, faba bean/wheat intercropped; C, control treatment; D, water deficit treatment. Harvest 1, beginning of water deficit; Harvest 2, during water deficit; Harvest 3, re-watering.

Table 4. Soluble sugar concentrations [in %] of glucose, fructose and sucrose in crop leaves.

	Glucose			Fructose			Sucrose		
Treatment	**Harvest 1**	**Harvest 2**	**Harvest 3**	**Harvest 1**	**Harvest 2**	**Harvest 3**	**Harvest 1**	**Harvest 2**	**Harvest 3**
Wheat_C	8.5±1.22AB	**6.4±0.55aA**	**9.5±0.60aB**	**10.5±1.19A**	**7.1±0.80B**	**16.4±0.95C**	35.5±5.90	51.6±7.40	43.2±9.62
Wheat_D	9.8±0.99	**8.9±0.93b**	**7.9±0.64b**	**10.5±0.51AB**	**9.1±0.85A**	**14.6±1.33B**	39.7±6.26	44.0±5.52	33.5±6.24
WIC_C	**7.1±1.54b**	**6.4±0.80ab**	8.7±0.59	**8.5±1.22b**	7.1±1.17	17.9±0.90	31.9±7.63	49.1±9.23	38.2±11.68
WIC_D	**11.7±0.72a**	**8.7±0.98b**	7.8±0.54	**9.9±0.36b**	9.0±0.77	16.8±1.19	31.5±3.91	41.8±5.94	23.4±4.14
WM_C	**11.3±1.21ab**	**6.7±0.53a**	11.3±0.85	**14.5±0.90a**	7.2±0.72	13.6±1.42	42.8±7.63	56.5±13.87	53.0±18.23
WM_D	**6.2±1.25b**	**9.3±2.21ab**	8.2±1.75	**11.8±1.18ab**	9.5±2.27	10.4±1.94	56.1±15.02	48.6±12.64	53.8±11.93
S4_C	**6.7±1.04aA**	**3.4±0.43B**	**4.9±0.37AB**	**4.0±0.95AB**	**2.2±0.23A**	**4.3±0.52B**	47.8±8.42	45.5±5.12	58.6±5.6
S4_D	**3.7±0.46b**	3.6±0.39	5.0±0.56	**3.1±0.28AB**	**2.2±0.20A**	**5.6±1.02B**	40.0±6.08	41.8±4.65	58.0±4.98
S4_FBIC_C	**8.8±0.95a**	2.8±0.35	4.9±0.55	4.2±1.85	1.8±0.12	**5.4±0.71ab**	37.5±10.33	48.6±8.55	56.8±8.87
S4_FBIC_D	**4.0± 0.74ab**	3.7±0.59	4.9±1.14	3.4±0.19	2.5±0.34	**7.6±1.52a**	41.7±8.28	37.9±7.55	54.4±8.93
S4_FBM_C	**4.7±1.13ab**	4.2±0.65	5.1±0.55	3.9±0.83	2.8±0.25	**3.3±0.21b**	58.0±12.44	42.3±6.53	60.5±8.29
S4_FBM_D	**3.4±0.62b**	3.5±0.58	5.1±0.41	3.0±0.55	2.1±0.21	**3.8±0.33ab**	38.4±10.10	45.7±5.82	61.5±5.23
S62_C	4.7±0.38	**5.5±1.05a**	5.1±0.36	4.4±0.33	**5.5±1.11a**	7.3±0.64	45.9±4.87	**39.9±5.45a**	39.0±4.95
S62_D	**4.9±0.52A**	**3.2±0.18bB**	**5.4±0.24A**	**4.4±0.45A**	**3.1±0.28bA**	**8.7±0.64B**	**38.2±5.64AB**	**25.7±2.23bA**	**47.1±3.73B**
S62_FBIC_C	5.3±0.59	**7.3±1.70a**	5.1±0.30	5.0±0.44	**7.5±1.72a**	8.0±0.41	50.0±6.84	46.0±9.59	38.9±10.36
S62_FBIC_D	5.2±0.91	**3.3±0.20ab**	5.1±0.19	4.2±0.66	**3.4±0.48ab**	8.0±0.95	42.3±10.65	24.2±2.03	55.1±4.07
S62_FBM_C	4.1±0.30	**3.8±0.37ab**	5.2±0.71	3.9±0.33	**3.5±0.38ab**	6.8±1.23	41.8±7.26	33.9±4.72	39.0±2.68
S62_FBM_D	4.7±0.63	**3.2±0.34b**	5.7±0.43	4.6±0.70	**2.9±0.32b**	9.6±0.77	34.0±4.83	27.2±4.19	39.1±2.41

Different small and large letters in columns and rows indicate statistically significant differences between treatments (Dunn's Test or Kruskal-Wallis-test; $p \leq 0.05$, means ± SE). Abbreviations: FBM/WM, faba bean/wheat grown in monoculture; FBIC/WIC, faba bean/wheat intercropped; C, control treatment; D, water deficit treatment. Harvest 1, beginning of water deficit; Harvest 2, during water deficit; Harvest 3, re-watering.

Table 5. pH value in the rhizosphere of wheat and faba bean genotypes.

Treatment	Harvest 1	Harvest 2	Harvest 3
Wheat_C	**7.10±0.06A**	**7.05±0.06A**	**7.39±0.01B**
Wheat_D	**7.11±0.06A**	**7.02±0.05B**	**7.43±0.02C**
WIC_C	7.05±0.08	**6.93±0.04a**	7.39±0.01
WIC_D	7.12±0.08	**6.91±0.02a**	7.42±0.03
WM_C	7.20±0.09	**7.30±0.06b**	7.40±0.02
WM_D	7.08±0.10	**7.23±0.02b**	7.44±0.04
S4_C	**7.12±0.09AB**	**7.11±0.04A**	**7.37±0.01B**
S4_D	**7.10±0.09A**	**7.00±0.05A**	**7.36±0.01B**
S4_FBIC_C	7.25±0.13	**7.03±0.06ab**	7.39±0.01
S4_FBIC_D	7.24±0.10	**6.94±0.03a**	7.37±0.01
S4_FBM_C	7.03±0.12	**7.20±0.01b**	7.35±0.00
S4_FBM_D	6.97±0.13	**7.07±0.08ab**	7.35±0.02
S62_C	**7.18±0.08AB**	**6.91±0.04A**	**7.36±0.01B**
S62_D	**7.02±0.07A**	**6.95±0.04A**	**7.36±0.01B**
S62_FBIC_C	7.03±0.12	**6.82±0.03a**	**7.38±0.00a**
S62_FBIC_D	7.02±0.10	**6.84±0.04ab**	**7.36±0.00ab**
S62_FBM_C	7.34±0.01	**7.01±0.01ab**	**7.34±0.01b**
S62_FBM_D	7.01±0.11	**7.03±0.01b**	**7.36±0.01ab**

Different small and large letters in columns and rows indicate statistically significant differences between treatments (Dunn'sTest or Kruskal-Wallis-test; $p \leq 0.05$, means ± SE). Abbreviations: FBM/WM, faba bean/wheat grown in monoculture; FBIC/WIC, faba bean/wheat intercropped; C, control treatment; D, water deficit treatment. Harvest 1, beginning of water deficit; Harvest 2, during water deficit; Harvest 3, re-watering.

Table 6. Total organic carbon and nitrogen [%] in the rhizosphere of wheat and faba bean genotypes.

	C:N ratio			C_{total} [%]			N_{total} [%]		
Treatment	Harvest 1	Harvest 2	Harvest 3	Harvest 1	Harvest 2	Harvest 3	Harvest 1	Harvest 2	Harvest 3
Wheat_C	11.12±0.21	10.99±0.21	10.78±0.34	**2.02±0.04A**	**2.03±0.04A**	**1.56±0.14B**	**0.18±0.00A**	**0.19±0.01A**	**0.14±0.01B**
Wheat_D	11.46±0.15	10.97±0.25	11.28±0.14	**2.04±0.04A**	**2.02±0.03A**	**1.74±0.08B**	**0.18±0.00A**	**0.19±0.00A**	**0.15±0.01B**
WIC_C	11.02±0.31	**10.59±0.14a**	10.67±0.51	2.03±0.03	2.05±0.02	**1.34±0.16a**	0.19±0.00	0.19±0.00	**0.12±0.01a**
WIC_D	11.32±0.17	**10.96±0.38a**	11.16±0.20	2.01±0.04	2.07±0.03	**1.60±0.07a**	0.18±0.01	0.19±0.01	**0.14±0.01a**
WM_C	11.34±0.21	**11.81±0.19b**	11.01±0.20	2.01±0.10	1.98±0.12	**2.02±0.02b**	0.18±0.01	0.17±0.01	**0.18±0.00b**
WM_D	11.75±0.25	**11.00±0.08ab**	11.53±0.10	2.12±0.05	1.92±0.02	**2.04±0.03b**	0.18±0.00	0.18±0.00	**0.18±0.00b**
S4_C	11.15±0.35	11.28±0.48	11.50±0.27	2.03±0.04	2.06±0.13	1.87±0.08	0.18±0.01	0.19±0.01	0.16±0.00
S4_D	11.43±0.15	11.26±0.33	11.47±0.31	2.10±0.03	2.07±0.16	1.85±0.09	**0.18±0.00A**	**0.19±0.02A**	**0.16±0.01B**
S4_FBIC_C	**10.32±0.23a**	**10.21±0.20a**	11.32±0.54	2.08±0.05	2.09±0.01	**1.79±0.12ab**	**0.20±0.00a**	0.21±0.01	0.16±0.01
S4_FBIC_D	**11.10±0.15ab**	**10.46±0.16ab**	10.88±0.05	2.08±0.02	2.05±0.02	**1.63±0.17a**	**0.19±0.00ab**	0.20±0.00	0.15±0.01
S4_FBM_C	**11.98±0.24b**	**12.35±0.51b**	11.67±0.13	1.98±0.07	2.03±0.28	**1.96±0.09ab**	**0.17±0.01b**	0.17±0.02	0.17±0.01
S4_FBM_D	**11.76±0.12b**	**12.06±0.23b**	12.05±0.46	2.13±0.05	2.09±0.35	**2.06±0.03b**	**0.18±0.00ab**	0.17±0.03	0.17±0.01
S62_C	11.47±0.10	11.63±0.26	11.34±0.33	1.96±0.04	1.88±0.10	1.79±0.09	0.17±0.00	0.16±0.01	0.16±0.01
S62_D	**12.11±0.37A**	**10.88±0.35B**	**11.41±0.19AB**	1.86±0.09	1.70±0.19	1.64±0.11	0.16±0.01	0.15±0.02	0.14±0.01
S62_FBIC_C	11.35±0.17	**11.06±0.29a**	11.16±0.57	2.01±0.03	**2.12±0.04a**	1.69±0.08	0.18±0.00	**0.19±0.00a**	0.15±0.01
S62_FBIC_D	11.53±0.19	**11.12±0.14ab**	11.41±0.39	1.93±0.07	**2.09±0.04a**	1.42±0.15	0.17±0.01	**0.19±0.01a**	0.13±0.01
S62_FBM_C	11.59±0.11	**12.19±0.12b**	11.53±0.41	1.90±0.08	**1.65±0.07ab**	1.90±0.17	0.16±0.01	**0.14±0.01b**	0.17±0.02
S62_FBM_D	12.68±0.62	**10.65±0.72ab**	11.41±0.14	1.79±0.16	**1.32±0.26b**	1.85±0.05	0.15±0.02	**0.12±0.02b**	0.16±0.00

Different small and large letters in columns and rows indicate statistically significant differences between treatments (Dunn's-test or Kruskal-Wallis-test; $p \leq 0.05$, means ± SE). Abbreviations: FBM/WM, faba bean/wheat grown in monoculture; FBIC/WIC, faba bean/wheat intercropped; C, control treatment; D, water deficit treatment. Harvest 1, beginning of water deficit; Harvest 2, during water deficit; Harvest 3, re-watering.

Table 7. Bacterial diversity and richness in the rhizosphere soil with regard to water treatment and cropping system.

	Richness			Diversity		
Treatment	1	2	3	1	2	3
Wheat_C	845.25±74.03	866.07±55.6	794.57±223.71	6.19±0.22	6.24±0.12	5.82±1.12
Wheat_D	841.90±48.62	818.37±105.68	691.02±341.75	6.18±0.13	6.03±0.59	5.35±1.8
WIC_C	826.56±87.46	859.44±68.18	845.96±67.03	6.14±0.26	6.22±0.14	6.10±0.33
WIC_D	848.04±56.29	802.91±115.8	609.23±433.16	6.20±0.14	5.94±0.71	4.87±2.26
WM_C	877.98±26.99	877.68±26.91	657.53±442.47	6.29±0.05	6.27±0.06	5.08±2.19
WM_D	831.15±36.04	849.28±88.03	813.70±64.94	6.14±0.12	6.21±0.22	6.09±0.24
S4_C	744.71±295.43	516.54±382.25	805.37±95.79	5.76±1.17	4.49±2.14	5.93±0.62
S4_D	795.20±40.3	792.59±113.95	751.91±145.78	6.04±0.14	5.85±0.86	5.77±0.78
S4_FBIC_C	806.70±118.42	470.05±384.88	753.83±108.6	6.09±0.21	4.50±1.9	5.58±0.88
S4_FBIC_D	821.95±31.61	833.45±70.86	630.77±129.17	6.14±0.06	6.17±0.2	5.19±0.9
S4_FBM_C	698.23±398.1	578.53±453.94	844.03±76.41	5.51±1.59	4.47±2.89	6.19±0.17
S4_FBM_D	759.53±5.82	738.10±153.72	842.78±75.47	5.91±0.1	5.43±1.29	6.21±0.26
S62_C	780.94±87.63	867.72±47.53	740.08±289.75	5.99±0.34	6.21±0.15	5.51±1.63
S62_D	835.76±39.45	614.13±368.85	820.79±90.57	6.15±0.18	4.94±1.94	6.15±0.26
S62_FBIC_C	771.65±117.25	859.73±64.46	816.08±136.46	6.02±0.33	6.16±0.21	5.91±0.79
S62_FBIC_D	846.80±32.91	562.87±420.59	834.40±75.33	6.18±0.15	4.71±1.86	6.17±0.24
S62_FBM_C	793.33±44.83	875.70±36.08	664.08±402.33	5.94±0.43	6.26±0.07	5.12±2.27
S62_FBM_D	824.73±47.16	652.58±386.75	802.63±123.42	6.11±0.23	5.11±2.26	6.12±0.34

Diversity is expressed as Shannon values (H') and richness is based on the number of unique sequences Different small and large letters in columns and rows indicate statistically significant differences between treatments (Dunn's Test or Kruskal-Wallis-test, $p \leq 0.05$, means ± SD). Abbreviations: FBM/WM, faba bean/wheat grown in monoculture; FBIC/WIC, faba bean/wheat intercropped; C, control treatment; D, water deficit treatment. Harvest 1, beginning of water deficit; Harvest 2, during water deficit; Harvest 3, re-watering.

Table 8. Bacterial diversity and richness in the leaf endosphere with regard to water treatment and cropping system.

	Richness			Diversity		
Treatment	**1**	**2**	**3**	**1**	**2**	**3**
Wheat_C	125.65± 168.19	70.35± 26.13	107.66± 78.49	2.62± 1.01	2.07± 0.53	2.60± 0.82
Wheat_D	68.05± 17.04	100.05± 172.80	122.65± 190.28	2.22± 0.55	2.05± 1.21	2.46± 1.28
WIC_C	143.13± 207.28	75.25± 30.63	118.01± 91.69	2.85± 1.16	1.97± 0.50	2.67±0.81
WIC_D	64.99± 18.91	49.94± 18.99	164.06± 233.55	2.05± 0.52	1.62± 0.35	2.89± 1.41
WM_C	90.70± 32.02	60.55± 11.10	86.95± 45.97	2.15± 0.40	2.27± 0.60	2.47± 0.96
WM_D	76.20± 7.97	200.28± 297.58	50.18± 24.54	2.68± 0.38	2.91± 1.89	1.69± 0.51
S4_C	139.64± 89.81	103.45± 53.81	118.65± 103.71	2.44± 0.60	2.20± 0.45	2.44± 0.52
S4_D	105.95± 53.26	102.79± 39.44	72.45± 29.70	2.22± 0.45	2.21± 0.33	1.88± 0.48
S4_FBIC_C	93.90± 50.16	94.93± 59.90	142.37± 151.43	2.17± 0.21	2.07± 0.59	2.49± 0.82
S4_FBIC_D	85.05± 33.80	93.90± 58.12	79.68± 19.92	2.09± 0.39	2.03± 0.29	2.18± 0.01
S4_FBM_C	185.38± 103.56	111.98± 54.53	94.93± 47.66	2.71± 0.77	2.33± 0.28	2.40± 0.39
S4_FBM_D	126.85± 65.66	111.68± 6.38	65.23± 39.01	2.36± 0.53	2.40± 0.28	1.58± 0.60
S62_C	107.70± 48.28	**117.55± 54.97b**	108.86± 71.58	2.36± 0.41	**2.26± 0.42a**	2.33± 0.56
S62_D	146.20± 91.11	**54.93± 12.60 a**	81.06± 41.94	**2.55± 0.54A**	**1.84± 0.18Bb**	**2.27± 0.45AB**
S62_FBIC_C	129.15± 42.91ab	107.68± 77.06	145.90± 86.89	2.67± 0.30ab	2.23± 0.59	2.68± 0.60
S62_FBIC_D	**224.83± 42.23b**	52.85± 12.37	95.28± 51.44	**3.01± 0.30a**	1.89± 0.06	2.20± 0.46
S62_FBM_C	86.25± 48.70 ab	127.43± 29.20	71.83± 27.33	**2.05± 0.24b**	2.30± 0.24	1.99± 0.23
S62_FBM_D	**87.23± 67.77a**	57.00± 14.36	62.10± 19.11	2.21± 0.41ab	1.80± 0.26	2.36± 0.51

Diversity is expressed as Shannon values (H') and richness is based on the number of unique sequences. Different small and large letters in columns and rows indicate statistically significant differences between treatments (Dunn's Test or Kruskal-Wallis-test, $p \leq 0.05$, means ± SD). Abbreviations: FBM/WM, faba bean/wheat grown in monoculture; FBIC/WIC, faba bean/wheat intercropped; C, control treatment/sufficiently irrigated; D, water deficit treatment. Harvest 1, beginning of water deficit; Harvest 2, during water deficit; Harvest 3, re-watering.

Table 9. Effect of the tested parameter on the bacterial community composition for each harvest.

Treatment	Rhizosphere Soil						Leaves					
	Harvest 1		Harvest 2		Harvest 3		Harvest 1		Harvest 2		Harvest 3	
	R^2 (%)	p	R^2 (%)	p	R^2 (%)	p	R^2 (%)	p	R^2 (%)	p	R^2 (%)	p
Cropping system	2.1	0.23	1.1	0.922	1.2	0.918	1.6	0.446	1.7	0.366	1.4	0.652
Crop species	**4.2**	**0.008**	**3.4**	**0.047**	2.0	0.347	**7.1**	**0.002**	**4.5**	**0.048**	**4.5**	**0.019**
Genotype	**6.1**	**0.016**	2.7	0.595	2.6	0.712	1.4	0.891	1.8	0.642	3.5	0.221
Water-deficit	2.5	0.12	2.2	0.206	1.6	0.577	1.3	0.637	0.8	0.832	1.7	0.553
Harvest	**1.6**	**0.003**					0.4	0.613				

Results of the permutational multivariate analysis of variance (PERMANOVA) with Bray-Curtis distances testing for the different treatments. Statistically significant differences ($p \leq 0.05$) between the treatments for each plant compartment are written in bold. Cropping system compares monoculture versus intercropping. Genotype compares S_004 versus S_062. Harvest was tested for all harvests together without strata.

Figures

Figure 1. Experimental design. Abbreviations: FBM/WM, faba bean/wheat monoculture; IC, intercropping; C, control (blue container); D, water deficit treatment (red container).

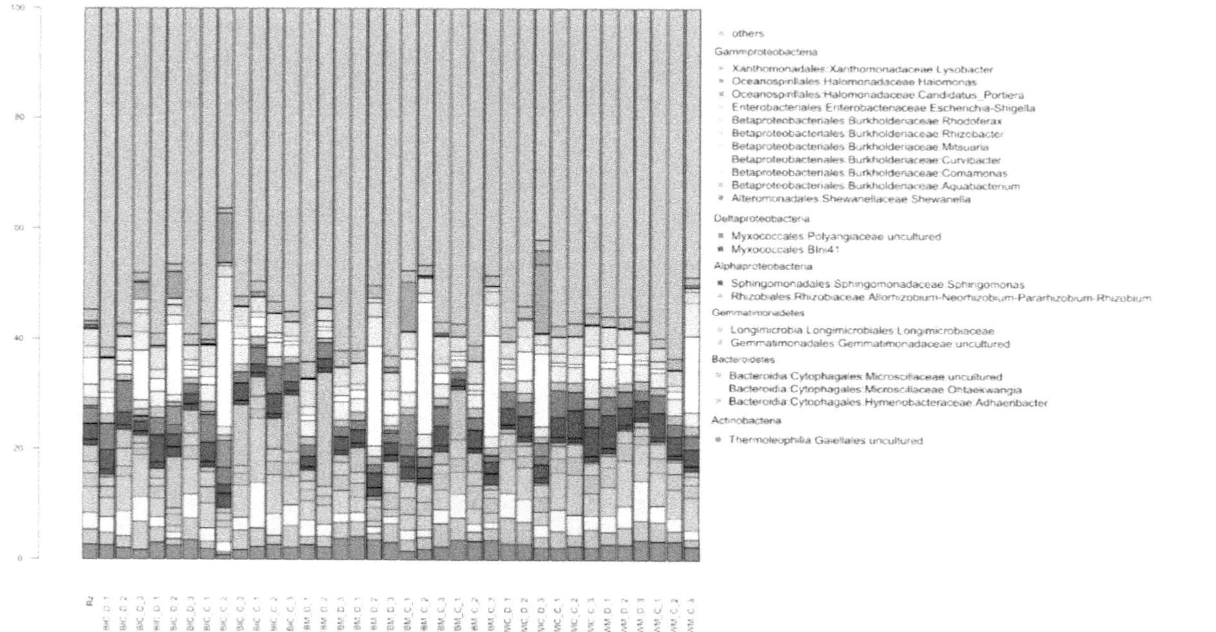

Figure 2. Abundant bacterial genera in the rhizosphere soil and the investigated cropping systems with regard to water treatment and harvest. Only genera with an abundance >1% in at least one of the investigated cropping systems are shown. Mean relative abundances of each taxon were calculated based on relative abundances calculated for each sample. Abbreviations: C, control treatment; D, water deficit treatment; S4/S62, faba bean genotype; FBM/WM, faba bean/ wheat monoculture, FBIC/WIC, faba bean/wheat intercropped.

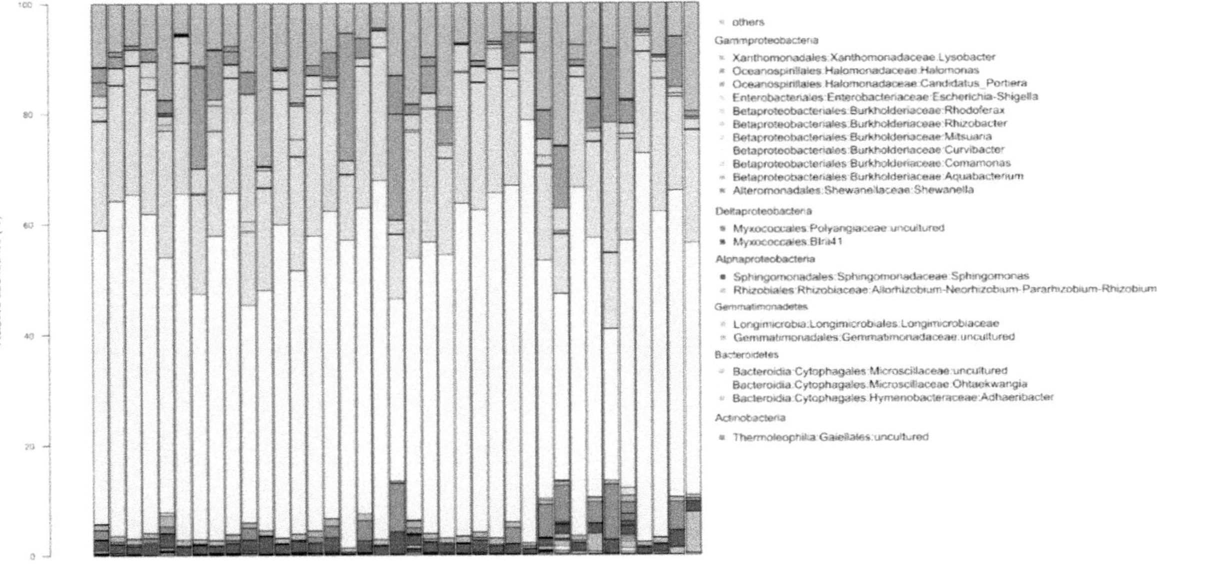

Figure 3. Abundant bacterial genera in the leaf endosphere and the investigated cropping systems with regard to water treatment and harvest. Only genera with an abundance >1% in at least one of the investigated cropping systems are shown. Mean relative abundances of each taxon were calculated based on relative abundances calculated for each sample. Abbreviations: C, control treatment; D, water deficit treatment; S4/S62, faba bean genotype; FBM/WM, faba bean/ wheat monoculture, FBIC/WIC, faba bean/wheat intercropped.

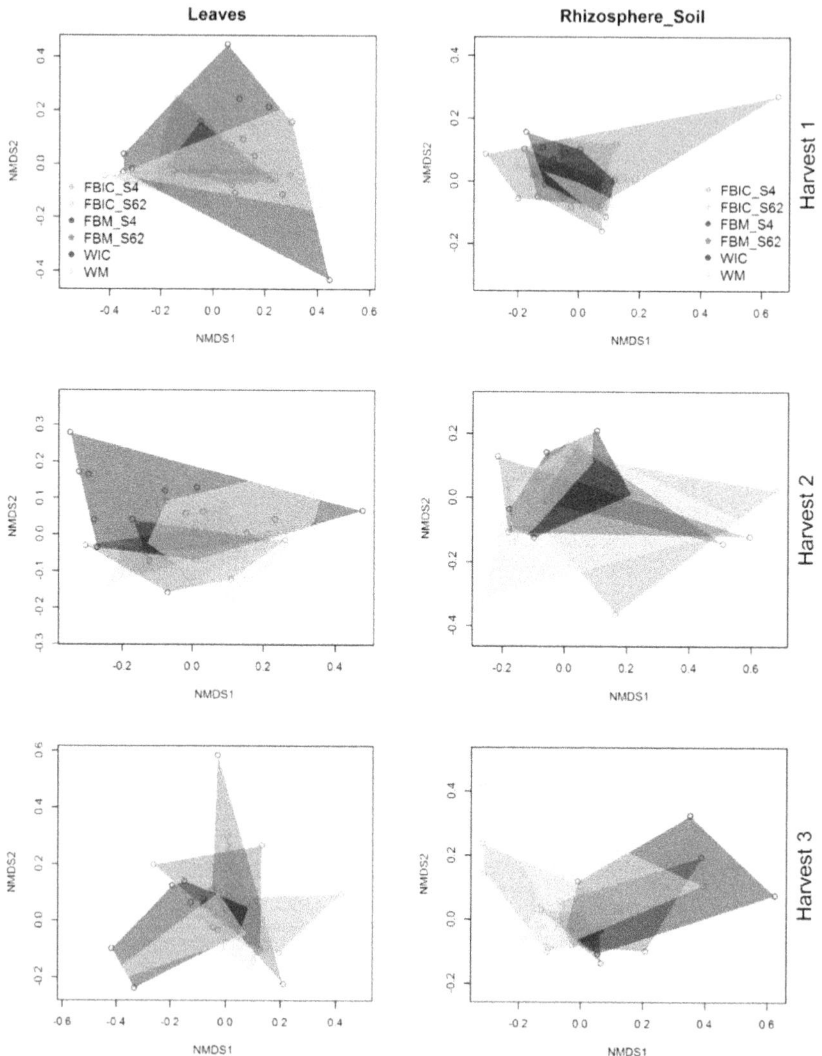

Figure 4. Response of bacterial communities in the leaf endosphere and rhizosphere soil towards cropping system. Ordination is based on Bray-Curtis dissimilarities between samples. NMDS ordination of microbial community is color-coded by the respective cropping system and genotype. Abbreviations: FBM/WM, faba bean/wheat monoculture; FBIC/WIC, faba bean/wheat intercropping system. S4/S62, faba bean genotype.

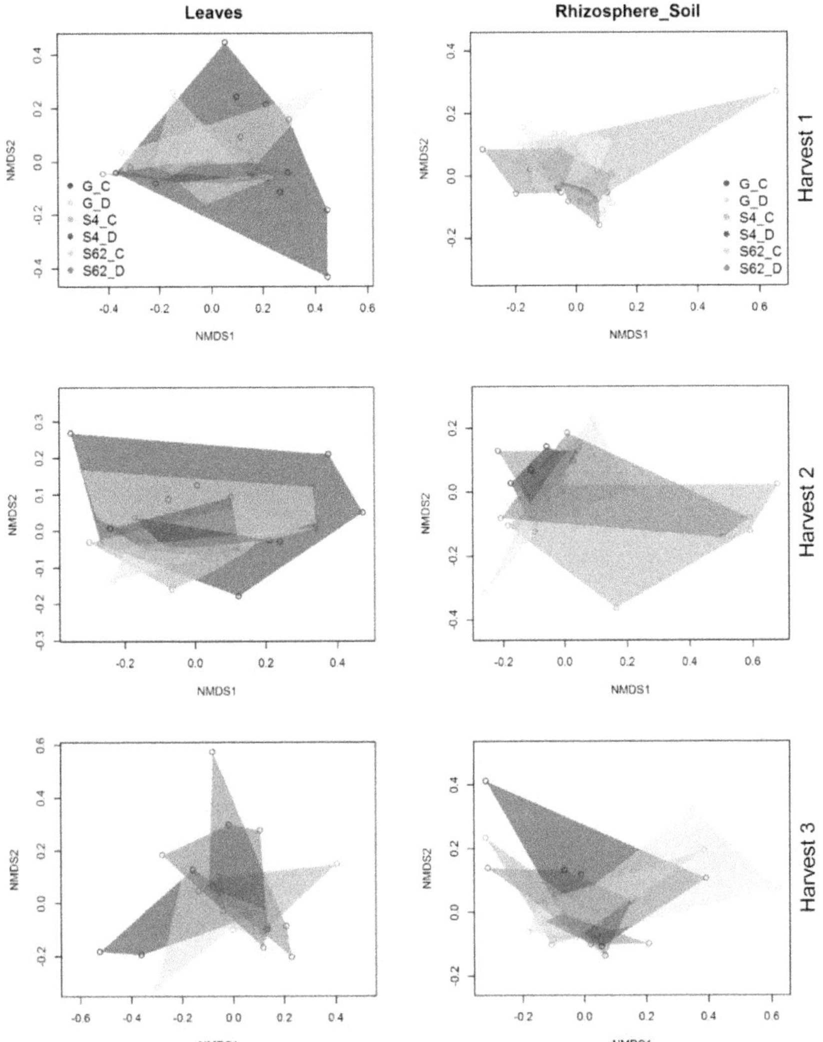

Figure 5. Response of bacterial communities in the leaf endosphere and rhizosphere soil towards water treatment regarding the different crop genotypes. Ordination is based on Bray-Curtis dissimilarities between samples. NMDS ordination of microbial community is color-coded by the respective water treatment and genotype. Abbreviations: S4/S62, faba bean genotype; G, Genius (wheat); C, control/well-watered conditions; D, water deficit.

Figure 6. Bipartite association network for bacterial taxa within different cropping regimes for the three harvests. Significant associated taxa are shown. Abbreviations: FBM/WM, faba bean/wheat monoculture; FBIC/WIC, faba bean/wheat intercropped; C, control/well-watered conditions; D, water deficit.

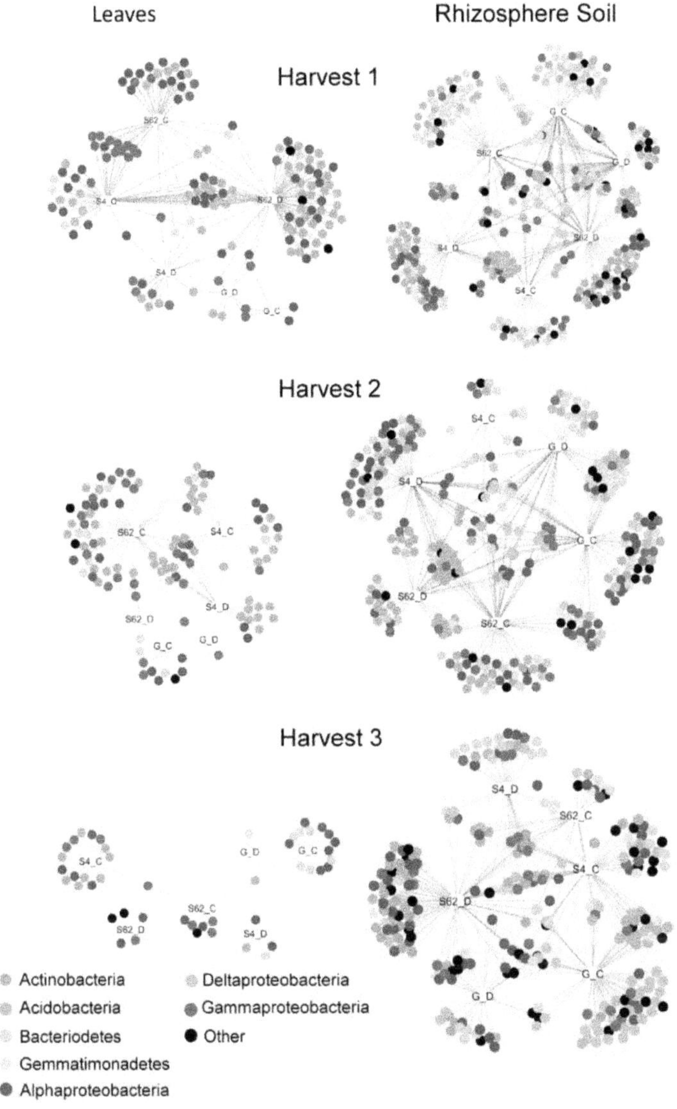

Figure 7. Bipartite association network for bacterial taxa within the different genotypes and water treatments. Significant associated taxa are shown. Abbreviations: G, Genius (wheat); S4/S62, faba bean genotypes; C, control/well-watered conditions; D, water deficit.

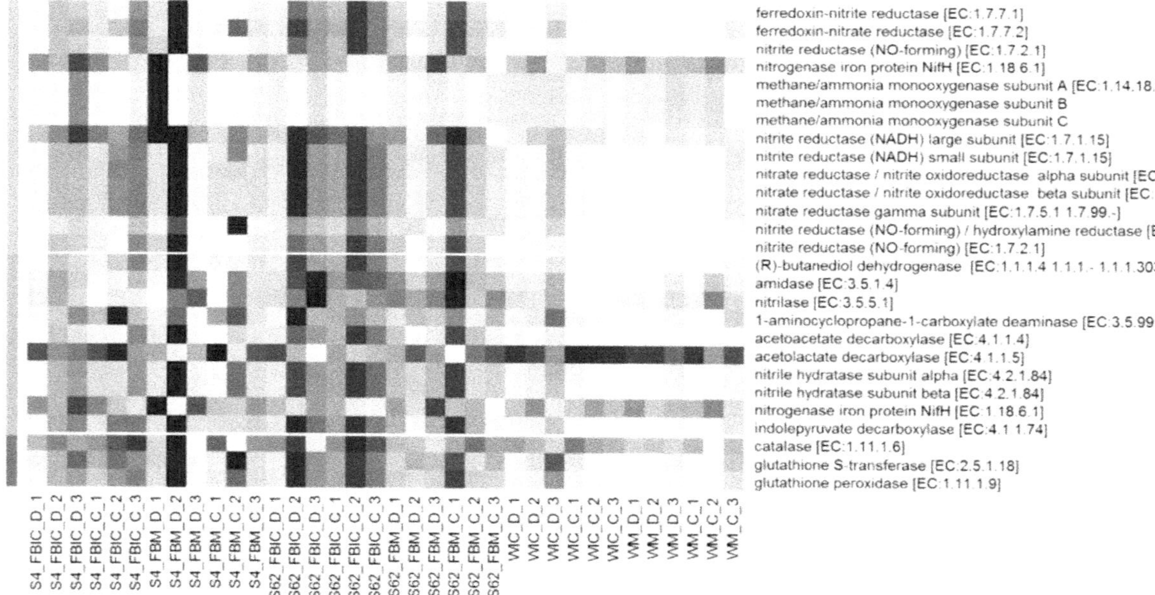

Figure 8. Predicted abundances of enzyme-encoding genes involved in nitrogen cycling (orange), plant growth promotion (green) and stress (blue) in the rhizosphere. Colour code of the heatmap refers to gene abundance, with high predicted abundances (dark blue) and low predicted abundances (white). Abbreviations: 1-3, sampling time; S4/S62, faba bean genotype; D/C, water deficit/control treatment; FBIC/WIC, faba bean/wheat intercropped; FBM/WM, faba bean/wheat monoculture.

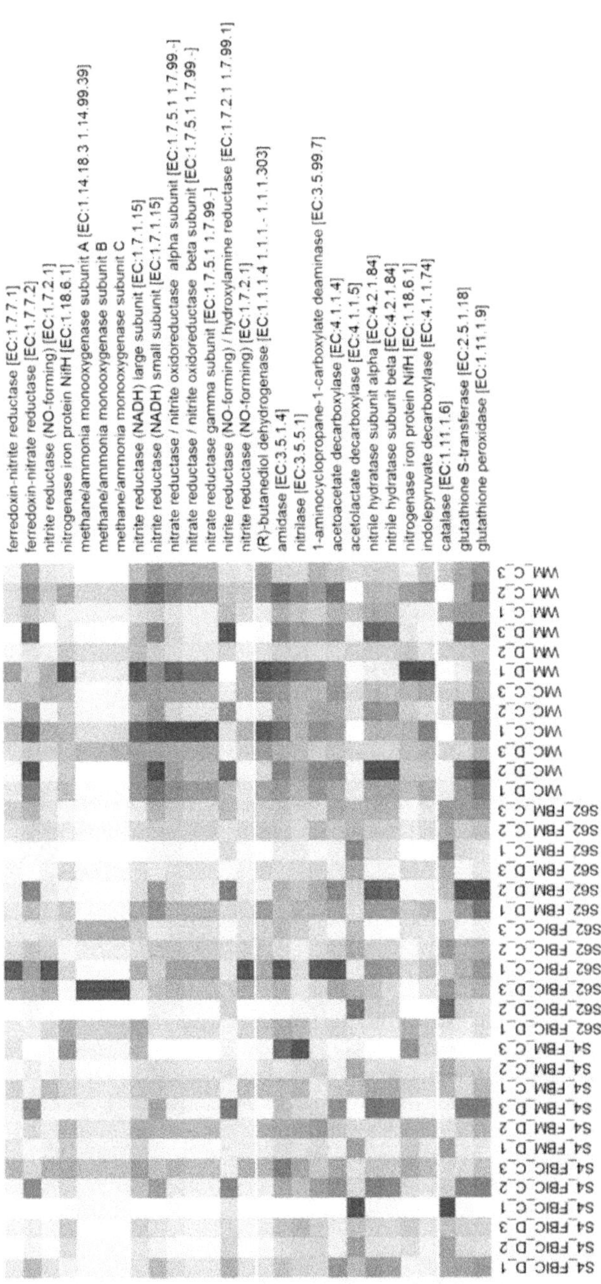

Figure 9. Predicted abundances of enzyme-encoding genes involved in nitrogen cylcing (orange), plant growth promotion (green) and stress (blue) in the leaf endosphere. Colour code of the heatmap refers to gene abundance, with high predicted abundances (dark green) and low predicted abundances (white). Abbreviations: 1-3, sampling time; S4, S62, faba bean genotype; D/C, water deficit/control treatment; FBIC/WIC, faba bean/wheat intercropped; FBM/WM, faba bean/wheat monoculture.

6 References

Abid G, Hessini K, Aouida M, Aroua I, Baudoin JP, Muhovski Y, Mergeai G, Sassi K, Machraoui M, Souissi F, et al. 2017. Agro-physiological and biochemical responses of faba bean (Vicia faba L. var. 'minor') genotypes to water deficit stress. Biotechnologie Agronomie Societe Et Environnement.21:146-159.

Anderson MJ. 2001. A new method for non-parametric multivariate analysis of variance. Austral Ecology. Feb;26:32-46.

Anderson MJ. 2006. Distance-based tests for homogeneity of multivariate dispersions. Biometrics. Mar;62:245-253.

Andreote FD, da Rocha UN, Araujo WL, Azevedo JL, van Overbeek LS. 2010. Effect of bacterial inoculation, plant genotype and developmental stage on root-associated and endophytic bacterial communities in potato (Solanum tuberosum). Antonie Van Leeuwenhoek International Journal of General and Molecular Microbiology. May;97:389-399.

Asshauer KP, Wemheuer B, Daniel R, Meinicke P. 2015. Tax4Fun: predicting functional profiles from metagenomic 16S rRNA data. Bioinformatics. Sep;31:2882-2884.

Barnard RL, Osborne CA, Firestone MK. 2013. Responses of soil bacterial and fungal communities to extreme desiccation and rewetting. Isme Journal. Nov;7:2229-2241.

Berg G. 2009. Plant-microbe interactions promoting plant growth and health: perspectives for controlled use of microorganisms in agriculture. Applied Microbiology and Biotechnology. Aug;84:11-18.

Blagodatskaya E, Kuzyakov Y. 2013. Active microorganisms in soil: Critical review of estimation criteria and approaches. Soil Biology & Biochemistry. Dec;67:192-211.

Bodenhausen N, Horton MW, Bergelson J. 2013. Bacterial Communities Associated with the Leaves and the Roots of Arabidopsis thaliana. Plos One. Feb;8:9.

Bray EA. 1997. Plant responses to water deficit. Trends in Plant Science. Feb;2:48-54.

Bulgarelli D, Rott M, Schlaeppi K, van Themaat EVL, Ahmadinejad N, Assenza F, Rauf P, Huettel B, Reinhardt R, Schmelzer E, et al. 2012. Revealing structure and assembly cues for Arabidopsis root-inhabiting bacterial microbiota. Nature. Aug;488:91-95.

Bull AT, Asenjo JA. 2013. Microbiology of hyper-arid environments: recent insights from the Atacama Desert, Chile. Antonie Van Leeuwenhoek International Journal of General and Molecular Microbiology. Jun;103:1173-1179.

Cardinale M, Ratering S, Suarez C, Montoya AMZ, Geissler-Plaum R, Schnell S. 2015. Paradox of plant growth promotion potential of rhizobacteria and their actual promotion effect on growth of barley (Hordeum vulgare L.) under salt stress. Microbiological Research.181:22-32.

Christensen JH, Christensen OB. 2007. A summary of the PRUDENCE model projections of changes in European climate by the end of this century. Climatic Change. May;81:7-30.

Coleman-Derr D, Tringe SG. 2014. Building the crops of tomorrow: advantages of symbiont-based approaches to improving abiotic stress tolerance. Frontiers in Microbiology. Jun;5.

De Caceres M, Legendre P. 2009. Associations between species and groups of sites: indices and statistical inference. Ecology. Dec;90:3566-3574.

de Zelicourt A, Al-Yousif M, Hirt H. 2013. Rhizosphere Microbes as Essential Partners for Plant Stress Tolerance. Molecular Plant. Mar;6:242-245.

Edgar RC. 2010. Search and clustering orders of magnitude faster than BLAST. Bioinformatics. Oct;26:2460-2461.

Edgar RC, Haas BJ, Clemente JC, Quince C, Knight R. 2011. UCHIME improves sensitivity and speed of chimera detection. Bioinformatics. Aug;27:2194-2200.

Fahad S, Bajwa AA, Nazir U, Anjum SA, Farooq A, Zohaib A, Sadia S, Nasim W, Adkins S, Saud S, et al. 2017. Crop Production under Drought and Heat Stress: Plant Responses and Management Options. Frontiers in Plant Science. Jun;8.

Fitzpatrick CR, Copeland J, Wang PW, Guttman DS, Kotanen PM, Johnson MTJ. 2018. Assembly and ecological function of the root microbiome across angiosperm plant species. Proceedings of the National Academy of Sciences of the United States of America. Feb;115:E1157-E1165.

Fox J, Weisberg S. 2011. An R companion to applied regression. Second Edition. Sage.

Gagne-Bourque F, Mayer BF, Charron JB, Vali H, Bertrand A, Jabaji S. 2015. Accelerated Growth Rate and Increased Drought Stress Resilience of the Model Grass Brachypodium distachyon Colonized by Bacillus subtilis B26. Plos One. Jun;10.

Gdanetz K, Trail F. 2017. The wheat microbiome under four management strategies, and potential for endophytes in disease protection. Phytobiomes. Oct; 158-168.

Goh CH, Vallejos DFV, Nicotra AB, Mathesius U. 2013. The Impact of Beneficial Plant-Associated Microbes on Plant Phenotypic Plasticity. Journal of Chemical Ecology. Jul;39:826-839.

Hamedi J, Mohammadipanah F. 2015. Biotechnological application and taxonomical distribution of plant growth promoting actinobacteria. Journal of Industrial Microbiology & Biotechnology. Feb;42:157-171.

Hardoim PR, van Overbeek LS, Berg G, Pirttila AM, Compant S, Campisano A, Doring M, Sessitsch A. 2015. The Hidden World within Plants: Ecological and Evolutionary Considerations for Defining Functioning of Microbial Endophytes. Microbiology and Molecular Biology Reviews. Sep;79:293-320.

Hartman K, van der Heijden MGA, Roussely-Provent V, Walser JC, Schlaeppi K. 2017. Deciphering composition and function of the root microbiome of a legume plant. Microbiome. Jan;5.

Henry A, Doucette W, Norton J, Bugbee B. 2007. Changes in crested wheatgrass root exudation caused by flood, drought, and nutrient stress. Journal of Environmental Quality. May-Jun;36:904-912.

Herzog S, Wemheuer F, Wemheuer B, Daniel R. 2015. Effects of Fertilization and Sampling Time on Composition and Diversity of Entire and Active Bacterial Communities in German Grassland Soils. Plos One. Dec;10.

Hu FL, Feng FX, Zhao C, Chai Q, Yu AZ, Yin W, Gan YT. 2017. Integration of wheat-maize intercropping with conservation practices reduces CO2 emissions and enhances water use in dry areas. Soil & Tillage Research. Jun;169:44-53.

Kaisermann A, de Vries FT, Griffiths RI, Bardgett RD. 2017. Legacy effects of drought on plant-soil feedbacks and plant-plant interactions. New Phytologist. Sep;215:1413-1424.

Kaisermann A, Maron PA, Beaumelle L, Lata JC. 2015. Fungal communities are more sensitive indicators to non-extreme soil moisture variations than bacterial communities. Applied Soil Ecology. Feb;86:158-164.

Kaurin A, Mihelic R, Kastelec D, Grcman H, Bru D, Philippot L, Suhadolc M. 2018. Resilience of bacteria, archaea, fungi and N-cycling microbial guilds under plough and conservation tillage, to agricultural drought. Soil Biology & Biochemistry. May;120:233-245.

Kavamura VN, Taketani RG, Lanconi MD, Andreote FD, Mendes R, de Melo IS. 2013. Water Regime Influences Bulk Soil and Rhizosphere of Cereus jamacaru Bacterial Communities in the Brazilian Caatinga Biome. Plos One. Sep;8.

Kembel SW, Cowan PD, Helmus MR, Cornwell WK, Morlon H, Ackerly DD, Blomberg SP, Webb CO. 2010. Picante: R tools for integrating phylogenies and ecology. Bioinformatics. Jun;26:1463-1464.

Kuchenbuch RO, Buczko U. 2011. Re-visiting potassium- and phosphate-fertilizer responses in field experiments and soil-test interpretations by means of data mining. Journal of Plant Nutrition and Soil Science. Apr;174:171-185.

Lancashire PD, Bleiholder H, Vandenboom T, Langeluddeke P, Stauss R, Weber E, Witzenberger A. 1991. A UNIFORM DECIMAL CODE FOR GROWTH-STAGES OF CROPS AND WEEDS. Annals of Applied Biology. Dec;119:561-601.

Li P, Ye SF, Liu H, Pen AH, Ming F, Tang XM. 2018. Cultivation of Drought-Tolerant and Insect-Resistant Rice Affects Soil Bacterial, but Not Fungal, Abundances and Community Structures. Frontiers in Microbiology. Jun;9.

Mahoney AK, Yin CT, Hulbert SH. 2017. Community Structure, Species Variation, and Potential Functions of Rhizosphere-Associated Bacteria of Different Winter Wheat (Triticum aestivum) Cultivars. Frontiers in Plant Science. Feb;8.

Martinez Arbizu P. 2017. Pairwiseadonis: Pairwise multilevel comparison using adonis. R Package Version 0.0.1.

McMurdie PJ, Holmes S. 2014. Waste Not, Want Not: Why Rarefying Microbiome Data Is Inadmissible. Plos Computational Biology. Apr;10.

Montanari-Coelho KK, Costa AT, Polonio JC, Azevedo JL, Marin SRR, Fuganti-Pagliarini R, Fujita Y, Yamaguchi-Shinozaki K, Nakashima K, Pamphile JA, et al. 2018. Endophytic bacterial microbiome associated with leaves of genetically modified (AtAREB1) and conventional (BR 16) soybean plants. World Journal of Microbiology & Biotechnology. Apr;34.

Mwanamwenge J, Loss SP, Siddique KHM, Cocks PS. 1999. Effect of water stress during floral initiation, flowering and podding on the growth and yield of faba bean (Vicia faba L.). European Journal of Agronomy. Jun;11:1-11.

Naveed M, Mitter B, Reichenauer TG, Wieczorek K, Sessitsch A. 2014. Increased drought stress resilience of maize through endophytic colonization by Burkholderia phytofirmans PsJN and Enterobacter sp FD17. Environmental and Experimental Botany. Jan;97:30-39.

Naylor D, DeGraaf S, Purdom E, Coleman-Derr D. 2017. Drought and host selection influence bacterial community dynamics in the grass root microbiome. Isme Journal. Dec;11:2691-2704.

Nguyen LTT, Osanai Y, Lai K, Anderson IC, Bange MP, Tissue DT, Singh BK. 2018. Responses of the soil microbial community to nitrogen fertilizer regimes and historical exposure to extreme weather events: Flooding or prolonged-drought. Soil Biology & Biochemistry. Mar;118:227-236.

Nuccio ML, Paul M, Bate NJ, Cohn J, Cutler SR. 2018. Where are the drought tolerant crops? An assessment of more than two decades of plant biotechnology effort in crop improvement. Aug; 273: 110-119.

Ogle DH. 2016. Introductory fisheries analyses with R. Chapman & Hall/CRC.

Oksanen O, Blanchet FG, Kindt R, Legendre P, Minchin PR, O'Hara RB, Simpson GL, Solymos P, Stevens MHH, Wagner H. 2016. Vegan: Community Ecology Package. R Package Version 2.3-5.

Osakabe Y, Osakabe K, Shinozaki K, Tran LSP. 2014. Response of plants to water stress. Frontiers in Plant Science. Mar;5.

Ouyang WJ, Struik PC, Yin XY, Yang JC. 2017. Stomatal conductance, mesophyll conductance, and transpiration efficiency in relation to leaf anatomy in rice and wheat genotypes under drought. Journal of Experimental Botany. Aug;68:5191-5205.

Pinheiro J, Bates D, DebRoy S, Sarkar D and R Core Team. 2017. nlme: Linear and Nonlinear Mixed Effects Models. R package version 3.1-131.

Preece C, Peñuelas J. 2016. Rhizodeposition under drought and consequences for soil communities and ecosystem resilience. Plant and Soil. Dec;409:1-17.

R Core Team. 2016. R: A Language and Environment for Statistical Computing. Vienna: R Foundation for Statistical Computing. Available online at: http://www.R-project.org

Ritz C, Streibig JC. 2016. Package 'drc' Analysis of dose-response curves. Version 3.0-1.

Ryan RP, Germaine K, Franks A, Ryan DJ, Dowling DN. 2008. Bacterial endophytes: recent developments and applications. Fems Microbiology Letters. Jan;278:1-9.

Santos-Medellin C, Edwards J, Liechty Z, Nguyen B, Sundaresan V. 2017. Drought Stress Results in a Compartment-Specific Restructuring of the Rice Root-Associated Microbiomes. Mbio. Jul-Aug;8.

Senbayram M, Trankner M, Dittert K, Bruck H. 2015. Daytime leaf water use efficiency does not explain the relationship between plant N status and biomass water-use efficiency of tobacco under non-limiting water supply. Journal of Plant Nutrition and Soil Science. Aug;178:682-692.

Shannon P, Markiel A, Ozier O, Baliga NS, Wang JT, Ramage D, Amin N, Schwikowski B, Ideker T. 2003. Cytoscape: A software environment for integrated models of biomolecular interaction networks. Genome Research. Nov;13:2498-2504.

Song YN, Zhang FS, Marschner P, Fan FL, Gao HM, Bao XG, Sun JH, Li L. 2007. Effect of intercropping on crop yield and chemical and microbiological properties in rhizosphere of wheat (Triticum aestivum L.), maize (Zea mays L.), and faba bean (Vicia faba L.). Biology and Fertility of Soils. Jun;43:565-574.

Spinoni J, Naumann G, Vogt J. 2015. Spatial patterns of European droughts under a moderate emission scenario. Adv. Sci. Res.12: 179–186.

Streit W, Daniel R. 2010. Metagenomics. Methods and Protocols. Springer Science+Business Media, LLC. Series Volume 668.

Tahtamouni ME, Khresat S, Lucero M, Sigala J, Unc A. 2016. Diversity of endophytes across the soil-plant continuum for Atriplex spp. in arid environments. Journal of Arid Land. Apr;8:241-253.

Taschen E, Amenc L, Tournier E, Malagoli P, Fustec J, Bru D, Philippot L, Bernard L. 2017. Cereal-legume intercropping modifies the dynamics of the active rhizospheric bacterial community. Rhizosphere 3: 191–195.

Tian BY, Zhang CJ, Ye Y, Wen JM, Wu YM, Wang HZ, Li HM, Cai SX, Cai WT, Cheng ZQ, et al. 2017. Beneficial traits of bacterial endophytes belonging to the core communities of the tomato root microbiome. Agriculture Ecosystems & Environment. Sep;247:149-156.

Tichy L, Chytry M. 2006. Statistical determination of diagnostic species for site groups of unequal size. Journal of Vegetation Science. Dec;17:809-818.

Vandenkoornhuyse P, Quaiser A, Duhamel M, Le Van A, Dufresne A. 2015. The importance of the microbiome of the plant holobiont. New Phytologist. Jun;206:1196-1206.

Vandermeer JH. 1992. The ecology of intercropping. New York, NY: Cambridge University Press.

Wagner MR, Lundberg DS, del Rio TG, Tringe SG, Dangl JL, Mitchell-Olds T. 2016. Host genotype and age shape the leaf and root microbiomes of a wild perennial plant. Nature Communications. Jul;7.

Wallace J, Laforest-Lapointe I, Kembel SW. 2018. Variation in the leaf and root microbiome of sugar maple (Acer saccharum) at an elevational range limit. Peerj. Aug;6.

Weinbauer MG, Fritz I, Wenderoth DF, Hofle MG. 2002. Simultaneous extraction from bacterioplankton of total RNA and DNA suitable for quantitative structure and function analyses. Applied and Environmental Microbiology. Mar;68:1082-1087.

Wemheuer F, Kaiser K, Karlovsky P, Daniel R, Vidal S, Wemheuer B. 2017. Bacterial endophyte communities of three agricultural important grass species differ in their response towards management regimes. Scientific Reports. Jan;7.

Wemheuer F, Wemheuer B, Kretzschmar D, Pfeiffer B, Herzog S, Daniel R, Vidal S. 2016. Impact of grassland management regimes on bacterial endophyte diversity differs with grass species. Letters in Applied Microbiology. Apr;62:323-329.

Wemheuer B, and Wemheuer F. 2017. "Assessing bacterial and fungal diversity in the plants endosphere," in Metagenomics - Methods and Protocols, Vol. 1539, edsW. Streit, and R. Daniel (New York, NY: Humana Press), 75–84.

Xu L, Naylor D, Dong ZB, Simmons T, Pierroz G, Hixson KK, Kim YM, Zink EM, Engbrecht KM, Wang Y, et al. 2018. Drought delays development of the sorghum root microbiome and enriches for monoderm bacteria. Proceedings of the National Academy of Sciences of the United States of America. May;115:E4284-E4293.

Yang CH, Huang GB, Chai Q, Luo ZX. 2011. Water use and yield of wheat/maize intercropping under alternate irrigation in the oasis field of northwest China. Field Crops Research. Dec;124:426-432.

Yang ZP, Yang WP, Li SC, Hao JM, Su ZF, Sun M, Gao ZQ, Zhang CL. 2016. Variation of Bacterial Community Diversity in Rhizosphere Soil of Sole-Cropped versus Intercropped Wheat Field after Harvest. Plos One. Mar;11.

Zampieri M, Ceglar A, Dentener F, Toreti A. 2017. Wheat yield loss attributable to heat waves, drought and water excess at the global, national and subnational scales. Environmental Research Letters. Jun;12.

Zhang NN, Sun YM, Li L, Wang ET, Chen WX, Yuan HL. 2010. Effects of intercropping and Rhizobium inoculation on yield and rhizosphere bacterial community of faba bean (Vicia faba L.). Biology and Fertility of Soils. Aug;46:625-639.

Zhang NN, Sun YM, Wang ET, Yang JS, Yuan HL, Scow KM. 2015. Effects of intercropping and Rhizobial inoculation on the ammonia-oxidizing microorganisms in rhizospheres of maize and faba bean plants. Applied Soil Ecology. Jan;85:76-85.

Zhang YZ, Wang ET, Li M, Li QQ, Zhang YM, Zhao SJ, Jia XL, Zhang LH, Chen WF, Chen WX. 2011. Effects of rhizobial inoculation, cropping systems and growth stages on endophytic bacterial community of soybean roots. Plant and Soil. Oct;347:147-161.

Zhou Y, Zhu HH, Fu SL, Yao Q. 2017. Variation in Soil Microbial Community Structure Associated with Different Legume Species Is Greater than that Associated with Different Grass Species. Frontiers in Microbiology. May;8.

Chapter 2.3:
Response of the active bacterial and fungal communities in the rhizosphere differ towards water deficit

Sandra Granzow, Annika Meißner, Birgit Pfeiffer, Rolf Daniel, Stefan Vidal, Franziska Wemheuer

Manuscript draft

Response of the active bacterial and fungal communities in the rhizosphere differ towards water deficit

Sandra Granzow [1,2*], Annika Meißner[2,3], Birgit Pfeiffer[3,4], Rolf Daniel[4], Stefan Vidal[1] and Franziska Wemheuer[1]

[1]Division of Agricultural Entomology, Department of Crop Sciences, University of Göttingen, Göttingen, Germany

[2]Center of Biodiversity and Sustainable Land Use, University of Göttingen, Göttingen, Germany

[3] Institute of Applied Plant Nutrition, University of Göttingen, Göttingen, Germany

[4]Genomic and Applied Microbiology and Göttingen Genomics Laboratory, Institute of Microbiology and Genetics, University of Göttingen, Göttingen, Germany

*** Correspondence:**
Dr. Sandra Granzow
sandra.granzow@agr.uni-goettingen.de

Keywords: active bacterial and fungal community, intercropping, water deficit, microbial interactions

Abstract

Drought limits plant growth and yield, but can also impact soil ecosystem functioning. Integrated soil management including new genotypes and intercropping of plants might improve the sustainability of agricultural production in a changing climate. As plant-associated microorganisms play key roles in enhancing plant tolerance to environmental stressors such as drought, it is of crucial interest to better understand how water deficit affects the active microbial community of important crops. In the present study, we investigated how water deficit changes the active bacterial and fungal community in the rhizosphere soil of winter wheat (genotype: Genius) and two winter faba bean genotypes (S_004; S_062) under different cropping systems. Our major results were that both bacterial and fungal communities were altered by water deficit; however, they responded differently towards drought. Changes of bacterial community composition were dependent on crop species and genotype, whereas alpha-diversity showed a marked resistance towards water deficit. In contrast, fungal community composition responded more sensitive, but the response of fungal alpha-diversity was dependent on crop genotype. Cropping system alone only influenced fungal community composition but not bacteria. Furthermore, we recorded complex microbial interactions dependent on cropping system and water deficit. For example, under water deficit the number of positive correlations in bacteria increased in wheat cropping systems compared to well-watered plants. For fungi, we observed an increase in positive correlations under intercropped wheat compared to monoculture under well-watered conditions. To our knowledge, this is the first study investigating the combined and separate effect of intercropping and water deficit on the metabolically active plant-associated bacterial and fungal communities of two important crop species. Obtained results highlight that the combination of crop species, genotype and cropping system play key roles in the response of the active microbiome in the rhizosphere soil towards drought. Further research on field-scale might deepen our understanding how sustainable agricultural practices and plant-associated microorganisms might mitigate future drought events.

1 Introduction

Drought is one of the key abiotic stressors that limit plant growth and yield worldwide (Fahad et al., 2017; Zampieri et al., 2017). Previous studies showed that drought events can also significantly influence soil ecosystem functioning, including biogeochemical cycling or soil organic matter dynamics (Austin et al., 2004; Preece and Peñuela, 2016). Drought is thus a serious threat for food security in agricultural production. The development of new genotypes and intercropping of plants are key elements of integrated soil management to improve the sustainability of agricultural production in a changing climate (Coleman-Derr and Tringe, 2014; Daryanto, Wang and Jacinthe, 2016). For example, intercropping of wheat and maize significantly increased water use and water use efficiency compared to sole cropping (Yang et al., 2011). On the other hand, Saharan et al. (2018) showed that the combination of intercropping of pigeon pea/finger millet and beneficial microorganisms such as arbuscular mycorrhizal fungi and rhizobacteria increased biomass production and nutrient uptake even under dry conditions. Soil-derived beneficial microorganism can form symbiotic and/or mutualistic associations with roots of host plants and can be important promoter of plant growth and health through better nutrient acquisition or alleviation of abiotic stressors (Coleman-Derr and Tringe, 2014; Vimal et al., 2017). Hence, changes in abundance or composition of root-associated microbial communities as response to environmental stressors might also impact plant performance (Berg et al., 2014; Ahkami et al., 2017). As consequence, it is of crucial interest to better understand how environmental changes such as drought alter the microbiome in the plant rhizosphere.

Previous studies on microbial responses towards drought and re-watering reported significant changes of bacterial and fungal communities (Kaisermann et al., 2015, Schmidt et al. 2017; Meisner et al., 2018). For example, Santos-Medellin et al. (2017) found that drought significantly changed bacterial and fungal community composition in different rice compartments. The authors concluded that the restructuring of the associated microbiome might contribute to plant survival under extreme environmental conditions. Recently, de Vriese et al. (2018) investigated in the response of soil bacterial and fungal communities over time towards drought in a field-based mesocosm experiment consisting of common grassland species. The authors showed that bacterial co-occurrence networks were characterised by properties that indicate low stability under disturbances, whereas fungal networks were more stable towards drought. In addition, they indicated that changes in bacterial communities were linked more strongly to soil functioning during drought recovery than do changes in fungi.

As most previous studies have focused on bacterial or fungal responses separately (Mahoney, Yin and Hulbert, 2017; Xue et al., 2018; but see Li and Wu, 2018; deVriese et al., 2018), bacterial-fungal interactions in the rhizosphere of intercropped plants under water deficit remain poorly understood. In a previous study, we found that plant compartment and plant species altered the effects of cropping systems on microbial communities and we observed different responses of fungal and bacterial communities towards cropping systems (Granzow et al., 2017). Moreover, the number of negative inter-domain interactions between fungi and bacteria decreased in bulk and rhizosphere soil in intercropping regimes compared to monoculture indicating beneficial effects (Granzow et al., 2017). So far, most studies investigating the response of microbial communities towards drought focused on the entire microbial community (Santos-Medellin et al., 2017; deVriese et al., 2018, but see Barnard et al., 2013). However, the potentially active microbial community might be more sensitive to abiotic stresses and thus is more closely related to ecosystem functionality (Blagodatskaya and Kuzyakov, 2013; Herzog et al., 2015).

Hence, the aim of the present study was to investigate the influence of water deficit and re-watering on the metabolically active fungal and bacterial communities and their interactions in the rhizosphere of two important crop species under different systems. For this purpose, winter wheat (*Triticum aestivum* L.; genotype: Genius) and two genotypes of winter faba bean (*Vicia faba* L.; S_004 and S_062) were grown in monoculture or in row intercropping with (water deficit treatments) or without drought stress (control treatments). Rhizosphere soil was collected at three time points: beginning of water deficit, during water deficit and after re-watering. Bacterial and fungal communities in rhizosphere were examined by iTag sequencing of bacterial 16S rRNA genes and the fungal internal transcribed spacer (ITS) region, respectively, amplified by two-step reverse transcriptase (RT) PCR.

We hypothesized that (i) water deficit changes microbial community composition and diversity and (ii) bacterial and fungal communities respond differently towards water deficit as they differ in their lifestyles in terms of colonization area in the rhizosphere soil (Deveau et al., 2018). We expected further that (iii) crop species, faba bean genotype and cropping system would alter the response of bacterial and fungal communities towards water deficit. Finally, we hypothesized that (iv) co-associations and microbial interactions are also influenced by water deficit and cropping systems.

2 Material and Methods

2.1 Plant material

To examine the combined influence of cropping system and water deficit on the active fungal and bacterial community in roots and attached soil (here regarded as rhizosphere soil) a greenhouse experiment was conducted in autumn 2016. Seeds of the two winter faba bean genotypes were provided by the Institute of Plant breeding of the University of Göttingen. The two winter faba bean genotypes (S_004 and S_062) were selected based on a previous field trial within the IMPAC[3] project (*Novel genotypes for mixed cropping allow for improved sustainable land use across arable land, grassland and woodland*). The genotype S_004 is characterized by medium height and leaf size, low tillering, late maturity and high yield. In contrast, genotype S_062 is very short with small leaflets, high tillering, and early maturing. Seeds of winter wheat (genotype: Genius) were provided by Norddeutsche Pflanzenzucht Hans-Georg Lembke KG. All seeds were surface-sterilized by serial washing according to Andreote et al. (2010) with one modification. Immersion in sterile distilled water was performed four times for 30 s. Surface sterilized seeds were placed on wet sterile tissues and germinated at 7 °C under dark conditions until seedlings developed roots with a length of approximately 4 cm.

2.2 Experimental design and soil substrate

Pre-germinated seeds of faba bean and wheat were sown in monoculture or as mixture in polypropylene containers (Sunware; 45.5 x 36 x 24 cm) in a randomized block design (day 0, DAO, days after onset of experiment). Twelve treatments were established: faba bean monoculture S_004 with or without water deficit (S4_FBM_D/C), faba bean monoculture S_062 with or without water deficit (S62_FBM_D/C), faba bean S_004 intercropped with wheat with or without water deficit (S4_FBIC_D/C; WIC_D/C), faba bean S_062 intercropped with wheat with or without water deficit (S62_FBIC_D/C; WIC_D/C) and wheat monoculture with or without water deficit (WM_D/C; Table 1). Each treatment was replicated four times, resulting in a total of 40 containers. We defined two different cropping systems (monoculture and intercropping), whereas cropping regimes compromised each treatment, e.g. WM_D and FBM_C.

For monocultures, 30 faba bean or 72 wheat seeds per container were sown in six rows. For intercropping systems, 15 faba bean and 36 wheat seeds were sown in alternate rows (Vandermeer, 1992). Each container was filled with air-dried, sieved (< 10 mm) and layered soil from the experimental study site in Reinshof (51.48 ° N, 9.92 ° E and 157 m asl.), Germany. The soil volume of each pot accounted for approximately 20 L with a dry weight of 18 kg. Filling of the pots was performed in layers adding distilled water to each layer to prevent soil compaction. After emergence of the seedlings, the soil was covered by gravel to minimize water loss by evaporation. The soil was classified as Gleyic Fluvisol according to the FAO classification system and contained 21 % clay, 68 % silt and 11 % sand, with pH 7.3 and 2.8 % humus. Nutrients such as phosphorus (50 mg P/kg dry soil) and potassium (140 mg K/kg dry soil) were in an optimal range according to the German nutrient-availability class system (Kuchenbuch and Buczko 2011).

2.3 Water management and growth conditions

During the experiment, photosynthetic photon flux density was 400 µmol m^{-2} s^{-1} at plant level with a 10/14 h day/night photoperiod. Furthermore, the CO_2 concentration reached around 450 ppm. There was a relative humidity of 50 % and an average air temperature of 23 °C. Water loss by transpiration was documented by placing the pots permanently on balances (TQ30, ATP Messtechnik, Germany). The weight reduction was measured every 30 minutes in order to constantly determine water consumption. This systems avoids hidden drought due to higher transpiration of increased biomasses (Senbayram et al., 2015). Plants of all treatments were irrigated with distilled water to 90 % field capacity. After a growing period of 24 days and a BBCH of 14/34 for faba bean and a BBCH of 14/15 of wheat plants (Lancashire et al., 1991), the amount of water in water deficit treatments was reduced to 75 % compared to control treatments (beginning of water deficit). At day 28, the amount of water in these treatments was further reduced to 25 % (during water deficit). Day 34, all water deficit treatments were re-watered with the adequate amount of water according to plant growth and water consumption. All control pots were sufficiently irrigated during the whole experimental duration (6 weeks).

2.4 Sampling

Soil and plant samples were collected from control and water deficit treatments at day 29 (beginning of water deficit), day 34 (during water deficit) and at day 38 (after re-watering of water deficit plants) (Figure 1). For microbial community analysis, one faba bean and two wheat plants per container and harvest were randomly sampled which showed no obvious sign of any disease infection. The roots were gently shaken to remove the non-rhizosphere soil. Rhizosphere soil for pH value and C/N was collected by carefully brushing the roots. Rhizosphere soil and roots of each plant species and each pot were pooled for molecular analysis. All samples for molecular analysis were immediately frozen in liquid nitrogen, transferred to the laboratory and stored at -80 °C. In total, 96 faba bean (48 plants of each genotype) and 144 wheat plants were collected. Rhizosphere and aerial plant parts of each crop species and container were pooled, resulting in a total of 96 faba bean and 72 wheat samples (Table 1).

2.5 Edaphic properties

For determination of edaphic properties such as total organic carbon and total organic nitrogen, subsamples of all rhizosphere samples were dried at 60 °C for two days and subsequently sieved to < 2 mm. Carbon and nitrogen concentrations from dried subsamples were determined using a NA-1500N analyser (Thermo Fisher Scientific, Waltham, USA). Afterwards, the carbon-to-nitrogen (C/N) ratio was calculated. The gravimetric soil water content (%) of all soil samples was calculated based on the fresh weight and the oven-dried weight. The pH values of all rhizosphere soil samples were measured as follows: 10 g of dried and sieved soil was added in a small beaker with 25 ml 0.01 M calcium chloride. Soil solution was homogenized after 30 min and 60 min, and subsequently soil pH_{CaCl} was measured. Details on soil properties are provided in Table S1.

2.6 RNA Extraction and Purification

Environmental RNA of the rhizosphere was extracted from 2 g soil per sample employing the RNA PowerSoil total RNA isolation kit as recommended by the manufacturer (MoBio Laboratories, Carlsbad, CA, USA, now Qiagen, Hilden, Germany). Residual DNA was removed with the TURBO DNA-free™ kit (Thermo Fisher Scientific, Waltham, USA) from the extracted RNA according to the manufacturer's protocol. In addition, 1/40 volume Ribolock RNase Inhibitor (40 U/µL) (Thermo Fisher Scientific, Waltham, USA) was added in the first step of the DNA digestion. The absence of DNA was confirmed by PCR using the internal transcribed spacer region as target gene for amplification of fungi. For details of the PCR reaction and cycling conditions as well as the primer see the first PCR according to Wemheuer and Wemheuer (2017). The DNA-free RNA was purified according to Streit and Daniel (2012). RNA concentrations were determined using a NanoDrop ND-1000 spectrophotometer (NanoDrop Technologies, Wilmington, DE, USA).

2.7 Synthesis of cDNA from total RNA

Purified RNA from 168 rhizosphere samples were converted to cDNA by employing the SuperScript™III reverse transcriptase Kit as recommended by the supplier (Invitrogen, Karlsruhe, Germany) with two modifications. Same reverse primer 1193r (20 µM) and ITS4 (20 µM) were used for the reaction as for the following PCR. After the last step, 0.5 µl RNase H (5 U/µl; Fermentas) was added, and samples were incubated for 15 min at 37 °C and subsequently for 10 min at 65 °C. The cDNA was stored at -20 °C.

2.8 Amplification of 16S rRNA gene

Bacterial community in the rhizosphere was assessed by PCR approach targeting the V5-V7 region of the 16S rRNA gene. The following primers were used: 799F (Chelius and Triplett, 2001) and 1193R (Bodenhausen et al., 2013; Hartman et al., 2017) containing MiSeq adaptors (underlined) Miseq-799F 5'-<u>TCGTCGGCAGCGTCAGATGTGTATAAGAGACAG</u>AACM GGATTAGATACCCKG-3'; MiSeq- 1193R 5'<u>GTCTCGTGGGCTCG GAGATGTGTATAA GAGACAG</u>ACGTCATCCCCACCTTCC-3'. The PCR mixture (25 µl) contained 5 µl of five-fold Phusion GC buffer, 200 µM of each of the four deoxynucleoside triphosphates, 4 µM of each primer, 0.5 U of Phusion high fidelity DNA polymerase (Thermo Scientific) and approximately 50 ng of cDNA as template. The following thermal cycling scheme was used: initial denaturation at 98 °C for 30 s, 30 cycles of denaturation at 98 °C for 15 s, annealing at 53 °C for 30 s, followed by extension at 72 °C for 30 s. The final extension was carried out at 72 °C for 2 min. Negative controls were performed using the reaction mixture without template. Genomic DNA of *Escherichia coli* strain DH5α was used as template in the positive control. Three independent PCRs were performed per sample. Obtained PCR products per sample were controlled for appropriate size, pooled in equal amounts, and purified using the NucleoMag NGS Clean up (Macherey-Nagel, Düren, Germany).

Quantification of the purified PCR products was performed using the Quant-iT dsDNA HS assay kit and a Qubit fluorometer (Thermo Scientific) as recommended by the manufacturer. Quantified PCR products were barcoded using the Nextera XT-Index kit (Illumina, San Diego, USA) and the Kapa HIFI Hot Start polymerase (Kapa Biosystems, Wilmington, USA). The Göttingen Genomics Laboratory determined the sequences of the partial 16S rRNA genes employing the MiSeq Sequencing platform and the MiSeq Reagent Kit v3 (2 x 300 cycles) as recommended by the manufacturer (Illumina).

2.9 Amplification of ITS region

The fungal community in the rhizosphere was assessed by PCR targeting the ITS2 region with the primers ITS3_KYO2 (Toju et al., 2012) and ITS4 (White et al., 1990) containing the MiSeq adaptors (underlined): MiSeq-ITS3_KYO2 (5′-TCGTCGGCAGCGTCAGATGTGTATAAGAGACAGGATGAAGAACGYAGYRAA-3′) and MiSeq-ITS4 (5′-GTCTCGTGGGCTCGGAGATGTGTATAAGAGACAGTCCTCCGCTTA TTGATATGC -3′). The PCR mixture (25 µl) contained: 5 µl of five-fold Phusion GC buffer, 200 µM of each of the four deoxynucleoside triphosphates, 4 µM of each primer, 5 % DMSO, 25 mM $MgCl_2$, 0.5 U of Phusion High Fidelity DNA polymerase (Thermo Scientific) and approximately 10 ng DNA sample as template. For details in the thermal cycling conditions see Wemheuer and Wemheuer (2017) and Granzow et al. (2017). Negative controls were performed using the reaction mixture without template. Genomic DNA of *Aspergillus nidulans* was used as template in the positive control. Three independent PCRs were performed per sample. Obtained PCR products per sample were controlled for appropriate size, pooled in equal amounts, and purified using the NucleoMag NGS Clean up (Macherey-Nagel). Quantification of the PCR products was performed using the Quant-iT dsDNA HS assay kit and a Qubit fluorometer (Thermo Scientific) as recommended by the manufacturer. Purified PCR products were barcoded using the Nextera XT-Index kit (Illumina) and the Kapa HIFI Hot Start polymerase (Kapa Biosystems). The Göttingen Genomics Laboratory determined the sequences of the the ITS2 region employing the MiSeq Sequencing platform and the MiSeq Reagent Kit v3 (2 x 300 cycles) as recommended by the manufacturer (Illumina).

2.10 Processing of microbial community dataset

Generated sequencing data were initially quality filtered with the Trimmomatic tool version 0.36 (Bolger et al., 2014). Low quality reads were truncated if the quality dropped below 15 in a sliding window of 4 bp. Subsequently, all reads shorter than 100 bp and orphan reads were removed. Remaining sequences were merged, quality-filtered and further processed with USEARCH version 10.0.240 (Edgar, 2010). Filtering included the removal of reads shorter than 350 (bacteria) and 100 (fungi) or longer than 450 bp (bacteria) and 586 (fungi) as well as the removal low quality reads (expected error > 1) and reads with more than one ambitious base.

Processed sequences of all samples were concatenated to one file and subsequently dereplicated into unique sequences. These sequences were denoised with the unoise3 algorithm implemented in USEARCH (Edgar, 2010). All OTUs consisting of one single sequence (singletons) were removed. Subsequently, remaining chimeric sequences were removed using UCHIME (Edgar et al., 2011) in reference mode with the QIIME release of the UNITE database version 7.2 (Kõljalg et al., 2013) for fungi. Filtered sequences were mapped on remaining unique sequences to determine the occurrence and abundance of each unique sequence in every sample. To assign taxonomy of fungal chimera-free sequences were classified by BLAST alignment against the most recent UNITE database (Kõljalg et al., 2013) with an e-value threshold of 1e-20. Concatenated sequences of all sequences were mapped on the final set of unique sequences to calculate the evenness and abundance of each unique sequence in all samples. All non-fungal or non-bacterial zOTUs were removed based on their taxonomic classification in the respective database. Final zOTU tables for bacteria and fungi are provided in Table S2 and S3. Only zOTUs occurring in more than one sample were considered for further statistical analysis. Samples with less than 445 (bacteria) and 63 (fungi) sequences per sample were removed prior statistical analysis, resulting in 160 and 126 samples for bacteria and fungi.

2.11 Statistical Analysis

All statistical analyses were performed using R version 3.4.0 (R Core Team, 2016). Differences in edaphic, plant and bacterial community data were considered as statistically significant with $P \leq 0.05$. Differences in alpha or beta diversity as well as sequencing depth with regard to cropping system and water treatment (yes/no) were tested by a Kruskal-Wallis test. There were no significant differences of the mean sequencing depths between the cropping systems and water treatments. In consequence, zOTU tables were not rarefied as recommended by McMurdie and Holmes (2014).

Alpha diversity indices (Richness, Shannon index of diversity and Michaelis Menten Fit) were calculated in the *vegan* package version 2.4.4 (Oksanen et al., 2016) and the *drc* package version 3.0-1 (Ritz and Streibig, 2016). OTU tables were rarefied using the *rrarefy* function in *vegan* and samples with less than 2060 (bacteria) and 213 (fungi) sequences were removed prior alpha diversity analysis. Sample coverage was estimated using the Michaelis-Menten Fit calculated in R. For this purpose, richness and rarefaction curves were calculated using the *picante* package version 1.6-2 (Kembel et al., 2010). Richness and diversity were calculated using the *specnumber* and *diversity* function, respectively. The Michaelis-Menten Fit was subsequently calculated from generated rarefaction curves using the *MM2* model within the *drc* package version 3.0-1 (Ritz and Streibig, 2016). All alpha diversity indices were calculated ten times. The average from each iteration was used for further statistical analysis. Final tables containing bacterial and fungal richness, diversity, Michaelis-Menten Fit and coverage are provided in Table S4 and S5.

Data were tested for normal distribution with *shapiro* and homogeneity of variance with *leveneTest* function with the package *car* version 2.1-5 (Fox and Weisberg, 2011). For global differences (for all three harvests) between measured edaphic properties and plant parameters were calculated with a linear mixed model with the function *lme* and the R package *nlme* version

3.1-131 (Pinheiro et al., 2017) with pot number as random factor. Data was log-transformed when not normal distributed. F-values were evaluated with ANOVA and type = "marginal". In addition, each harvest was tested separately with a post hoc test using Dunn's test with *p*-value adjustment "BH" and the function *dunnTest* in the R package *FSA* version 0.8.17 (Ogle, 2016). Alpha-diversity was evaluated with Kruskal-Wallis test or post hoc test using *dunnTest*. Differences in community composition were investigated by permutational multivariate analysis of variance (PERMANOVA; Anderson, 2001) based on Bray-Curtis distance matrices using 999 random permutations. Bacterial and fungal communities were tested separately. OTU tables were subsampled ten times and all tables were summed up to account for low abundant species. Global effects (calculated for all three sampling times together) for crop species on fungal and bacterial communities were tested with strata = pot, as we had pseudoreplicated data. A significant *p*-value in PERMANOVA for beta-diversity can be driven by true biological differences, differences within group (variance) or both (Anderson, 2001). In case of significant *p*-values in PERMANOVA, we tested for differences in homogeneity using permutational analysis of multivariate dispersions (PERMDISP, Anderson, 2006) with 999 permuations. NMDS, PERMANOVA and PERMDISP were run using functions; *metaMDS*, *adonis* and *betadisper*, respectively, in the R package *vegan* (Oksanen et al., 2016). Differences in community composition were visualized using the *metaMDS* function within the *vegan* package (Oksanen et al., 2016). To investigate in differences between cropping regimes, pairwise Adonis with *p*-value adjustment "BH" based on Bray-Curtis distances were used (Martinez Arbizu, 2017).

To identify zOTUs highly associated to cropping regime, multipattern analyses were applied. For that purpose, bacteria and fungi were investigated using the *multipatt* function from the *IndicSpecies* package version 1.7.6 (DeCáceres and Legendre, 2009). Only bacterial and fungal zOTUs found in at least three samples were used. The biserial coefficients (R) with a particular cropping regime were corrected for unequal sample size using the function *r.g* (Tichy and Chytry, 2006). For visualization, a bipartite network was generated using the treatment as source nodes and the taxa as target nodes. Network generation was performed using the edge-weighted spring embedded layout algorithm in Cytoscape version 3.3.0 (Shannon et al., 2003). The results of the multipattern analyses are provided in Table S6.

Correlation-based co-occurrence patterns were calculated with respect to cropping regime to investigate the interactions between bacteria and fungi in the rhizosphere soil. Therefore, bacterial and fungal zOTU tables were combined resulting in 126 samples for each kingdom. To enhance reliability of the co-occurrence patterns, only zOTUs with an average abundance of more than 0.01 % were considered. Additionally, zOTUs present in at least three samples were taken into account. Pairwise correlation based on Spearman's rho was calculated using the *cor.test* function in R and the numbers of significant positive and significant negative correlations were counted. Positive correlations were considered as two taxa co-occurring or cooperation between the two taxa. Negative correlations were considered as two taxa avoiding each other or competition between the two taxa.

3 Results and Discussion

3.1 Edaphic properties

We investigated in several edaphic properties including pH value, total organic carbon and nitrogen, as previous studies have been shown that cropping system or drought can change chemical characteristics in the rhizosphere soil (Song et al., 2007b; Preece and Peñuelas, 2016). Partly in line, we found that cropping system was the most influencing factor on edaphic properties compared to water treatment. Results of linear mixed effect model showed that pH value was significantly influenced by cropping system in the rhizosphere of wheat (LMM, df=21, F=5.72, p=0.0026). This was mainly observed for harvest 2, with lowest pH values under WIC (Table 2). In addition, we observed that pH value was significantly lower in FBIC compared to FBM for both genotypes specific for harvest 2 (Kruskal-Wallis (KW)-test, S_004 x^2=6.37, df=1, p=0.012; S_062: x^2=10.63, df=1, p=0.001). C:N ratio as well as carbon were significantly affected by cropping system in the wheat rhizosphere (LMM, C:N ratio: df=21, F=5.96, p=0.023; carbon: df= 21, F=4.47, p=0.046). Total nitrogen and carbon had significant lower values under WIC compared to WM for harvest 3 whereas the opposite was observed for harvest 2. Cropping system also significantly influenced C:N ratio in S_004 (S_004; LMM, df=13, F=50.54, p<0.0001) and highest C:N ratio was found under FBIC compared to FBM (Table 3).

3.2 Overall microbial community

The response of the bacterial and fungal communities of faba bean and wheat towards water deficit under different cropping systems were assessed by Illumina (MiSeq) sequencing targeting the bacterial 16S rRNA gene and internal transcribed spacer region. After removal of low-quality reads, PCR artefacts (chimeras) and non-target contaminations, a total of 1,309,304 and 860,402 high quality reads were obtained for bacteria and fungi (Table S4, S5). Sequence numbers per sample varied between 445 to 71,551 (average 8,182.2) for bacteria and 63 to 104,940 for fungi. Obtained sequences were grouped into 5,809 bacterial and 1,073 fungal zOTUs (Table S2, S3). Calculated Michaelis-Menten Fit and rarefaction curves confirmed that the majority of microbial communities were recovered by the surveying effort (Figure S1, S2; Table S4, S5).

Bacteria were dominated by eight phyla (> 0.5 % of all sequences across all samples): Proteobacteria (53.71 %), Bacteroidetes (17.30 %), Actinobacteria (16.94 %), Gemmatimonadetes (5.50 %), Acidobacteria (2.28 %), Chloroflexi (1.36 %), Entotheonellaeota (1.12 %) and Fibrobacteres (0.55 %). The Proteobacteria were dominated by Gammaproteobacteria (29.95 %), followed by Alpha- (12.40 %) and Deltaproteobacteria (11.35 %). The abundant bacterial phyla were present in all samples and accounted for 98.81 %, of all sequences analysed in this study. At family level, *Burkolderiaceae* (19.94 %), *Microscillaceae* (7.89 %) and *Polyangiceae* (5.01 %) dominated the bacterial dataset. The most frequent bacterial genera were *Curvibacter* (4.79 %), *Rhizobacter* (3.14 %), *Comamonas* (3.02 %), *Blrii*41 (3.07 %) and *Ohtaekwangia* (3.02 %) (Figure 2).

Fungi were dominated by six abundant phyla: Ascomycota (30.77 %), Glomeromycota (12.98 %), Basidiomycota (8.70 %), Mucoromycota (1.06 %), Zoopagomycota (0.62 %) and Mortierellomycota (0.54 %). Approximately 45.16 % of all sequences belonged to unidentified fungi. At family level, *Mycosphaerellaceae* (15.78 %), *Gigasporaceae* (6.08 %) and *Glomeraceae* (5.45 %) dominated the fungal dataset. The most frequent fungal genera were *Polythrincium* (15.67 %), *Dentiscutata* (6.08 %), *Cladosporium* (1.58 %), *Periconia* (1.18 %) and *Rhizopus* (1.01 %) (Figure 3). Abundant bacterial and fungal taxa were also found in previous studies investigating in microbial communities in the rhizosphere soil (Zhou et al., 2017; Li et al., 2018).

3.3 Fungal community in the rhizosphere soil was more sensitive towards water deficit than bacteria

According to our first hypothesis that water deficit affects microbial community diversity and composition, we calculated diversity (represented by the Shannon index H') and richness (number of observed unique sequences) with regard to harvest. A general influence of water deficit on bacterial and fungal alpha-diversity was not found. Furthermore, differences between water treatments on beta-diversity were not immediately evident with the NMDS (non-metric multidimensional scaling) analysis based on Bray-Curtis dissimilarities (Figure 4). However, PERMANOVA found that water deficit significantly influenced fungal community composition in harvest 1 and 2 which explained 7.7 % and 6.8 % of the variance (PERMANOVA, p=0.006 (H1); p=0.049 (H2)), whereas bacteria were not influenced (Table 6). In accordance with this observation was the study by Schmidt et al. (2017). They investigated how reduced moisture conditions impacted soil fungal communities from temperate grassland over the course of an entire season. As a result, they reported that fungal diversity was not different between the experimental moisture levels, whereas fungi changed in their composition, in both abundances and presence/absence of species (Schmidt et al., 2017). In contrast to our results, Naylor et al. (2017) showed that bacterial diversity and composition was significantly influenced by drought in the root endosphere, rhizosphere and bulk soil of different grasses. In our study, results indicate that fungal communities were more sensitive towards water deficit compared to bacteria in the rhizosphere. In accordance with this assumption were previous findings (Kaisermann et al., 2015; He et al., 2017). On the other hand, studies have shown that fungi were more resistant towards drought compared to bacteria (Barnard et al., 2013; Meisner et al., 2018) or that both exhibited a similar response (Sayer et al., 2017; Kaurin et al., 2018). For example, Barnard et al. (2013) investigated in active (RNA-based) and entire (DNA-based) bacterial and fungal communities in grassland bulk soil. They found that fungal community composition exhibited a marked resistance to changes in water availability, whereas only the active bacterial community responded towards desiccation (Barnard et al., 2013). These contradictory findings between studies might be attributed to different investigated compartments (bulk vs. rhizosphere soil), differences in experimental settings such as drought intensities or soil characteristics, and precipitation history in soil which has been shown to affect microbial communities and their response towards drought (Santos-Medellin et al., 2017; Kaisermann et al., 2015; Kaisermann et al. 2017).

3.4 Crop species and genotype influenced response of bacteria and fungi towards water deficit

We further evaluated whether crop species and genotypes had an influence on microbial communities. We found a significantly higher fungal diversity in wheat rhizosphere compared to faba bean specific for harvest 2 (KW-test, x^2= 4.3, df=1, p=0.038), whereas bacterial diversity was unaffected by crop species. A significant difference between genotypes on bacterial and fungal alpha-diversity was not observed. In general, crop species explained 2.1 % (PERMANOVA, p=0.003; PERMDISP, F=4.73, p=0.029) and 2.0 % (PERMANOVA, p=0.001) of the variance in the bacterial and fungal dataset. Faba bean genotype significantly influenced bacterial community composition specific for harvest 1 but fungi were completely unaffected. Here, genotype explained 6.1 % of the variance (PERMANOVA, p=0.01). Moreover, several taxa were more abundant in one of the two crop species or faba bean genotypes (Figure 2, 3). Higher relative abundances of the bacterial genus *Rhizobium* were more often found in faba bean rhizosphere (4.12 %) compared to wheat (0.6 %). In addition, the fungal genus *Polythrincium* was frequent in faba bean rhizosphere especially in genotype S_004 with 27.8 % relative abundance compared to S_062 (19.04 %) and wheat (3.34 %).

In line with our results, previous studies reported that plant identity is one of the important factors shaping the microbial community in the rhizosphere soil (Dawson et al., 2017; Zhou et al., 2017). For example, Zhou et al. (2017) investigated in three legume and grass species grown in a mesocosms and demonstrated that legume and grass differentially shaped the bacterial and fungal community composition and diversity in the soil. They also found that fungal diversity was significantly higher in grass compared to different legume species, whereas bacteria showed the opposite effect, indicating that bacteria and fungi respond differently towards plant identity (Zhou et al., 2017). In accordance to our results, Li et al. (2018) found that bacterial community composition in the rhizosphere was substantially different between two rice cultivars. However, bacterial alpha-diversity and fungi displayed no responsiveness. In addition, they indicated that cultivar dependent effects were stronger for bacteria than for fungi (Li et al., 2018) which was similar to our observation. They explained that observed changes in bacterial community composition were related to alterations in pH and Bt protein concentration in the soil (Li et al., 2018). Similar, legume and grass species not only differently affect edaphic properties such as pH but also differ in their quantity and quality of root exudates (Siczek et al., 2018; Zhou et al., 2017) which might also explain observed changes towards crop species.

We further hypothesized that crop species and genotype would alter the response of bacterial and fungal communities towards water deficit. In the rhizosphere of S_004, we recorded significantly lower fungal diversity and richness in water deficit compared to well-watered plants specific for harvest 1 (KW-test, shannon, x^2=5.0, df=1, p=0.025; richness, x^2=5.0, df=1, p=0.025) (Table 5).

Evaluation of microbial community with NMDS showed that only bacterial community composition for faba bean genotype S_062 exhibited a distinct clustering between control and water deficit treatment for harvest 1 (Figure 5). PERMANOVA also confirmed that water deficit significantly influenced bacterial community composition for genotype S_062 and explained 21.1 % of the variance (PERMANOVA, p=0.002). However, dispersion among water treatments was not homogenous (PERMDISP, F=18.63, p=0.002).

Previous studies indicated that indirect environmental factors such as plant species which are also influenced by drought might play a larger role in altering microbial communities than direct effects of desiccation (Kaisermann et al., 2017; deVries et al., 2018). Water deficit can lead to plant stress that changes plant metabolism including the composition and quality of plant residuals such as root exudates (Henry et al., 2007; Preece and Peñuelas, 2016). As crop species but also genotypes differ in their susceptibility towards water deficit, changes in root exudation might also be different and thus, the response of the microbial community (Preece and Peñuelas, 2016). Similarly, Santos-Medellin and coworkers (2017) investigated in four different rice cultivars and plant compartments and recorded compartment-specific cultivar effects on drought response for the bacterial community composition. However, they only found few individual OTUs which showed differential responses to drought based on genotype, indicating that communities assembled in each cultivar responded similar towards drought. Partly in line with this study, we observed that response of bacterial community composition towards water deficit was dependent on faba bean genotype.

3.5 Cropping system influenced fungal community but not bacteria in the rhizosphere soil

We further evaluated the influence of cropping system on the microbial community composition and diversity. We found no significant effect of cropping system on bacterial and fungal diversity and richness. NMDS analysis showed for fungi a clustering between the cropping system WIC and WM especially for harvest 1 and 2 (Figure 6). However, PERMANOVA found that cropping system only significantly affected fungal community composition in wheat for harvest 2. Here, cropping system explained 9.8 % of the variance in the fungal dataset (PERMANOVA, p=0.024). In contrast, bacterial community composition was completely unaffected by cropping system. In accordance to this result, Wang et al. (2012) showed that fungal community composition in the rhizosphere of wheat in monoculture was significantly different compared to wheat in intercropping system, whereas bacteria were not influenced. They also explained that soil type and crop species were the main effects which influenced microbial communities in the rhizosphere soil (Wang et al., 2012). Other studies reported that intercropping and monoculture significantly affected bacterial and/or fungal diversity which was in contrast to our results (Yang et al., 2016; Li and Wu, 2018). Li and Wu (2018) showed that from seven intercropping systems only the combination of cucumber/mustard and cucumber/trifolium increased bacterial and fungal diversity in bulk soil compared to cucumber monoculture, indicating that crop species exhibited a strong influence on microbial communities.

Moreover, we found that differences between water treatments were pronounced for a specific cropping system. For example, fungal diversity and richness was significantly lower in the rhizosphere of S_004 in FBM_D compared to FBM_C for harvest 1 (KW-test, shannon, x^2=4.5, df=1, p=0.033; richness, x^2=4.5, df=1, p=0.033). In the cropping system WM, we observed significantly higher fungal richness (KW-test, H2, x^2=3.85, df=1, p=0.049) under water deficit, whereas bacterial diversity (KW-test, H1, x^2=4.08, df=1, p=0.043) showed significantly lower diversity compared to the control treatment. In contrast, microbial community composition was not influenced by the combination of water deficit and cropping system. We speculate that the response of bacterial and fungal diversity and/or richness towards water deficit under a specific cropping system might be attributed to intraspecific below-ground water competition which was increased under monoculture resulting in a more sensitive microbial community.

3.6 Associated bacterial and fungal taxa as well as microbial interactions are altered by water deficit and cropping system

To identify bacterial and fungal taxa responsible for the observed differences among water deficit and cropping system, we performed a multipattern analysis to investigate which microorganisms are significantly associated with those treatments (Table S6). In general, the wheat cropping regimes harbored the highest number of associated bacterial and fungal taxa, whereas faba bean cropping regimes the least number (Figure 7). Most significant associated bacterial taxa were shared between cropping regimes and we found that the cropping regimes WM_C and WM_D had the most uniquely associated bacterial taxa for all sampling times. However, identity of associated bacterial taxa changed over time and between cropping regimes. For example, drought cropping regimes especially from faba bean plants showed more associated bacterial taxa from the phylum Actinobacteria than well-watered plants which was most pronounced for harvest 1. In addition, Bacteroidetes was associated more often with FBIC_D for harvest 3 than in harvest 1 or 2. In contrast, number of associated fungal taxa varied between the three harvests (Figure 7). For example, most unique associated fungal taxa were found in the cropping regime WM_D for harvest 1 and 2, whereas for harvest 3 most unique associated fungi were found in WM_C. The main fungal classes associated with drought especially in WM were assigned to *Agaricomycetes* and *Dothideomycetes* for harvest 1 and 2.

In accordance to our results, previous studies observed an enrichment of the bacterial phylum Actinobacteria under drought stress in root endosphere, bulk as well as rhizosphere soil (Kavamura et al., 2013; Naylor et al., 2017; Santos-Medellin et al., 2017). As Actinobacteria are well-known to be highly tolerant for life in arid environments, they might increase in abundance under drought, whereas sensitive taxa diminish (Bull and Asenjo, 2013; Kavamura et al., 2013). Similarly, Kavamura et al. (2013) found that the phylum Bacteroidetes strongly correlated with rainy season in soil, whereas Actinobacteria with dry season. Moreover, Meisner and coworkers (2018) showed that the bacterial phylum Bacteriodetes was enriched when soil had a drought history which might additionally explain the increased number of associated bacterial taxa in the water deficit treatment for the re-watering phase. They also indicated that fungal OTUs belonging to *Dothideomycetes* but also to *Agaricomycetes* responded sensitive towards drought which is in accordance to our result.

Observed taxa are frequently described in plant microbiome surveys (Gdanetz et al., 2017; Naylor et al., 2017) but their specific roles in association with plants under water deficit remains unclear. However, we speculate that crops under water deficit selected competent microorganisms which provide the crops some degree of tolerance or assist in their development through growth promotion (Goh et al., 2013; Coleman-Derr and Tringe, 2014).

We further investigated the effect of cropping regimes on inter-and intra-domain interactions of fungi and bacteria. We calculated the number of significant correlations between OTUs for each harvest. Positive interactions (indicating species co-occurrence) are regarded indicative for cooperation, whereas negative interactions indicate avoidance or competition. In general, bacteria had more total significant interactions but less positive interactions than fungi (Table 7). Inter-domain interactions displayed less positive interactions than bacteria or fungi. In addition, faba bean rhizosphere had more positive intra-and inter-domain interactions than wheat. We observed a marked increase of positive intra-domain interactions in the fungal community in WIC_C compared to WM_C in each harvest. For bacteria, we recorded more positive correlations under water deficit in wheat compared to well-watered conditions. For example, the cropping regime WIC_D (67.58 %) and WM_D (64.34 %) showed more positive bacterial intra-domain interactions compared to WIC_C (57.19 %) or WM_C (56.25 %) for harvest 2. For inter-domain interactions between bacteria and fungi, we observed no consistent pattern. For example, number of positive inter-domain correlations decreased in the cropping regime FBM_D (25.25 %) and FBIC_D (38.49 %) compared to FBIC_C (77.75 %) in the re-watering phase. In contrast, higher abundance of positive inter-domain interactions was observed in WM_D (43.08 %) and WIC_C (55.99 %) compared to WM_C (18.19 %) for harvest 3.

Similar to our results, deVries et al. (2018) found that in general, fungal networks contained fewer negative correlations than bacterial networks in grassland bulk soil. Furthermore, deVries et al. (2018) showed that drought reduced the proportion of negative correlations in bacteria, which was in accordance with our observations but specific in the wheat rhizosphere. Another study by Li and Wu (2018) reported that only a specific crop species combination from seven intercropping systems showed an increase of positive bacterial and/or fungal correlations compared to monoculture which was in line with our observations in fungi under WIC. For our findings, we speculate that changes in interactions might be related to shifts in water availability that might reduce competitive ability of dominant microbial taxa towards other taxa which are better adapted to the current moisture content (Kaisermann et al., 2015). As mentioned above, different crop species differ in their root exudation profile which might also affect interactions within the plant microbiome (Zhou et al., 2017). As indicated by previous research (Granzow et al., 2017; Kaisermann et al., 2017), we further assume that inter- and intraspecific competition between plants for water (or nutrients) in the specific cropping system had different effects on each crop species and thus on their associated microbial communities. Bacteria and fungi co-occur in the same habitat, the rhizosphere; however, they differ in their lifestyle in terms of colonization area which might further explain differences in the observed results towards water deficit and cropping systems. For example, bacterial habitats are reduced to soil particle of few mm^3 or in specific zones in a biofilm on roots (Deveau et al., 2018). In contrast, fungi have an extended and exploratory hyphal network with which they locally interact with other plants, microorganisms and microfauna (Deveau et al., 2018).

4 Conclusion

Our study provides novel findings of the response of the active microbial communities in the rhizosphere soil towards water deficit and cropping system in two important agricultural crops using Illumina MiSeq sequencing. In accordance to our hypotheses, we found that both bacterial and fungal communities were altered by water deficit; however, they responded differently towards drought. Changes of bacterial community composition were dependent on crop species and genotype, whereas alpha-diversity showed a marked resistance towards water deficit. In contrast, fungal community composition responded more sensitive towards water deficit but fungal alpa-diversity was altered dependent on crop genotype. Cropping system alone changed only fungal community composition but not bacteria. However, we recorded complex changes in microbial interactions when considering water deficit and cropping system. Obtained results highlight that the combination of crop species, genotype and cropping system play key roles in the response of the active microbiome in the rhizosphere soil towards drought. Further research on field-scale might deepen our understanding how sustainable agricultural practices and plant-associated microorganisms might mitigate future drought events.

5 Acknowledgment

This study is part of the project IMPAC[3] and was funded by the Federal Ministry of Education and Research (FKZ 031A351A). The authors thank Prof. Dr. Wolfgang Link from the Working Group "Breeding Research Faba Bean" (Division of Plant Breeding at the University of Göttingen) for providing the seed material.

Tables

Table 1. Sampling numbers for each container and harvest.

Treatments /Compartments	ID	Rhizosphere	Plants/treatment
	Harvest 1		
Faba bean monoculture S_004	S4_FBM	1 (8/7)	8
Faba bean monoculture S_062	S62_FBM	1 (8/7)	8
Faba bean intercropping S_004	S4_FBIC	1 (8/8)	8
Faba bean intercropping S_062	S62_FBIC	1 (8/4)	8
Wheat monoculture	WM	2 (8/5)	16
Wheat intercropped	WIC	2 (15/12)	32
	Harvest 2		
Faba bean monoculture S_004	S4_FBM	1 (7/6)	8
Faba bean monoculture S_062	S62_FBM	1 (7/6)	8
Faba bean intercropping S_004	S4_FBIC	1 (8/8)	8
Faba bean intercropping S_062	S62_FBIC	1 (8/8)	8
Wheat monoculture	WM	2 (8/6)	16
Wheat intercropped	WIC	2 (15/11)	32
	Harvest 3		
Faba bean monoculture S_004	S4_FBM	1 (8/6)	8
Faba bean monoculture S_062	S62_FBM	1 (7/4)	8
Faba bean intercropping S_004	S4_FBIC	1 (8/6)	8
Faba bean intercropping S_062	S62_FBIC	1 (8/6)	8
Wheat monoculture	WM	2 (7/6)	16
Wheat intercropped	WIC	2 (14/10)	32
Total (for each harvest)		64	32(FB), 48(W)
Total (all)		192	240

WM, wheat in monoculture; FBM, faba bean in monoculture, FBIC, faba bean samples in intercropping; WIC, wheat samples in intercropping. Numbers before brackets refer to sampled plants per pot. Numbers in brackets refer to the number of samples left after removal of samples with too low sequencing numbers. First number in brackets refers to bacteria, second to fungi. Harvest 1 refers to "beginning of water deficit", harvest 2 refers to "during water deficit" and harvest 3 refers to "re-watering". Sample size (n) for the cropping system WIC was 16 and for the other cropping systems, n=8.

Table 2. pH value in the rhizosphere of wheat and faba bean genotypes.

Treatment	Harvest 1	Harvest 2	Harvest 3
Wheat_C	**7.10±0.06A**	**7.05±0.06A**	**7.39±0.01B**
Wheat_D	**7.11±0.06A**	**7.02±0.05B**	**7.43±0.02C**
WIC_C	7.05±0.08	**6.93±0.04a**	7.39±0.01
WIC_D	7.12±0.08	**6.91±0.02a**	7.42±0.03
WM_C	7.20±0.09	**7.30±0.06b**	7.40±0.02
WM_D	7.08±0.10	**7.23±0.02b**	7.44±0.04
S4_C	**7.12±0.09AB**	**7.11±0.04A**	**7.37±0.01B**
S4_D	**7.10±0.09A**	**7.00±0.05A**	**7.36±0.01B**
S4_FBIC_C	7.25±0.13	**7.03±0.06ab**	7.39±0.01
S4_FBIC_D	7.24±0.10	**6.94±0.03a**	7.37±0.01
S4_FBM_C	7.03±0.12	**7.20±0.01b**	7.35±0.00
S4_FBM_D	6.97±0.13	**7.07±0.08ab**	7.35±0.02
S62_C	**7.18±0.08AB**	**6.91±0.04A**	**7.36±0.01B**
S62_D	**7.02±0.07A**	**6.95±0.04A**	**7.36±0.01B**
S62_FBIC_C	7.03±0.12	**6.82±0.03a**	**7.38±0.00a**
S62_FBIC_D	7.02±0.10	**6.84±0.04ab**	**7.36±0.00ab**
S62_FBM_C	7.34±0.01	**7.01±0.01ab**	**7.34±0.01b**
S62_FBM_D	7.01±0.11	**7.03±0.01b**	**7.36±0.01ab**

Different small and large letters in columns and rows indicate statistically significant differences between treatments (Dunn's-test or Kruskal-Wallis-test, $p \leq 0.05$, means ± SE). Abbreviations: FBM/WM, faba bean/wheat grown in monoculture; FBIC/WIC, faba bean/wheat intercropped; C, control treatment; D, water deficit treatment. Harvest 1, beginning of water deficit; Harvest 2, during water deficit; Harvest 3, re-watering.

Table 3. Carbon and nitrogen [%] in the rhizosphere of wheat and faba bean genotypes.

	C:N ratio			C_{total} [%]			N_{total} [%]		
Treatment	**Harvest 1**	**Harvest 2**	**Harvest 3**	**Harvest 1**	**Harvest 2**	**Harvest 3**	**Harvest 1**	**Harvest 2**	**Harvest 3**
Wheat_C	11.12±0.21	10.99±0.21	10.78±0.34	**2.02±0.04A**	**2.03±0.04A**	**1.56±0.14B**	**0.18±0.00A**	**0.19±0.01A**	**0.14±0.01B**
Wheat_D	11.46±0.15	10.97±0.25	11.28±0.14	**2.04±0.04A**	**2.02±0.03A**	**1.74±0.08B**	**0.18±0.00A**	**0.19±0.00A**	**0.15±0.01B**
WIC_C	11.02±0.31	**10.59±0.14a**	10.67±0.51	2.03±0.03	2.05±0.02	**1.34±0.16a**	0.19±0.00	0.19±0.00	**0.12±0.01a**
WIC_D	11.32±0.17	**10.96±0.38a**	11.16±0.20	2.01±0.04	2.07±0.03	**1.60±0.07a**	0.18±0.01	0.19±0.01	**0.14±0.01a**
WM_C	11.34±0.21	**11.81±0.19b**	11.01±0.20	2.01±0.10	1.98±0.12	**2.02±0.02b**	0.18±0.01	0.17±0.01	**0.18±0.00b**
WM_D	11.75±0.25	**11.00±0.08ab**	11.53±0.10	2.12±0.05	1.92±0.02	**2.04±0.03b**	0.18±0.00	0.18±0.00	**0.18±0.00b**
S4_C	11.15±0.35	11.28±0.48	11.50±0.27	2.03±0.04	2.06±0.13	1.87±0.08	0.18±0.01	0.19±0.01	0.16±0.00
S4_D	11.43±0.15	11.26±0.33	11.47±0.31	2.10±0.03	2.07±0.16	1.85±0.09	**0.18±0.00A**	**0.19±0.02A**	**0.16±0.01B**
S4_FBIC_C	**10.32±0.23a**	**10.21±0.20a**	11.32±0.54	2.08±0.05	2.09±0.01	**1.79±0.12ab**	**0.20±0.00a**	0.21±0.01	0.16±0.01
S4_FBIC_D	**11.10±0.15ab**	**10.46±0.16ab**	10.88±0.05	2.08±0.02	2.05±0.02	**1.63±0.17a**	**0.19±0.00ab**	0.20±0.00	0.15±0.01
S4_FBM_C	**11.98±0.24b**	**12.35±0.51b**	11.67±0.13	1.98±0.07	2.03±0.28	**1.96±0.09ab**	**0.17±0.01b**	0.17±0.02	0.17±0.01
S4_FBM_D	**11.76±0.12b**	**12.06±0.23b**	12.05±0.46	2.13±0.05	2.09±0.35	**2.06±0.03b**	**0.18±0.00ab**	0.17±0.03	0.17±0.01
S62_C	11.47±0.10	11.63±0.26	11.34±0.33	1.96±0.04	1.88±0.10	1.79±0.09	0.17±0.00	0.16±0.01	0.16±0.01
S62_D	**12.11±0.37A**	**10.88±0.35B**	**11.41±0.19AB**	1.86±0.09	1.70±0.19	1.64±0.11	0.16±0.01	0.15±0.02	0.14±0.01
S62_FBIC_C	11.35±0.17	**11.06±0.29a**	11.16±0.57	2.01±0.03	**2.12±0.04a**	1.69±0.08	0.18±0.00	**0.19±0.00a**	0.15±0.01
S62_FBIC_D	11.53±0.19	**11.12±0.14ab**	11.41±0.39	1.93±0.07	**2.09±0.04a**	1.42±0.15	0.17±0.01	**0.19±0.01a**	0.13±0.01
S62_FBM_C	11.59±0.11	**12.19±0.12b**	11.53±0.41	1.90±0.08	**1.65±0.07ab**	1.90±0.17	0.16±0.01	**0.14±0.01b**	0.17±0.02
S62_FBM_D	12.68±0.62	**10.65±0.72ab**	11.41±0.14	1.79±0.16	**1.32±0.26b**	1.85±0.05	0.15±0.02	**0.12±0.02b**	0.16±0.00

Different small and large letters in columns and rows indicate statistically significant differences between treatments (Dunn's-test or Kruskal-Wallis-test, $p \leq 0.05$, means ± SE). Abbreviations: FBM/WM, faba bean/wheat grown in monoculture; FBIC/WIC, faba bean/wheat intercropped; C, control treatment; D, water deficit treatment. Harvest 1, beginning of water deficit; Harvest 2, during water deficit; Harvest 3, re-watering.

Table 4. Bacterial richness and diversity in the rhizosphere soil with regard to water treatments and cropping systems.

	Richness			Diversity		
Treatment	**1**	**2**	**3**	**1**	**2**	**3**
Wheat_C	845.25±74.03	866.07±55.6	794.57±223.71	6.19±0.22	6.24±0.12	5.82±1.12
Wheat_D	841.90±48.62	818.37±105.68	691.02±341.75	6.18±0.13	6.03±0.59	5.35±1.8
WIC_C	826.56±87.46	859.44±68.18	845.96±67.03	6.14±0.26	6.22±0.14	6.10±0.33
WIC_D	848.04±56.29	802.91±115.8	609.23±433.16	6.20±0.14	5.94±0.71	4.87±2.26
WM_C	877.98±26.99	877.68±26.91	657.53±442.47	6.29±0.05	6.27±0.06	5.08±2.19
WM_D	831.15±36.04	849.28±88.03	813.70±64.94	6.14±0.12	6.21±0.22	6.09±0.24
S4_C	744.71±295.43	516.54±382.25	805.37±95.79	5.76±1.17	4.49±2.14	5.93±0.62
S4_D	795.20±40.3	792.59±113.95	751.91±145.78	6.04±0.14	5.85±0.86	5.77±0.78
S4_FBIC_C	806.70±118.42	470.05±384.88	753.83±108.6	6.09±0.21	4.50±1.9	5.58±0.88
S4_FBIC_D	821.95±31.61	833.45±70.86	630.77±129.17	6.14±0.06	6.17±0.2	5.19±0.9
S4_FBM_C	698.23±398.1	578.53±453.94	844.03±76.41	5.51±1.59	4.47±2.89	6.19±0.17
S4_FBM_D	759.53±5.82	738.10±153.72	842.78±75.47	5.91±0.1	5.43±1.29	6.21±0.26
S62_C	780.94±87.63	867.72±47.53	740.08±289.75	5.99±0.34	6.21±0.15	5.51±1.63
S62_D	835.76±39.45	614.13±368.85	820.79±90.57	6.15±0.18	4.94±1.94	6.15±0.26
S62_FBIC_C	771.65±117.25	859.73±64.46	816.08±136.46	6.02±0.33	6.16±0.21	5.91±0.79
S62_FBIC_D	846.80±32.91	562.87±420.59	834.40±75.33	6.18±0.15	4.71±1.86	6.17±0.24
S62_FBM_C	793.33±44.83	875.70±36.08	664.08±402.33	5.94±0.43	6.26±0.07	5.12±2.27
S62_FBM_D	824.73±47.16	652.58±386.75	802.63±123.42	6.11±0.23	5.11±2.26	6.12±0.34

Diversity is expressed as Shannon values (H') and richness is based on the number of unique sequences. Different small and large letters in columns and rows indicate statistically significant differences between treatments (Dunn's-test or Kruskal-Wallis-test, $p \leq 0.05$, means ± SD). Abbreviations: FBM/WM, faba bean/wheat grown in monoculture; FBIC/WIC, faba bean/wheat intercropped; C, control treatment; D, water deficit treatment. Harvest 1, beginning of water deficit; Harvest 2, during water deficit; Harvest 3, re-watering.

Table 5. Fungal richness and diversity in the rhizosphere soil with regard to water treatments and cropping systems.

	Richness			Diversity		
Treatment	**1**	**2**	**3**	**1**	**2**	**3**
Wheat_C	69.09±25.61	66.66±17.21	57.80±23.82	3.29±0.94	3.22±0.75	3.05±0.71
Wheat_D	49.07±32.26	58.94±29.1	48.44±34.02	2.50±1.3	3.22±0.98	2.24±1.65
WIC_C	66.34±28.71	66.48±21.59	49.57±23.93	3.19±1.06	3.06±0.9	2.78±0.67
WIC_D	42.54±36.48	43.98±31.06	50.43±42.9	2.29±1.52	2.80±1.16	2.40±2.08
WM_C	78.70±8.2	67.00±4.42	77.00±6.72	3.64±0.19	3.53±0.15	3.67±0.27
WM_D	65.40±13.15	78.90±6.89	45.45±30.33	3.04±0.39	3.79±0.12	2.00±1.44
S4_C	**73.90±24.99a**	49.20±30.59	59.80±14.86	**3.49±0.86a**	2.59±1.26	3.21±0.63
S4_D	**47.11±27.59b**	50.93±30.07	42.78±22.46	**2.30±1.18b**	2.58±1.16	2.54±0.95
S4_FBIC_C	65.65±32.2	60.65±19.38	64.00±13.3	3.26±1.14	3.18±0.6	3.34±0.66
S4_FBIC_D	46.97±28.04	46.15±37.35	28.30±22.2	2.36±1.24	2.29±1.35	2.20±0.55
S4_FBM_C	84.90±0.5	37.75±38.19	43.00±NA	3.80±0.15	2.01±1.57	2.70±NA
S4_FBM_D	47.23±31.6	60.50±7.78	50.03±21.6	2.26±1.32	3.16±0.44	2.71±1.14
S62_C	21.38±22.65	58.14±25.79	30.20±39.29	1.57±1.37	2.86±0.79	1.62±1.69
S62_D	57.18±23.84	36.77±30.85	27.06±31.05	2.82±1.01	2.32±1.41	1.77±1.37
S62_FBIC_C	34.10±NA	44.75±27.04	47.97±43.63	2.97±NA	2.45±0.77	2.36±1.88
S62_FBIC_D	61.90±23.96	29.08±17.46	25.63±37.18	2.92±1.11	2.45±1.04	1.55±1.65
S62_FBM_C	17.13±25.72	76.00±7.85	3.55±0.07	1.11±1.23	3.41±0.44	0.51±0.45
S62_FBM_D	52.47±27.93	47.03±46.05	29.20±32.81	2.71±1.13	2.13±2.07	2.10±1.29

Diversity is expressed as Shannon values (H') and richness is based on the number of unique sequences. Different small and large letters in columns and rows indicate statistically significant differences between treatments (Dunn's-test or Kruskal-Wallis-test, $p \leq 0.05$, means ± SD). Abbreviations: FBM/WM, faba bean/wheat grown in monoculture; FBIC/WIC, faba bean/wheat intercropped; C, control treatment; D, water deficit treatment. Harvest 1, beginning of water deficit; Harvest 2, during water deficit; Harvest 3, re-watering.

Table 6. Effects of the tested parameters on bacterial and fungal community composition for each harvest.

	Bacteria						Fungi					
	Harvest 1		Harvest 2		Harvest 3		Harvest 1		Harvest 2		Harvest 3	
Treatment	R^2 (%)	*p*	R^2 (%)	*p*	R^2 (%)	*p*	R^2 (%)	*p*	R^2 (%)	*p*	R^2 (%)	*p*
Cropping system	2.1	0.23	1.1	0.916	1.3	0.826	1.7	0.812	1.7	0.832	2.7	0.382
Crop species	**4.2**	**0.005**	**3.4**	**0.049**	2.1	0.319	**5.8**	**0.001**	**4.9**	**0.001**	**4.7**	**0.003**
Genotype	**6.1**	**0.01**	2.7	0.649	2.6	0.724	3.2	0.735	2.9	0.823	5.4	0.221
Water deficit	2.5	0.125	2.2	0.227	1.7	0.559	**4.1**	**0.017**	**3.2**	**0.049**	2.9	0.243
Harvest	**2.2**	**0.015**					**2.0**	**0.033**				

Results of the permutational multivariate analysis of variance (PERMANOVA) with Bray-Curtis distances testing for the different treatments. Statistically significant differences ($p \leq 0.05$) between the treatments for each plant compartment are written in bold. Cropping systems compares monoculture versus intercropping. Genotype compares S_004 versus S_062. Harvest was tested for all harvests together without strata.

Table 7. Positive (+) and negative (-) relative interactions with regard to cropping regimes.

		Harvest 1			Harvest 2			Harvest 3		
	Treatment	**Total**	**+ (%)**	**- (%)**	**Total**	**+ (%)**	**- (%)**	**Total**	**+ (%)**	**- (%)**
B:B	**FBIC_D**	67155	56.16	43.84	182063	97.49	2.51	89622	80.96	19.04
	FBIC_C	26456	58.88	41.12	79873	98.29	1.71	105301	87.93	12.07
	FBM_D	94036	76.59	23.41	49951	64.31	35.69	151945	58.16	41.84
	FBM_C	59102	54.93	45.07	99886	98.09	1.91	75704	54.82	45.18
	WIC_D	47469	62.88	37.12	81741	67.58	32.42	8	62.50	37.50
	WIC_C	91633	57.54	42.46	109847	57.19	42.81	173569	65.40	34.60
	WM_D	0	NA	NA	49164	64.34	35.66	49641	53.36	46.64
	WM_C	119471	52.52	47.48	48673	56.26	43.74	4251	85.42	14.58
F:F	**FBIC_D**	950	87.68	12.32	545	100	0	1	100	0
	FBIC_C	550	91.09	8.91	3326	99.10	0.90	1812	100	0
	FBM_D	8672	97.52	2.48	190	100	0	305	100	0
	FBM_C	668	100.00	0.00	1771	99.94	0.06	0	NA	NA
	WIC_D	4996	99.98	0.02	66	100	0	0	NA	NA
	WIC_C	5419	98.41	1.59	1915	72.17	27.83	4240	99.88	0.12
	WM_D	0	NA	NA	178	56.74	43.26	22	100	0
	WM_C	31	87.10	12.90	479	52.40	47.60	2556	71.48	28.52
B:F	**FBIC_D**	13123	56.15	43.85	7629	68.62	31.38	317	38.49	61.51
	FBIC_C	4691	59.41	40.59	10866	64.64	35.36	19946	77.49	22.51
	FBM_D	22273	73.05	26.95	2471	54.11	45.89	10765	25.25	74.75
	FBM_C	7747	62.99	37.01	15845	71.65	28.35	71	45.07	54.93
	WIC_D	7182	44.46	55.54	2025	41.73	58.27	0	NA	NA
	WIC_C	39402	52.97	47.03	23299	53.80	46.20	32121	55.99	44.01
	WM_D	0	NA	NA	5213	52.31	47.69	1706	43.08	56.92
	WM_C	3738	48.50	51.50	9400	53.59	46.41	5482	18.19	81.81

Total refers to total number of significant interactions. Abbreviations: C, control treatment/sufficiently irrigated; D, water deficit, drought treatment; FBM/WM, faba bean/wheat monoculture; FBIC/WIC, faba bean/ wheat intercropped; B:B, bacterial intra-domain interactions; F:F, fungal intra-domain interations; B:F, bacterial and fungal inter-domain interactions.

Figures

Figure 1. Experimental design. Abbreviations: FBM/WM, faba bean/wheat monoculture; IC, intercropping; C, control (blue container); D, water deficit treatment (red container).

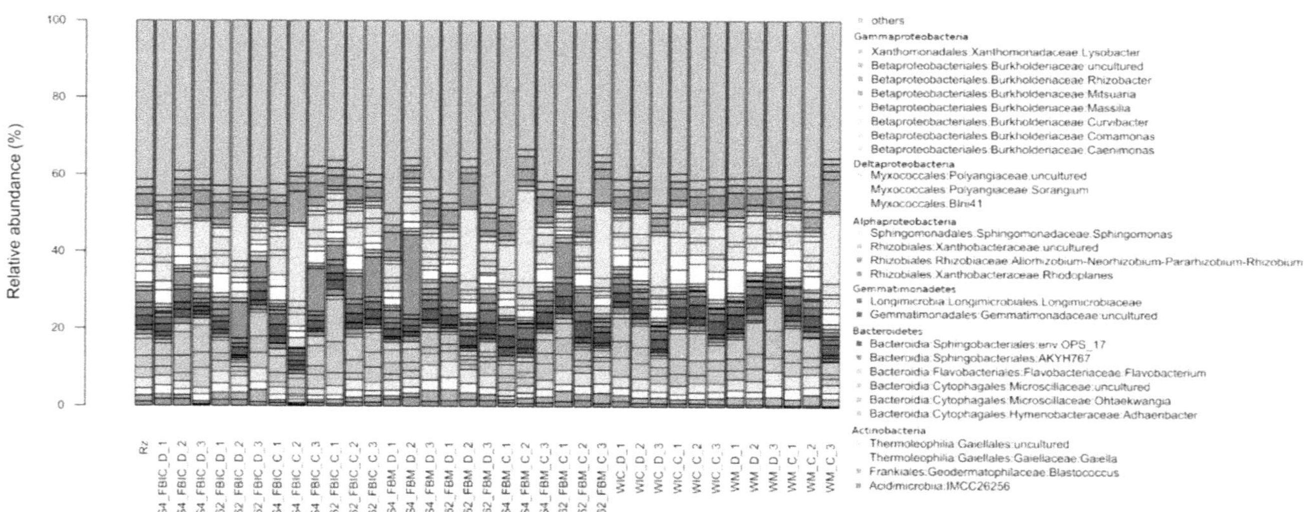

Figure 2. Abundant bacterial genera in the rhizosphere soil and the investigated cropping systems with regard to water treatment and harvest. Only genera with an abundance >1% in at least one of the investigated cropping systems are shown. Mean relative abundances of each taxon were calculated based on relative abundances calculated for each sample. Abbreviations: C, control treatment; D, water deficit treatment; S4/S62, faba bean genotype; FBM/WM, faba bean/ wheat monoculture, FBIC/WIC, faba bean/wheat intercropped.

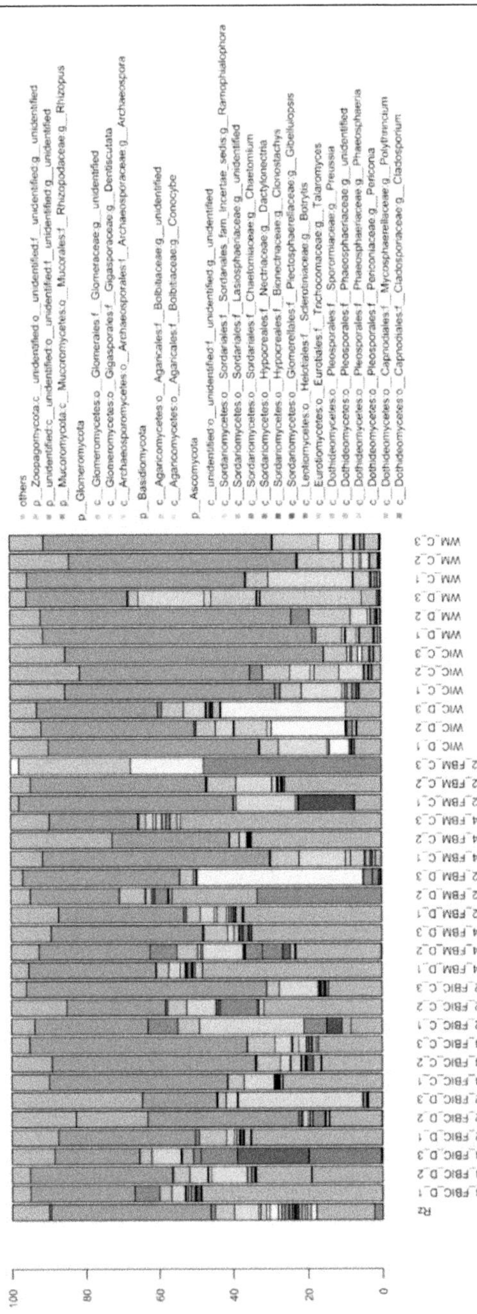

Figure 3. Abundant fungal genera in the rhizosphere soil and the investigated cropping systems with regard to water treatment and harvest. Only genera with an abundance >0.05% in at least one of the investigated cropping systems are shown. Mean relative abundances of each taxon were calculated based on relative abundances calculated for each sample. Abbreviations: C, control treatment; D, water deficit treatment; S4/S62, faba bean genotype; FBM/WM, faba bean/ wheat monoculture, FBIC/WIC, faba bean/wheat intercropped.

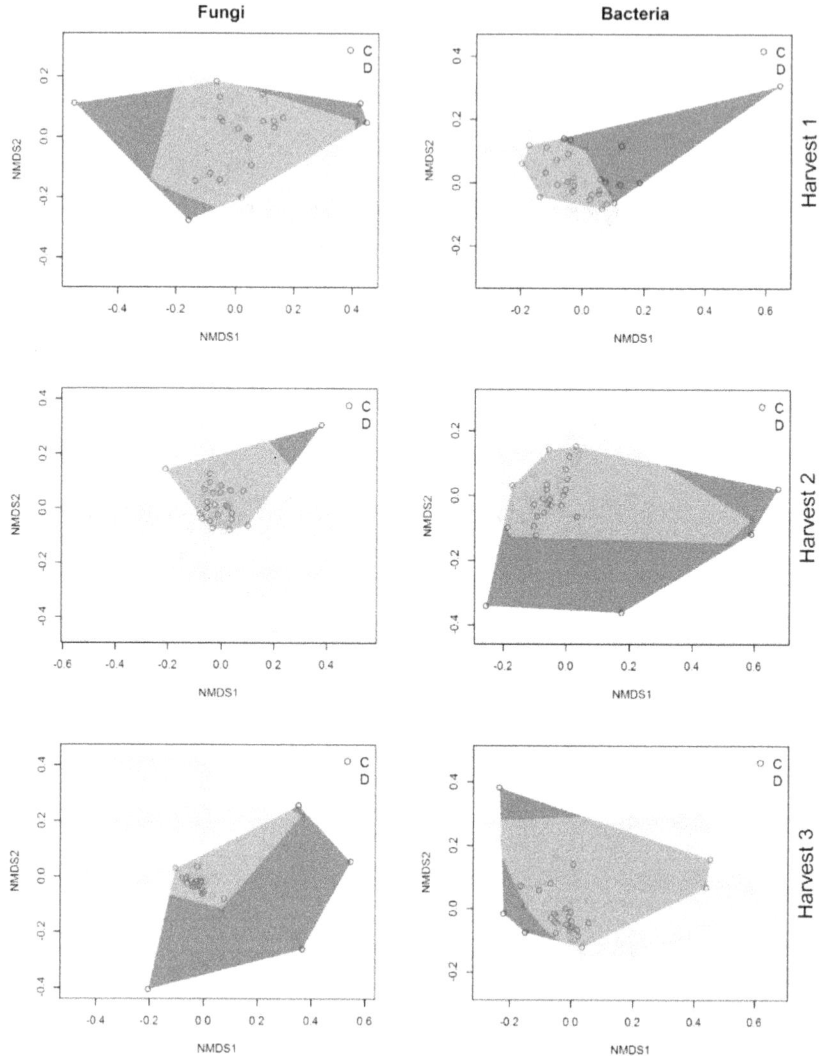

Figure 4. Response of bacterial and fungal communities in the rhizosphere soil towards water treatment. Ordination is based on Bray-Curtis dissimilarities between samples. NMDS ordination of microbial community is color-coded by the respective water treatment. Abbreviations: C, control treatment; D, water deficit treatment.

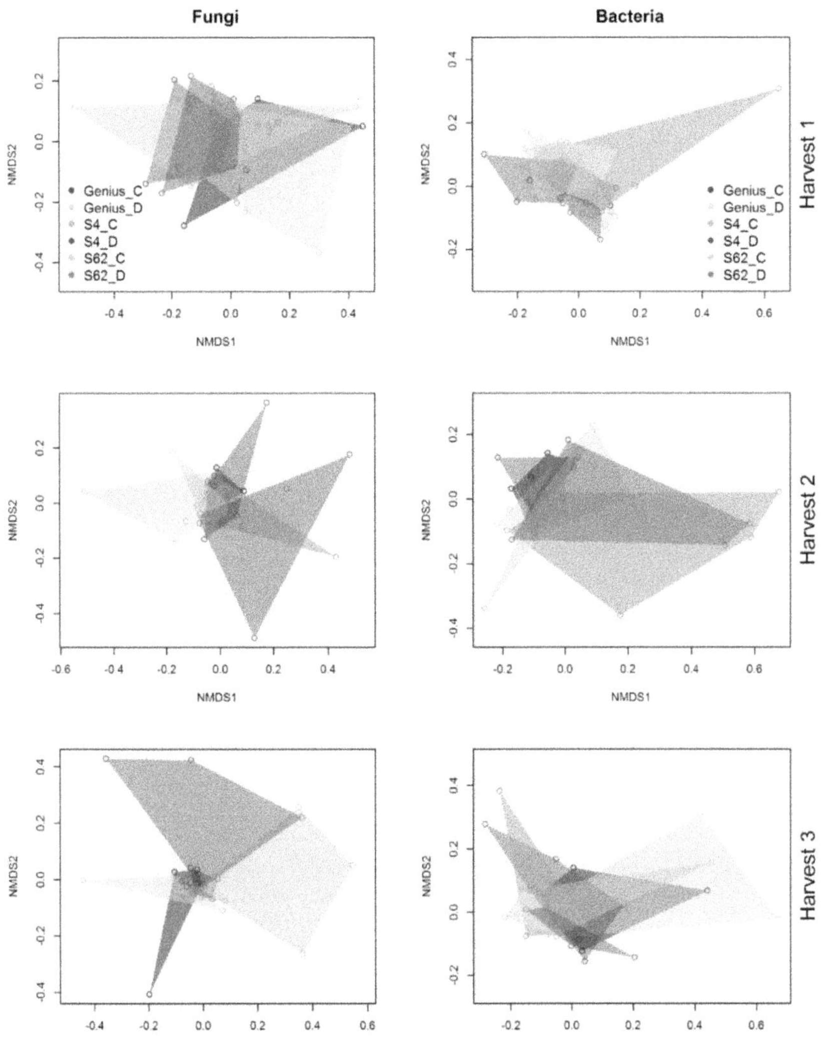

Figure 5. Response of bacterial and fungal communities in the rhizosphere soil towards water treatment regarding the different crop genotypes. Ordination is based on Bray-Curtis dissimilarities between samples. NMDS ordination of microbial community is color-coded by the respective water treatment and genotype. Abbreviations: S4/S62, faba bean genotype; C, control; D, water deficit.

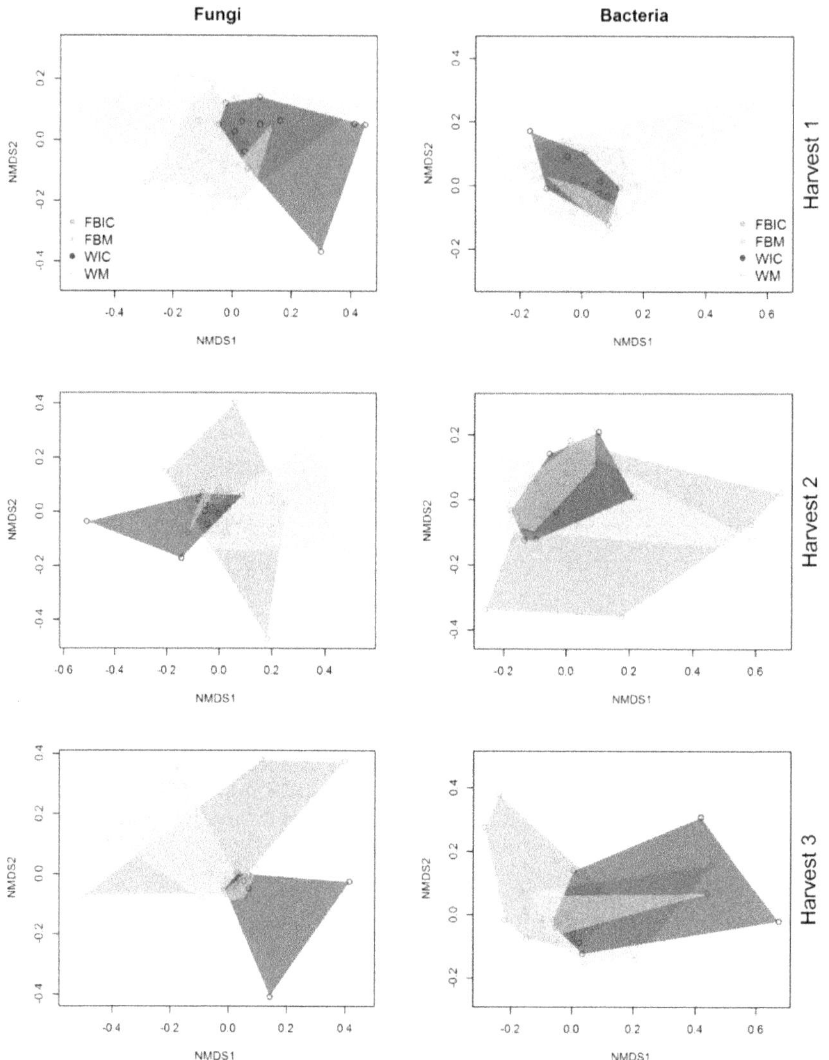

Figure 6. Response of bacterial and fungal communities in the rhizosphere soil towards cropping system. Ordination is based on Bray-Curtis dissimilarities between samples. NMDS ordination of microbial community is color-coded by the respective cropping system. Abbreviations: FBM/WM, faba bean/wheat monoculture; FBIC/WIC, faba bean/wheat intercropping.

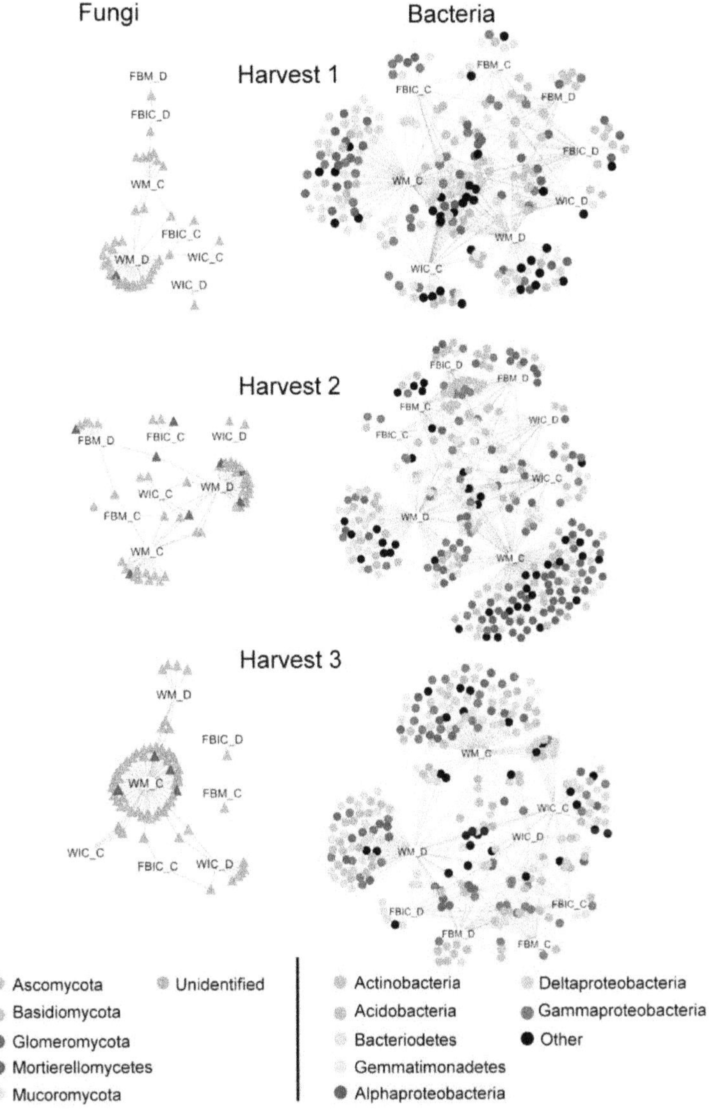

Figure 7. Bipartite association network for bacterial and fungal taxa within different cropping regimes for the three harvests. Significant associated taxa are shown. Abbreviations: FBM/WM, faba bean/wheat monoculture; FBIC/WIC, faba bean/wheat intercropped; C, control; D, water deficit.

6 References

Ahkami A, White RA, Handakumbura P, Jansson C. 2017. Rhizosphere Engineering: Enhancing Sustainable Plant Ecosystem Productivity in a Challenging Climate. Rhizosphere. Jun;3:233-243.

Anderson MJ. 2001. A new method for non-parametric multivariate analysis of variance. Austral Ecology. Feb;26:32-46.

Anderson MJ. 2006. Distance-based tests for homogeneity of multivariate dispersions. Biometrics. Mar;62:245-253.

Andreote FD, da Rocha UN, Araujo WL, Azevedo JL, van Overbeek LS. 2010. Effect of bacterial inoculation, plant genotype and developmental stage on root-associated and endophytic bacterial communities in potato (Solanum tuberosum). Antonie Van Leeuwenhoek International Journal of General and Molecular Microbiology. May;97:389-399.

Austin AT, Yahdjian L, Stark JM, Belnap J, Porporato A, Norton U, Ravetta DA, Schaeffer SM. 2004. Water pulses and biogeochemical cycles in arid and semiarid ecosystems. Oct; 141: 221-235.

Barnard RL, Osborne CA, Firestone MK. 2013. Responses of soil bacterial and fungal communities to extreme desiccation and rewetting. Isme Journal. Nov;7:2229-2241.

Berg G, Grube M, Schloter M, Smalla K. 2014. Unraveling the plant nnicrobiome: looking back and future perspectives. Frontiers in Microbiology. Jun;5.

Blagodatskaya E, Kuzyakov Y. 2013. Active microorganisms in soil: Critical review of estimation criteria and approaches. Soil Biology & Biochemistry. Dec;67:192-211.

Bodenhausen N, Horton MW, Bergelson J. 2013. Bacterial Communities Associated with the Leaves and the Roots of Arabidopsis thaliana. Plos One. Feb;8:9.

Bolger AM, Lohse M, Usadel B. 2014. Trimmomatic: a flexible trimmer for Illumina sequence data. Bioinformatics. Aug;30:2114-2120.

Bull AT, Asenjo JA. 2013. Microbiology of hyper-arid environments: recent insights from the Atacama Desert, Chile. Antonie Van Leeuwenhoek International Journal of General and Molecular Microbiology. Jun;103:1173-1179.

Chelius MK, Triplett EW. 2001. The diversity of archaea and bacteria in association with the roots of Zea mays L. Microbial Ecology. Apr;41:252-263.

Coleman-Derr D, Tringe SG. 2014. Building the crops of tomorrow: advantages of symbiont-based approaches to improving abiotic stress tolerance. Frontiers in Microbiology. Jun;5.

Daryanto S, Wang LX, Jacinthe PA. 2017. Global synthesis of drought effects on cereal, legume, tuber and root crops production: A review. Agricultural Water Management. Jan;179:18-33.

Dawson W, Hor J, Egert M, van Kleunen M, Pester M. 2017. A Small Number of Low-abundance Bacteria Dominate Plant Species-specific Responses during Rhizosphere Colonization. Frontiers in Microbiology. May;8.

de Caceres M, Legendre P. 2009. Associations between species and groups of sites: indices and statistical inference. Ecology. Dec;90:3566-3574.

Deveau A. Bonito G, Uehling J, Paoletti M, Becker M, Bindschedler S, Hacquard S, Herve V, Labbe J, Lastovetsky OA, Mieszkin S, Millet LJ, Vajna B, Junier P, Bonfante P, Krom BP, Olsson S, Elsas JD, Wick LY. 2018.Bacterial–fungal interactions: ecology, mechanisms and challenges. FEMS Microbiol Rev. May; 42: 335-352.

de Vries FT, Griffiths RI, Bailey M, Craig H, Girlanda M, Gweon HS, Hallin S, Kaisermann A, Keith AM, Kretzschmar M, et al. 2018. Soil bacterial networks are less stable under drought than fungal networks. Nature Communications. Aug;9.

Edgar RC. 2010. Search and clustering orders of magnitude faster than BLAST. Bioinformatics. Oct;26:2460-2461.

Edgar RC, Haas BJ, Clemente JC, Quince C, Knight R. 2011. UCHIME improves sensitivity and speed of chimera detection. Bioinformatics. Aug;27:2194-2200.

Fahad S, Bajwa AA, Nazir U, Anjum SA, Farooq A, Zohaib A, Sadia S, Nasim W, Adkins S, Saud S, et al. 2017. Crop Production under Drought and Heat Stress: Plant Responses and Management Options. Frontiers in Plant Science. Jun;8.

Fox J, Weisberg S. 2011. An R companion to applied regression. Second Edition. Sage.

Goh CH, Vallejos DFV, Nicotra AB, Mathesius U. 2013. The Impact of Beneficial Plant-Associated Microbes on Plant Phenotypic Plasticity. Journal of Chemical Ecology. Jul;39:826-839.

Granzow S, Kaiser K, Wemheuer B, Pfeiffer B, Daniel R, Vidal S, Wemheuer F. 2017. The Effects of Cropping Regimes on Fungal and Bacterial Communities of Wheat and Faba Bean in a Greenhouse Pot Experiment Differ between Plant Species and Compartment. Frontiers in Microbiology. May;8.

Hartman K, van der Heijden MGA, Roussely-Provent V, Walser JC, Schlaeppi K. 2017. Deciphering composition and function of the root microbiome of a legume plant. Microbiome. Jan;5.

He D, Shen WJ, Eberwein J, Zhao Q, Ren LJ, Wu QLL. 2017. Diversity and co-occurrence network of soil fungi are more responsive than those of bacteria to shifts in precipitation seasonality in a subtropical forest. Soil Biology & Biochemistry. Dec;115:499-510.

Henry A, Doucette W, Norton J, Bugbee B. 2007. Changes in crested wheatgrass root exudation caused by flood, drought, and nutrient stress. Journal of Environmental Quality. May-Jun;36:904-912.

Herzog S, Wemheuer F, Wemheuer B, Daniel R. 2015. Effects of Fertilization and Sampling Time on Composition and Diversity of Entire and Active Bacterial Communities in German Grassland Soils. Plos One. Dec;10.

Kaisermann A, de Vries FT, Griffiths RI, Bardgett RD. 2017. Legacy effects of drought on plant-soil feedbacks and plant-plant interactions. New Phytologist. Sep;215:1413-1424.

Kaisermann A, Maron PA, Beaumelle L, Lata JC. 2015. Fungal communities are more sensitive indicators to non-extreme soil moisture variations than bacterial communities. Applied Soil Ecology. Feb;86:158-164.

Kaurin A, Mihelic R, Kastelec D, Grcman H, Bru D, Philippot L, Suhadolc M. 2018. Resilience of bacteria, archaea, fungi and N-cycling microbial guilds under plough and conservation tillage, to agricultural drought. Soil Biology & Biochemistry. May;120:233-245.

Kavamura VN, Taketani RG, Lanconi MD, Andreote FD, Mendes R, de Melo IS. 2013. Water Regime Influences Bulk Soil and Rhizosphere of Cereus jamacaru Bacterial Communities in the Brazilian Caatinga Biome. Plos One. Sep;8.

Kembel SW, Cowan PD, Helmus MR, Cornwell WK, Morlon H, Ackerly DD, Blomberg SP, Webb CO. 2010. Picante: R tools for integrating phylogenies and ecology. Bioinformatics. Jun;26:1463-1464.

Kõljalg U, Nilsson RH, Abarenkov K, Tedersoo L, Taylor AFS, Bahram M, Bates ST, Bruns TD, Bengtsson-Palme J, Callaghan TM, et al. 2013. Towards a unified paradigm for sequence-based identification of fungi. Molecular Ecology. Nov;22:5271-5277.

Kuchenbuch RO, Buczko U. 2011. Re-visiting potassium- and phosphate-fertilizer responses in field experiments and soil-test interpretations by means of data mining. Journal of Plant Nutrition and Soil Science. Apr;174:171-185.

Lancashire PD, Bleiholder H, Vandenboom T, Langeluddeke P, Stauss R, Weber E, Witzenberger A. 1991. A UNIFORM DECIMAL CODE FOR GROWTH-STAGES OF CROPS AND WEEDS. Annals of Applied Biology. Dec;119:561-601.

Li P, Ye SF, Liu H, Pen AH, Ming F, Tang XM. 2018. Cultivation of Drought-Tolerant and Insect-Resistant Rice Affects Soil Bacterial, but Not Fungal, Abundances and Community Structures. Frontiers in Microbiology. Jun;9.

Li S, Wu FZ. 2018. Diversity and Co-occurrence Patterns of Soil Bacterial and Fungal Communities in Seven Intercropping Systems. Frontiers in Microbiology. Jul;9.

Mahoney AK, Yin CT, Hulbert SH. 2017. Community Structure, Species Variation, and Potential Functions of Rhizosphere-Associated Bacteria of Different Winter Wheat (Triticum aestivum) Cultivars. Frontiers in Plant Science. Feb;8.

Martinez Arbizu P. 2017. Pairwiseadonis: Pairwise multilevel comparison using adonis. R Package Version 0.0.1.

McMurdie PJ, Holmes S. 2014. Waste Not, Want Not: Why Rarefying Microbiome Data Is Inadmissible. Plos Computational Biology. Apr;10.

Meisner A, Jacquiod S, Snoek BL, ten Hooven FC, van der Putten WH. 2018. Drought Legacy Effects on the Composition of Soil Fungal and Prokaryote Communities. Frontiers in Microbiology. Mar;9.

Naylor D, DeGraaf S, Purdom E, Coleman-Derr D. 2017. Drought and host selection influence bacterial community dynamics in the grass root microbiome. Isme Journal. Dec;11:2691-2704.

Ogle DH. 2016. Introductory fisheries analyses with R. Chapman & Hall/CRC.

Oksanen O, Blanchet FG, Kindt R, Legendre P, Minchin PR, O'Hara RB, Simpson GL, Solymos P, Stevens MHH, Wagner H. 2016. Vegan: Community Ecology Package. R Package Version 2.3-5.

Pinheiro J, Bates D, DebRoy S, Sarkar D and R Core Team. 2017. nlme: Linear and Nonlinear Mixed Effects Models. R package version 3.1-131.

Preece C, Peñuelas J. 2016. Rhizodeposition under drought and consequences for soil communities and ecosystem resilience. Plant and Soil. Dec;409:1-17.

Quast C, Pruesse E, Yilmaz P, Gerken J, Schweer T, Yarza P, Peplies J, Glockner FO. 2013. The SILVA ribosomal RNA gene database project: improved data processing and web-based tools. Nucleic Acids Research. Jan;41:D590-D596.

R Core Team. 2016. R: A Language and Environment for Statistical Computing. Vienna: R Foundation for Statistical Computing. Available online at: http://www.R-project.org

Ritz C, Streibig JC. 2016. Package 'drc' Analysis of dose-response curves. Version 3.0-1.

Saharan K, Schütz L, Kahmen A, Wiemken A, Boller T, Mathimaran N. 2018. Finger Millet Growth and Nutrient Uptake Is Improved in Intercropping With Pigeon Pea Through "Biofertilization" and "Bioirrigation" Mediated by Arbuscular Mycorrhizal Fungi and Plant Growth Promoting Rhizobacteria. Front. Environ. Sci. Jun; 11.

Santos-Medellin C, Edwards J, Liechty Z, Nguyen B, Sundaresan V. 2017. Drought Stress Results in a Compartment-Specific Restructuring of the Rice Root-Associated Microbiomes. Mbio. Jul-Aug;8.

Sayer EJ, Oliver AE, Fridley JD, Askew AP, Mills RTE, Grime JP. 2017. Links between soil microbial communities and plant traits in a species-rich grassland under long-term climate change. Ecology and Evolution. Feb;7:855-862.

Schimel J, Balser TC, Wallenstein M. 2007. Microbial stress-response physiology and its implications for ecosystem function. Ecology. Jun;88:1386-1394.

Senbayram M, Trankner M, Dittert K, Bruck H. 2015. Daytime leaf water use efficiency does not explain the relationship between plant N status and biomass water-use efficiency of tobacco under non-limiting water supply. Journal of Plant Nutrition and Soil Science. Aug;178:682-692.

Shannon P, Markiel A, Ozier O, Baliga NS, Wang JT, Ramage D, Amin N, Schwikowski B, Ideker T. 2003. Cytoscape: A software environment for integrated models of biomolecular interaction networks. Genome Research. Nov;13:2498-2504.

Siczek A, Frac M, Kalembasa S, Kalembasa D. 2018. Soil microbial activity of faba bean (Vicia faba L.) and wheat (Triticum aestivum L.) rhizosphere during growing season. Applied Soil Ecology. Sep;130:34-39.

Song YN, Zhang FS, Marschner P, Fan FL, Gao HM, Bao XG, Sun JH, Li L. 2007. Effect of intercropping on crop yield and chemical and microbiological properties in rhizosphere of wheat (Triticum aestivum L.), maize (Zea mays L.), and faba bean (Vicia faba L.). Biology and Fertility of Soils. Jun;43:565-574.

Streit W, Daniel R. 2010. Metagenomics. Methods and Protocols. Springer Science+Business Media, LLC. Series Volume 668.

Tichy L, Chytry M. 2006. Statistical determination of diagnostic species for site groups of unequal size. Journal of Vegetation Science. Dec;17:809-818.

Toju H, Tanabe AS, Yamamoto S, Sato H. 2012. High-Coverage ITS Primers for the DNA-Based Identification of Ascomycetes and Basidiomycetes in Environmental Samples. Plos One. Jul;7.

Vandermeer JH. 1992. The ecology of intercropping. New York, NY: Cambridge University Press.

Vimal SR, Singh JS, Arora NK, Singh S. 2017. Soil-Plant-Microbe Interactions in Stressed Agriculture Management: A Review. Pedosphere. Apr;27:177-192.

Wan XH, Huang ZQ, He ZM, Yu ZP, Wang MH, Davis MR, Yang YS. 2015. Soil C:N ratio is the major determinant of soil microbial community structure in subtropical coniferous and broadleaf forest plantations. Plant and Soil. Feb;387:103-116.

Wang Y, Marschner P, Zhang FS. 2012. Phosphorus pools and other soil properties in the rhizosphere of wheat and legumes growing in three soils in monoculture or as a mixture of wheat and legume. Plant and Soil. May;354:283-298.

Wemheuer F, Kaiser K, Karlovsky P, Daniel R, Vidal S, Wemheuer B. 2017. Bacterial endophyte communities of three agricultural important grass species differ in their response towards management regimes. Scientific Reports. Jan;7.

Wemheuer F, Wemheuer B, Kretzschmar D, Pfeiffer B, Herzog S, Daniel R, Vidal S. 2016. Impact of grassland management regimes on bacterial endophyte diversity differs with grass species. Letters in Applied Microbiology. Apr;62:323-329.

Wemheuer B, and Wemheuer F. 2017. "Assessing bacterial and fungal diversity in the plants endosphere," in Metagenomics - Methods and Protocols, Vol. 1539, edsW. Streit, and R. Daniel (New York, NY: Humana Press), 75–84.

White TJ, Bruns T, Lee S, and Taylor J. 1990. "Amplification and direct sequencing of fungal ribosomal RNA genes for phylogenetics," in PCR Protocols: a Guide to Methods and Applications, eds M. A. Innis, D. H. Gelfand, J. J. Sninsky, and T. J.White (New York, NY: Academic Press). 18, 315–322.

Xue C, Penton CR, Zhu C, Chen H, Duan YH, Peng C, Guo SW, Ling N, Shen QR. 2018. Alterations in soil fungal community composition and network assemblage structure by different long-term fertilization regimes are correlated to the soil ionome. Biology and Fertility of Soils. Jan;54:95-106.

Yang CH, Huang GB, Chai Q, Luo ZX. 2011. Water use and yield of wheat/maize intercropping under alternate irrigation in the oasis field of northwest China. Field Crops Research. Dec;124:426-432.

Yang ZP, Yang WP, Li SC, Hao JM, Su ZF, Sun M, Gao ZQ, Zhang CL. 2016. Variation of Bacterial Community Diversity in Rhizosphere Soil of Sole-Cropped versus Intercropped Wheat Field after Harvest. Plos One. Mar;11.

Zampieri M, Ceglar A, Dentener F, Toreti A. 2017. Wheat yield loss attributable to heat waves, drought and water excess at the global, national and subnational scales. Environmental Research Letters. Jun;12.

Zhou Y, Zhu HH, Fu SL, Yao Q. 2017. Variation in Soil Microbial Community Structure Associated with Different Legume Species Is Greater than that Associated with Different Grass Species. Frontiers in Microbiology. May;8.

Chapter 2.4:
Crop species and cropping system alter the effect of *Metarhizium brunneum* seed application on plant-associated bacterial and fungal communities

Sandra Granzow, Hadis Jayanti, Annika Meißner, Birgit Pfeiffer, Rolf Daniel, Stefan Vidal, Franziska Wemheuer

Manuscript draft

Crop species and cropping system alter the effect of *Metarhizium brunneum* seed application on plant-associated bacterial and fungal communities

Granzow Sandra[1,2*], Jayanti Hadis[1], Meißner Annika[2,3], Pfeiffer Birgit[3,4], Daniel Rolf[4], Vidal Stefan[1] and Wemheuer Franziska[1]

[1]Division of Agricultural Entomology, Department of Crop Sciences, University of Göttingen, Göttingen, Germany

[2]Center of Biodiversity and Sustainable Land Use, University of Göttingen, Göttingen, Germany

[3] Institute of Applied Plant Nutrition, University of Göttingen, Göttingen, Germany

[4]Genomic and Applied Microbiology and Göttingen Genomics Laboratory, Institute of Microbiology and Genetics, University of Göttingen, Göttingen, Germany

*** Correspondence:**
Dr. Sandra Granzow
sandra.granzow@agr.uni-goettingen.de

Keywords: microbial communities, intercropping, *Metarhizium brunneum*, seed inoculation

Abstract

Entomopathogenic fungi are frequently used as biocontrol agents in a sustainable agriculture. In the last years, they received more attention due to their various ecological functions as endophytes such as plant growth promotion. To date, our knowledge how the application of entomopathogenic fungi influences microbial communities in soil and plant endosphere in different cropping systems is still limited. Hence, we investigated the separate and combined effect of seed inoculation with *Metarhizium brunneum* Cb15-III and cropping system on bacterial and fungal communities in the rhizosphere soil, root as well as leaf endosphere of winter faba bean (*Vicia faba* L.) and winter wheat (*Triticum aestivum* L.) using large-scale metabarcoding. For this purpose, faba bean and wheat were grown in monoculture and in row intercropping under greenhouse conditions. Plant and soil samples were collected after five (harvest 1) and seven weeks (harvest 2). In accordance to our first hypothesis we found that crop species and plant compartment exhibited a strong influence on the plant microbiome. Furthermore, *M. brunneum* application altered the fungal and bacterial community composition in the rhizosphere soil and bacterial community composition in the leaf endosphere for both sampling times. In addition, microbial diversity and richness showed sampling time- and kingdom-specific responses towards *M. brunneum* application. For example, a significantly lower fungal diversity and richness in leaf endosphere and rhizosphere soil of inoculated wheat compared to control plants was observed at harvest 2 only. Moreover, cropping system alone but also the combination of cropping system and application significantly affected plant microbiome. Fungal diversity and richness in root endosphere were significantly higher in intercropped wheat compared to monoculture and these differences were most pronounced between inoculated wheat cropping systems for harvest 2. For bacteria, we observed the same trend, but, in the rhizosphere, bacterial diversity and richness was significantly higher in intercropped wheat compared to monoculture which was most pronounced in inoculated wheat cropping systems for harvest 1. In addition, fungal application induced proliferation of specific bacteria in the rhizosphere, namely *Shewanella* spp. and *Halomonas* spp., which might be putatively important in plant growth promoting, biological control or in bioremediation. Overall, our findings highlight the importance to investigate the separate and combined effect of cropping system and application of entomopathogenic fungi on plant-associated microbial communities. The present findings increase our understanding of how the application of an entomopathogenic fungus affects microbial community which might gain further importance for biological control strategies in the future.

1 Introduction

Entomopathogenic fungi (EPF) are frequently used as biocontrol agents in sustainable agriculture worldwide, mainly applied as spore suspension to the aerial plant parts (Bing and Lewis, 1991; Batta et al., 2013). Batta et al. (2013) showed that leaf inoculation of *Metarhizium anisopliae* on *Brassica napus* plants resulted in an increased mortality rate of *Plutella xylostella* larvae. Results indicated that the fungus colonized the internal tissue and acted as endophyte antagonistic against *P. xylustella* larvae (Batta et al., 2013). In the last years, EPF have received more attention due to their various ecological functions as endophytes (Lacey et al., 2015; Vidal and Jaber, 2015). EPF as endophytes have been shown to promote plant growth and yield (Sasan and Bidochka, 2012; Jaber and Enkerli, 2016, Sánchez-Rodriguez et al., 2018) and improve plant nutrition (Raya-Diaz et al., 2017; Krell et al., 2018). Recently, Krell et al. (2018) demonstrated that endophytism of *M. brunneum* through encapsulated application significantly enhanced biomass as well as nitrogen and phosphorus content in potato. Consequently, the application of EPF may contribute to more sustainable production systems in certain crop plants (de Faria and Wraight, 2007; Ortiz-Urquiza, Luo and Keyhani, 2015).

While progress has been made in understanding the effects of entomopathogenic fungal application on the plant host and/or plant herbivore pest interactions (e.g., Raya-Diaz et al., 2017; Clifton et al., 2018), our knowledge how the application of EPF influences microbial communities is still limited as most of these studies focused on the soil microbiome (Hirsch et al., 2013; Mayerhofer et al., 2017; McKinnon et al., 2018, but see

Hong et al., 2017). Recently, Mayerhofer et al. (2017) examined the potential effect of *M. brunneum* application on fungal and prokaryotic communities in the bulk soil using high-throughput sequencing. They observed that fungal application did not affect the indigenous microbial community under field conditions, whereas smaller shifts in the soil fungal community were observed under greenhouse conditions. In a study by Hong et al. (2017), the application of *M. anisopliae* strain CQMa421 slightly affected bacterial diversity and community composition in the rice phyllosphere during the first six days in the booting stage of rice growth, while no significant changes in fungal diversity were observed during this period. Other studies investigating the effects of agricultural practices on soil-borne EPF reported that pesticides (Hummel et al., 2002) or cropping practices (Kepler et al., 2015; Clifton et al.,2015) significantly affected EPF such as *Beauveria bassiana* or *M. anisopliae*. However, it remains largely unknown how different cropping systems might influence the effect of EPF application on plant-associated microbial communities.

Hence, the aim of the present study was to investigate the combined effect of *M. brunneum* Cb15-III seed inoculation and a specific cropping system on plant-associated bacterial and fungal communities of two important crop species. For this purpose, winter wheat (*Triticum aestivum* L.) and winter faba bean (*Vicia faba* L.*)* were grown in monoculture and in row intercropping in a greenhouse pot experiment. Half of the plant seeds of both crops were treated with *M. brunneum* (inoculated plants), the other half was left untreated (control plants). *M. brunneum* was chosen because it has been reported that this species is able to transfer nutrients to their host plants (Behie and Bidochka, 2014; Krell et al., 2018). Plant and soil samples were collected after five and seven weeks of plant growing. Bacterial and fungal communities in rhizosphere and unplanted soil as well as in root and leaf endosphere were examined using Illumina Miseq sequencing targeting the bacterial 16S rRNA gene and the fungal internal transcribed spacer (ITS) region, respectively. We focused on four main hypotheses: (i) bacterial and fungal community composition and diversity depends on crop species and plant compartment. (ii) The application and establishment of an endophytic fungus in plant tissues creates a long-lasting effect on these bacterial and fungal communities (thus: no resilience). We further expected that (iii) cropping system influences effects of fungal application on plant microbiome and that (iv) responses of bacterial and fungal communities differ between the different plant compartments.

2 Material and Methods

2.1 Plant material

To examine the influence of cropping systems and seed inoculation with *M. brunneum* strain Cb15-III on the entire fungal and bacterial community in the soil and the plant endosphere, a greenhouse experiment was conducted in autumn 2016. This experiment was part of the IMPAC³-project (*Novel genotypes for mixed cropping allow for improved sustainable land use across arable land, grassland and woodland*). Seeding material from winter faba bean (genotype: S_062) was selected from field trial-tested inbred lines used within the IMPAC³ project and was provided by the Institute of Plant breeding at the University of Göttingen. Seeds of winter wheat (genotype: Genius) were provided by Norddeutsche Pflanzenzucht

Hans-Georg Lembke KG. Seeds of the two crop species were surface-sterilized by serial washing according to Andreote et al. (2010), with one modification: after immersion in sterile, distilled water for two times and 30 s, seeds were additionally washed in sterile, diethylpyrocarbonate (DEPC)-treated water. Surface-sterilized seeds germinated on a moistened filter paper with sterile distilled water at 4 °C in darkness for 72 hours. To determine how the host plant itself alters the response of plant-associated fungal and bacterial communities towards seed inoculation, sterile glass beads (Merck, Germany; Ø 6 mm) were used as controls in unplanted containers. Glass beads were treated in the same way as described above for plant seeds.

2.2 Fungal material

The entomopathogenic fungus *M. brunneum* strain Cb15-III was obtained from the fungal collection of the Agricultural Entomology Laboratory at the University of Göttingen, Germany, originally isolated from a luvisoil arable field. Fresh cultures of this strain were prepared prior experimental start. Therefore, *M. brunneum* was cultivated on Potato Dextrose Agar (PDA; Carl Roth, Karlsruhe, Germany) at 24 °C for 10 to 14 days. Fungal conidia were harvested with a sterile object plate under sterile conditions and suspended in autoclaved 0.9 % NaCl bidest water according to Gu et al. (2016). The conidial suspension was filtered through one layer of sterile gauze to remove hyphae and agar remnants. Subsequently, the conidia suspension was carefully mixed in a flask. Conidial concentrations were determined under a microscope using a haemocytometer (Thoma) and adjusted to a concentration of $6 * 10^7$ conidia ml^{-1}. The viability of spore suspension was controlled on PDA plates. Seeds and glass beads were placed in the spore suspension (inoculated treatments) or in 0.9 % NaCl (none-inoculated control treatments) for 16 hours.

2.3 Experimental design

Inoculated and non-inoculated seeds of faba bean and wheat were sown as monoculture or mixture in polypropylene containers (Sunware; 455 x 360 x 240 mm). Each container was filled with air-dried, sieved (< 10 mm) and layered soil from the experimental study site in Reinshof (51.48 ° N, 9.92 ° E and 157 m asl.), Germany. The soil was classified as Gleyic Fluvisol according to the FAO classification system and contained 21 % clay, 68 % silt and 11 % sand. The soil volume of each container accounted for approximately 20 L with a dry weight of 18 kg. Filling of the container was performed in layers adding distilled water to each layer to prevent soil compaction. For monocultures, 30 faba bean or 72 wheat seeds per container were sown in rows. For the intercropping treatment, 15 faba bean and 36 wheat seeds were sown in alternate rows. Unplanted container with inoculated or non-inoculated (control) glass beads were treated in the same way as described for the monoculture treatment.

In total, there were ten different treatments: faba bean monoculture non-inoculated or inoculated with the fungus *M. brunneum* strain Cb15-III (MB/MBF), wheat monoculture non-inoculated or inoculated (MW/MWF), faba bean intercropped non-inoculated or inoculated (XB/XBF), wheat intercropped non-inoculated or inoculated (XW/XWF) and unplanted soil non-inoculated or inoculated (C/CF) (Figure 1). Each treatment was replicated five times in a randomized block design. We defined two different cropping systems (monoculture and intercropping), whereas cropping regimes compromise each treatment, e.g. XB and XBF. All plants were cultured under light (12:12 h light/dark regime) and irrigated regularly for a growing period of seven weeks. The position of all containers was changed weekly to avoid spatial effects. To increase nutrient-limitation as well as intra- and interspecies interactions between the plants, no fertilizer treatments were applied.

2.4 Sampling

To investigate the effect of *M. brunneum* application on the plant microbiome over time, we collected samples from all treatments after a growing period of five weeks (harvest 1; H1) and seven weeks (harvest 2; H2). Fungal and bacterial communities in three different plant compartments were studied: the rhizosphere soil as well as the root and aerial (here regarded as leaf) endosphere. Moreover, fungal and bacterial communities in the soil of unplanted containers treated with inoculated and non-inoculated glass beads were analysed to investigate how the two crop plant species alter the response of these communities towards seed inoculation. To gain insights into the starting soil microbial community (CS), three composite soil samples of all unplanted soil containers were collected prior to the experimental start.

At harvest 1, plants reached BBCH stage 14-18 (wheat) and 14-16 (faba bean). The BBCH-scale describes the developmental stages of Mono- and Dicotyledonous weed species (Hess et al., 1997). For molecular analysis, one faba bean plant as well as two wheat plants were randomly collected from each container (Table 1). At harvest 2, plants reached a BBCH stage 15-21 (wheat) and 18-21 (faba bean). For molecular analysis, two faba bean plants and three wheat plants were harvested as described for harvest 1. All collected plants showed no obvious sign of any disease infection. In addition, rhizosphere soil samples (the soil tightly attached to the roots) were taken. For this purpose, the roots were gently shaken to remove the non-rhizosphere soil and the rhizosphere soil was collected by carefully brushing the roots. Rhizosphere samples were pooled per container and plant species. In total, 60 faba bean plants and 100 wheat plants were collected for molecular analysis, resulting in 160 plant and 80 rhizosphere samples (Table 1). All samples were immediately stored at -20 °C until further analyses. Additional rhizosphere and soil samples of the unplanted containers were collected at both harvests for determination of edaphic properties including the soil organic C and N content. For determination of plant properties such as the water content, or organic C and N in roots and leaves were sampled at harvest 2. For a detailed description of the sampling procedure see Table S1.

2.5 Edaphic and plant parameters

As temperature of leaf canopy can be a stress indicator for fungal infection (Yao et al., 2018), we evaluated the transpiration of the plant canopy. Thermal images were taken weekly from the 4th to the 7th week using a T640 infrared camera (FLIR Systems, OR, USA). The thermal images were analyzed with the software FLIR ResearchIR version 3.3.12277.1002 (FLIR

Systems, OR, USA). In addition, the height of five faba bean and wheat plants per container was measured weekly. The fresh biomass of plants (including roots and aerial parts) collected for molecular analysis was measured separately for below and aerial parts at both harvest times. Additionally, at harvest 1, pH values and water content in unplanted and rhizosphere soils were measured. Total nitrogen (N) and carbon (C) in roots and leaves as well as the water content from roots and aerial parts of ten faba bean and 20 wheat plants per container were collected at harvest 2. Finally, relative water content (RWC) in leaves from two plants of each crop species and container were determined at harvest 2.

For the determination of edaphic and plant properties, subsamples of homogenized and mixed plant and soil material were dried at 60 °C for two days and sieved to < 2 mm. Soil pH values were measured as follows: 2 g rhizosphere soil or unplanted soil of each container was mixed with 5 ml PCR grade water. After incubation for 24 h, pH_{bidest} was measured in the supernatant with a glass electrode (WTW, inoLab). Finally, 0.37 g KCl was added and pH_{KCl} was measured. Soil and plant organic C and N concentrations were determined using a LECO TruSpec CN analyser (Leco Corp., St. Joseph, MI). The gravimetric soil and plant water content (%) was calculated from oven-dried subsamples. Relative water content (RWC) of leaves was determined according to Barrs and Weatherley (1962). In brief, the youngest fully expanded leaf was taken around solar noon and the fresh weight (FW) was measured. Afterwards, the leaf samples were incubated in distilled water in closed boxes at approximately 23 °C for three hours. Then, the turgid weight (TW) was determined and the leaf samples were dried at 60 °C for 24 h to examine the dry weight (DW). RWC was calculated as follows: $RWC [\%] = [(FW - DW) / (TW - DW)] * 100$. Details on edaphic and plant parameters are provided in Table S2 and S3.

2.6 Surface sterilization of plant material

Foliar plant material was surface-sterilized according to Wemheuer and Wemheuer (2017). Surface sterilization of plant roots was performed as described in Granzow et al. (2017). The effectiveness of the sterilization process used was controlled as described previously (Wemheuer et al., 2016). In brief, aliquots of the water used in the final wash step were plated on common laboratory media plates, i.e. Luria-Bertani agar and potato dextrose agar. The plates were incubated in the dark at 25 °C for at least one week. No growth of bacteria or fungi was observed. Moreover, water from the same aliquots was subjected to PCR targeting the bacterial 16S rRNA gene and the ITS region of fungi as described below. No PCR products were detected. Surface sterilized plant samples were ground using a sterile mortar and pestle with liquid nitrogen and subsequently stored at -20 °C until further analyses.

2.7 Extraction of total microbial community DNA

Total DNA of aerial plant parts and roots was extracted employing the peqGOLD Plant DNA Mini kit (Peqlab, Erlangen, Germany) according to the manufacturer's instructions with two modifications described previously (Wemheuer et al., 2016). In brief, glass beads (Ø 3-6 mm) were used in the first step and 10 µl Proteinase K (20 mg ml^{-1}; Carl Roth,

Karlsruhe, Germany) was added. Total environmental DNA of rhizosphere as well as unplanted soil was extracted employing the PowerSoil® DNA Isolation kit (Qiagen, Venlo, Netherlands) according to the manufacturer's protocol. DNA concentrations of DNA extracts were quantified using a NanoDrop ND-1000 spectrophotometer (NanoDrop Technologies, Wilmington, DE, USA). In total, 253 samples (Table 1) were subjected to PCR targeting the bacterial 16S rRNA gene and the fungal ITS region.

2.8 Amplification of the 16S rRNA gene

Bacterial endophyte and soil communities were assessed by a nested PCR approach targeting the 16S rRNA gene as described in Wemheuer and Wemheuer (2017). For details of the first PCR reaction mixture and the thermal cycling scheme see Wemheuer et al. (2016). Briefly, the primers 799f (5′-AACMGGATTAGATACCCKG-3′) (Chelius & Triplett, 2001) and 1492R (5′-GCYTACCTTGTTACGACTT-3′) (Lane, 1991) were used in the first PCR to suppress co-amplification of chloroplast-derived 16S rRNA genes. Obtained PCR products were subjected to nested PCR. The V6-V8 region of the 16S rRNA gene was amplified with primers 968F and 1401R (Nübel et al., 1996) containing MiSeq adaptors (underlined) (MiSeq-968F 5′-<u>TCGTCGGCAGCGTCAGATGTGTATAAGAGACAG</u>AACGCGAAGAACCTTAC-3′; MiSeq- 1401R 5′-<u>GTCTCGTGGGCTCGGAGATGTGTATAAGAGACAG</u>CGGTGTGTACAAGACCC-3′) as described previously (Wemheuer and Wemheuer, 2017) with one modification: 0.5 U of Phusion high fidelity DNA polymerase (Thermo Scientific, Waltham, MA, USA) was used. Three independent PCRs were performed per sample. Genomic DNA of *Escherichia coli* was used as control. Negative controls were performed using the reaction mixture without template. Obtained PCR products were pooled in equal amounts and purified using the NucleoMag NGS Clean up (Macherey-Nagel). Quantification of the PCR products was performed using the Quant-iT dsDNA HS assay kit and a Qubit fluorometer (Thermo Scientific) as recommended by the manufacturer. PCR products were barcoded using the Nextera XT-Index kit (Illumina, San Diego, USA) and the Kapa HIFI Hot Start polymerase (Kapa Biosystems, Wilmington, USA). The Göttingen Genomics Laboratory determined the sequences of the partial 16S rRNA employing the MiSeq Sequencing platform and the MiSeq Reagent Kit v3 (2 x 300 cycles) as recommended by the manufacturer (Illumina, San Diego, USA).

2.9 Amplification of the ITS region

Fungal communities in soil and endosphere were assessed by a nested PCR approach targeting the ITS region as described previously (Granzow et al., 2017; Wemheuer and Wemheuer, 2017). In the first PCR, the primers ITS1-F_KYO2 (5′-TAGAGGAAGTAAAAGTCGTAA-3′) (Toju et al., 2012) and ITS4 (5′- TCCTCCGCTTATTGATATGC-3′) (White et al., 1990) were used to suppress co-amplification of plant-derived ITS regions. Obtained PCR products were subjected to nested PCR. The ITS2 region was subsequently amplified as described for the first PCR using approximately 50 ng product of the first PCR and the primers ITS3_KYO2 (Toju et al., 2012) and ITS4 (White et al., 1990) containing the MiSeq adaptors (underlined): MiSeq-ITS3_KYO2 (5′-<u>TCGTCGGCAGCGTCAGATGTGTATAA</u>

GAGACAGGATGAAGAACGYAGYRAA-3′) and MiSeq-ITS4 (5′-GTCTCGTGGGCTC GGAGATGTGTATAAGAGACAGTCCTCCGCTTATTGATATGC -3′). Genomic DNA of *Aspergillus nidulans* was used as template in the positive control. Negative controls were performed using the reaction mixture without template. Three independent PCRs were performed per sample. Obtained PCR products were pooled in equal amounts and quantified as described for the bacterial PCR products. Pooled PCR products were barcoded using the Nextera XT-Index kit (Illumina, San Diego, USA) and the Kapa HIFI Hot Start polymerase (Kapa Biosystems, Wilmington, USA). The Göttingen Genomics Laboratory determined the sequences of the ITS2 region employing the MiSeq Sequencing platform and the MiSeq Reagent Kit v3 (2 x 300 cycles) as recommended by the manufacturer (Illumina, San Diego, USA).

2.10 Detection of *Metarhizium brunneum* Cb15-III with conventional PCR

For the identification of the inoculated *Metarhizium* isolate in the soil and the endosphere, a nested PCR approach was applied. *Metarhizium* clade 1 specific primers, namely Ma1763 (5'-CCAACTCCCAACCCCTGTGAAT-3') and Ma2079 (5'-AAAACCAGCAGCCTCGC CGAT-3') were used to amplify an approximately 320-bp region of the internal transcribed spacers (ITS) (Schneider et al., 2012b). Reaction mixture (25 µl) of the first PCR contained: 2.5 µl of 10-fold Mg-free *Taq*-polymerase buffer (Thermo Fisher Scientific, Braunschweig, Germany), 200 µM of each of the four desoxynucleoside triphosphates, 25 mM $MgCl_2$, 4 µM of each primer, 5 % DMSO, 1 U/µl of Taq DNA polymerase (Thermo Fisher Scientific) and approximately 25 ng DNA samples as template. The following thermal cycling scheme was utilized: initial denaturation at 95 °C for 2 min followed by 40 cycles of denaturation at 95 °C for 1 min, annealing at 55 °C for 1 min, extension at 72 °C for 1.5 min and followed by final extension at 72 °C for 5 min. Obtained PCR products were subjected to nested PCR with the same primer pair. Reaction mixture (25 µl) of the second PCR contained: 5 µl of 5-fold Phusion GC buffer, 200 µM of each of the four desoxynucleoside triphosphates, 4 µM of each primer, 5 % DMSO, 25 mM $MgCl_2$, 0.5 U of Phusion High Fidelity DNA polymerase (Thermo Fisher Scientific). The following thermal cycling scheme was utilized: initial denaturation at 98 °C for 30 sec followed by 30 cycles of denaturation at 98 °C for 15 sec, annealing at 56 °C for 30 sec, extension at 72 °C for 30 sec and followed by final extension at 72 °C for 2 min. Negative controls were performed using the reaction mixture without template. Extracted DNA of *Metarhizium brunneum* Cb15-III was used as a template in the positive control. Presence of *Metarhizium* was confirmed on a 2 % agarose gel.

2.11 Processing of microbial community datasets

Generated sequencing data were initially quality filtered with the Trimmomatic tool version 0.36 (Bolger et al., 2014). Low quality reads were truncated if the quality dropped below 15 in a sliding window of 4 bp. Subsequently, all reads shorter than 100 bp and orphan reads were removed. Remaining sequences were merged, quality-filtered and further processed with USEARCH version 10.0.240 (Edgar, 2010). Filtering included the removal of reads shorter than 250 bp or longer than 490 (fungi) and shorter than 400 or longer than 470 bp (bacteria) as well as the removal of low-quality reads (expected error > 1) and reads with more than one ambitious base.

Processed sequences of all samples were dereplicated, concatenated, and obtained unique sequences were denoised and clustered into zero-radius operational taxonomic units (zOTUs) with the unoise3 algorithm implemented in USEARCH version 10.0.240 (Edgar, 2010). All OTUs consisting of one single sequence (singletons) were removed. Subsequently, remaining chimeric sequences were removed using UCHIME (Edgar et al. 2011) in reference mode with the SILVA SSU Ref NR 99 132 database (Quast et al., 2013) as reference data set for bacteria and the QIIME release of the UNITE database version 7.2 (Kõljalg et al., 2013) for fungi. Filtered sequences were mapped on remaining unique sequences to determine the occurrence and abundance of each unique sequence in every sample. To assign taxonomy of bacteria and fungi, chimera-free sequences were classified by BLAST alignment against the most recent SILVA database (Quast et al., 2013) and the most recent UNITE database (Kõljalg et al., 2013), repectively, with an e-value threshold of 1e-20. Concatenated sequences of all sequences were mapped on the final set of unique sequences to calculate the evenness and abundance of each unique sequence in all samples. All non-bacterial or non-fungal zOTUs were removed based on their taxonomic classification in the respective database. Final zOTU tables for bacteria and fungi are provided in Tables S4 and S5, respectively. Only zOTUs occurring in more than one sample were considered for further statistical analysis. Samples with less than 22 (bacteria) and 16 (fungi) sequences per sample were removed prior statistical analysis, resulting in 229 samples for bacteria, and 231 samples for fungi.

2.12 Statistical Analysis

All statistical analyses were performed using R version 3.4.0 (R Core Team, 2016) and the packages therein. Differences were considered as statistically significant with $p \leq 0.05$. Differences in alpha or beta diversity as well as sequencing depth with regard to cropping system, treatment and inoculation (yes/no) were tested by a Kruskal-Wallis test. There were no significant differences of the mean sequencing depths between the intercropping or monocropping systems, treatments or inoculation. In consequence, zOTU tables were not rarefied as recommended by McMurdie and Holmes (2014).

A variety of alpha diversity indices (Richness, Shannon index of diversity and Michaelis-Menten Fit) were calculated using the R-packages *picante* version 1.6-2 (Kembel et al., 2010) and *drc* version 3.0-1 (Ritz and Streigbig, 2016). Sample coverage was estimated using the Michaelis-Menten Fit calculated in R. For this purpose, richness and rarefaction curves were calculated utilizing the *picante* package (Kembel et al., 2010). OTU tables were rarefied using the *rrarefy* function in *vegan* version 2.4.4 and samples with less than 17,177 (soil bacteria), 4,172 (rhizosphere soil bacteria), 508 (root bacteria), 22 (leaves bacteria), 6,105 (soil, fungi), 596 (rhizosphere soil, fungi), 61 (root, fungi) and 16 (leaves, fungi) sequences were removed prior alpha diversity analysis. Richness and diversity were calculated using the *specnumber* and *diversity* functions, respectively. The Michaelis-Menten Fit was subsequently calculated from generated rarefaction curves using the MM2 model within the *drc* package (Ritz and Streibig, 2016). All alpha diversity indices were calculated ten times. The average from each iteration was used for further statistical analysis. Final tables containing bacterial and fungal richness and diversity are provided in Tables S6 and S7, respectively.

Differences in community composition were investigated by permutational multivariate analysis of variance (PERMANOVA; Anderson, 2001) based on Bray-Curtis distance matrices using 999 permutations. A significant *p*-value in PERMANOVA for beta-diversity can be driven by true biological differences, differences within treatment (variance) or both (Anderson, 2001). In case of significant *p*-values in PERMANOVA, we tested for differences in homogeneity using permutational analysis of multivariate dispersions (PERMDISP, Anderson, 2006) with 999 permutations. NMDS, PERMANOVA and PERMDISP were run using functions; *metaMDS*, *adonis* and *betadisper*, respectively, in the R package *vegan* (Oksanen et al., 2016). To investigate in differences between cropping cropping regimes, pairwise Adonis with *p*-value adjustment "BH" based on Bray-Curtis distances were used (Martinez Arbizu, 2017). Bacterial and fungal communities were tested separately. Differences with regard to crop species were tested after exclusion of unplanted soil samples. The effect of cropping systems and inoculation on diversity and richness of fungi and bacteria in all investigated compartments were analyzed separately to avoid spatial pseudoreplication. Global effects (calculated for both sampling times together) of plant compartment and crop species on fungal and bacterial communities were tested with strata = pot, as we had pseudoreplicated data. The two sampling dates were also analyzed separately as to assess whether the observed effects at harvest 1 would be maintained at harvest 2.

Data (including plant and soil parameter as well as alpha-diversity) were tested for normal distribution with *shapiro.test* Test and homogeneity of variance with *leveneTest* function with the package *car* version 2.1-5 (Fox and Weisberg, 2011). Differences between measured environmental and plant parameters were calculated with Kruskal-Wallis test followed by multiple comparing using *dunnTest* with Benjamini-Hochberg *p*-value adjustment or Tukey's post hoc test using the *HSD.test* function in the R package *FSA* version 0.8.17 (Ogle, 2016) and *agricolae* version 1.2-8 (De Mendiburu, 2014), respectively.

To identify zOTUs highly associated to application with respect to plant compartment and harvest, multipattern analyses were applied. For that purpose, bacteria and fungi were investigated using the *multipatt* function from the *IndicSpecies* package version 1.7.6 (DeCáceres and Legendre, 2009). Only fungal and bacterial zOTUs found in at least three samples were used. The biserial coefficients (*R*) with a particular plant species and treatment were corrected for unequal sample size using the function *r.g* (Tichy and Chytry, 2006).

3 Results and Discussion

3.1 Edaphic properties and plant parameters

We investigated in several edaphic properties such as total organic carbon and nitrogen, pH value and soil moisture, as previous studies have been shown that cropping system or application of microorganisms can change soil chemical characteristics (Xiao et al., 2004; Raya-Diaz et al., 2017). In agreement with these previous studies, we found a significantly higher pH value in MW compared to XW (Kruskal-Wallis-test (KW-test), with bidest $x^2=3.88$, df=1, $p=0.048$; with KCl, $x^2=5.24$, df=1, $p=0.022$), whereas inoculation did not influence pH value after five weeks of plant growth (Table 3). In contrast, soil moisture was not influenced when comparing these treatments (27.65 ± 1.18 %). Total nitrogen as well as C:N ratio in the faba bean rhizosphere was significantly influenced by application for harvest 2, showing lowest nitrogen values under application (KW-test, nitrogen, $x^2= 9.07$, df=1, $p=0.002$; C:N, $x^2=4.03$, df=1, $p=0.044$; Table 2). Similar, the inoculated wheat rhizosphere had significantly lower nitrogen content compared to the control treatment (KW-test, H1, $x^2=5.29$, df=1, $p=0.021$; H2, $x^2=10.83$, $p=0.001$). In addition, the combination of cropping system and inoculation significantly affected nitrogen content in the wheat and faba bean rhizosphere (KW-test, wheat, $x^2=12.38$, df=3, $p=0.01$; faba bean, $x^2=14.08$, df=3, $p<0.001$). Nitrogen content was significantly lower in the cropping regimes XBF/XWF compared to XB/XW and MB/MW but only at harvest 2. Previously, it has been reported that intercropping can increase wheat nitrogen accumulation in soil but decrease this parameter in faba bean which is in contrast to our observation (Xiao et al., 2004). We speculate that the significant decrease in nitrogen in the intercropping treatment could be partly explained by the limited available space for roots. This was much reduced in the containers after a growing period of seven weeks and interspecific competition for nutrients may thus have increased between the two crop plants. We further speculate that our applied fungus played an important role in altering nutrient availability in the rhizosphere between the two crop species as previously assumed (Behie and Bidochka, 2014).

Intercropping or the application of *M. brunneum* have been reported to enhance plant growth and biomass production (Xiao et al., 2004; Zhang et al., 2010; Krell et al., 2018). In line with these results, we found that plant height of wheat was significantly higher under intercropping compared to monoculture (Table 4). Faba bean plants were taller when inoculated with the entomopathogenic fungal isolate, but significantly only for the 4th week of plant growing (KW-test, $x^2=4.48$, df=1, $p=0.034$). After seven weeks of growing (H2), wheat aerial plant biomass (KW-test $x^2=7.77$, df=1, $p=0.005$), leaf nitrogen (KW-test, $x^2=6.62$, df=1, $p=0.01$) and water content (KW-test, $x^2=9.28$, df=1, p=0.002) were significantly higher in XW compared to MW (Table 3, 5). In addition, the C:N ratio in leaves was significantly lower in XW compared to MW (KW-test, $x^2=7.71$, df=1, $p=0.005$). Combination of application and cropping system showed that the C:N ratio in leaves were significantly lower in the cropping regime XWF compared to MW. In faba bean roots, total carbon had significantly lower values in the cropping regime XBF compared to MBF. In contrast to the rhizosphere soil, plants especially wheat were mainly affected by differences between cropping systems than fungal application. However, the combination of both treatments showed a significant influence, similar to the rhizosphere. We speculate that fungal application might play a direct role in shifting the balance of inter- and intraspecific competition between plants as already indicated in arbuscular mycorrhizal fungal inoculation experiments (Moora and Zobel, 1995; Hodge, 2003).

3.2 Overall bacterial and fungal community

The response of bacterial and fungal communities of faba bean and wheat towards inoculation with *Metarhizium* under different cropping systems was assessed by Illumina (MiSeq) sequencing targeting the bacterial 16S rRNA gene and the fungal internal transcribed spacer (ITS) region, respectively. After removal of low-quality reads, PCR artefacts (chimeras), non-target sequences and plant-derived contaminations, a total of 4,790,788 (bacteria) and 4,271,395 (fungi) high-quality reads were obtained (Tables S4 and S5). Prior to analyses, samples with less than 16 (fungi) or 22 (bacteria) as well as singletons were removed, resulting in 231 and 227 samples for fungi and bacteria, respectively. Sequence numbers per sample varied between 22 to 89,115 (average 20,916.7) for bacteria and between 16 to 112,646 (average 18,490.2) for fungi. Although samples were rarefied to low sequencing numbers, rarefaction curves (Figures S1-S8) and calculated Michaelis-Menten Fit confirmed that the majority of bacterial and fungal diversity was recovered by the surveying effort (Tables S6 and S7).

The eight dominant bacterial phyla (> 0.5 % of all sequences across all samples) were Proteobacteria (66.64 %), Actinobacteria (19.11 %), Acidobacteria (3.49 %), Firmicutes (2.58 %), Chloroflexi (2.64 %), Gemmatimonadetes (2.05 %), Verrrucomicrobia (1.02 %) and Bacteroides (0.75 %). The Proteobacteria were dominated by Gammaproteobacteria (37.32 %), followed by Alphaproteobacteria (27.7 %). The abundant bacterial phyla were present in all samples and accounted for 98.28 %, of all sequences analysed in this study. At family level, *Rhizobiaceae* (21.21 %) and *Burkholderiaceae* (18.44 %) were predominant (Figure 2). Abundant genera were, for example, *Allorhizobium-Neorhizobium-Pararhizobium-Rhizobium* (20.91 %), *Alicycliphilus* (12.38 %) *Halomonas* (3.84 %), *Lysobacter* (2.7 %), *Sphingomonas* (2.32 %) and *Shewanella* (2.05 %).

Fungi were represented by four abundant phyla (> 0.5 % of all sequences across all samples): Ascomycota (43.78 %), Basidiomycota (6.53 %), Mortierellomycota (4.36 %) and Mucoromycota (2.25 %). Approximately 43 % of all sequences belonged to unidentified fungi. The dominant fungal families were *Nectriaceae* (8.56 %), *Mortierellaceae* (3.94 %), *Phaeosphaeriaceae* (3.41 %) and *Pleosporaceae* (3.24 %) (Figure 3). At genus level: *Dactylonectria* (3.94 %), *Mortierella* (3.08 %), *Chaetomium* (2.43 %), *Rhizopus* (2.22 %), *Gibellulopsis* (2.19 %), *Alternaria* (2.18 %), *Gibberella* (2.15 %) and *Cladosporium* (2.11 %) were more frequent. Members of the genus *Metarhizium* accounted for 0.08 % of all sequences. Abundant bacterial and fungal taxa were also found in previous studies investigating in plant-associated microbial communities (Gdanetz et al., 2017; Zhou et al., 2017; Li et al., 2018).

3.3 Spatial and temporal distribution of *Metarhizium* species in soil and plants after application

We found six zOTUs: two belonging to *M. anisopliae* (0.077 %), one *M. marquandii* (0.0024 %), one *M. carneum* (0.002 %) and two belonging to *M. flavoviride* (0.0015 %). The identity of all *Metarhizium* zOTUs were higher than 99.721 %. Both *M. flavoviride* zOTUs occurred mainly in unplanted control soil (Table 6). zOTUs belonging to *M. marquandii* as well as *M. carneum* were detected in unplanted soil, rhizosphere soil and within roots in non-inoculated and inoculated treatments for both harvests. *M. anisopliae* was found more often in inoculated roots and leaves, especially in the treatment XBF and XWF in roots.

Conventional PCR approach with *Metarhizium* clade 1 specific primers also confirmed presence of *Metarhizium* spp. in the samples of harvest 1 (Figure S9-S16). However, we also observed positive bands with the correct amplicon length in several of the control samples. Previous studies reported that different *Metarhizium* species can be found naturally in agricultural soils including *M. brunneum*, *M. carneum* or *M. flavoviride* (Steinwender et al., 2014; Kepler et al., 2015; Mayerhofer et al., 2017). As we used agricultural soil from an arable field, we hypothesize that some of the different *Metarhizium* species might origin from it. We further speculate that one of the *M. anisopliae* zOTU is our inoculated fungus because the main occurrence was in inoculated planted treatments and relative abundance in the dataset was highest compared to the other *Metarhizium* zOTUs. However, we cannot confirm this assumption with our approach.

3.4 Bacterial and fungal communities are strongly affected by crop species and plant compartment

According to our first hypothesis that cop species and plant compartment affect microbial community composition and diversity, we calculated diversity (represented by the Shannon index H') and richness (number of observed unique sequences). Bacterial diversity and richness were significantly influenced by the crop plant species grown. The wheat root endosphere had a significantly higher bacterial diversity (KW-test, H1, x^2=13.73, df=1, p<0.001; H2, x^2=29.27, p<0.001) and richness (KW-test, H1, x^2=18.73, df=1, p<0.001; H2, x^2=29.27, p<0.001) compared to faba bean. Similarly, wheat rhizosphere showed higher diversity compared to faba bean but specific for harvest 2 (KW-test, x^2=6.79, df=1, p=0.009). In contrast, leaf endosphere showed higher bacterial diversity in faba bean compared to wheat for harvest 2 (KW-test, H2, x^2=5.0, df=1, p=0.025). Similar to bacteria, fungi showed significantly higher diversity (KW-test, x^2=10.19, df=1, p=0.001) and richness (KW-test, x^2=6.35, df=1, p=0.012) in the wheat rhizosphere compared to faba bean in harvest 2, whereas diversity (KW-test, x^2=6.35, df=1, p=0.012) and richness (KW-test, x^2=4.86, df=1, p=0.02) in the faba bean leaf endosphere was higher compared to wheat.

In addition to differences in richness and diversity, crop species and plant compartment significantly influenced microbial community composition in soil and endosphere. Compartment explained 32.7 % and 23.1 % of the variance in the bacterial and fungal dataset (PERMANOVA, p=0.001, p=0.001) and dispersion among compartments was heterogeneous (PERMDISP, F=5.60, p=0.001; F=39.79, p=0.001). Crop species explained most of the variance in the root and leaf endosphere of the bacterial and fungal dataset (Table 7). Several taxa were more abundant in one of the two crop species (Figure 2, 3). Higher relative abundances of several bacterial genera including *Rhizobium* (66.60 %), *Alicycliphilus* (7.48 %) and *Halomonas* (7.31 %) were recorded in faba bean compared to wheat plants, whereas the opposite was observed for *Alicycliphilus*

(30.58 %), *Rhizobium* (9.06 %) and *Sphingomonas* (5.41 %). The fungal genera *Dactylonectria* (11.78 %), *Cladosporium* (4.65 %) and *Alternaria* (4.37 %) were more abundant in faba bean plants. As plant species select microbial communities, each plant species can harbor specific microbial members as shown previously (Wemheuer et al., 2017; Zhou et al., 2017). Similarly, each plant compartment represents distinct niches in terms of nutrient availability, or exposure to abiotic or biotic factors which selects for a specific microbial assembly (Gdanetz et al., 2017; Wallace et al., 2018). The rhizosphere is strongly influenced by plant metabolism and differences in composition and quantity of root exudates between plant species such as legumes and grasses can influence soil microbial communities (Liu et al., 2017; Siczek et al., 2018). Differences in plant physiology such as different rooting structure or chemical composition between plants (Roumet et al., 2008) might also affect endophytic community.

3.5 *M. brunneum* application significantly affected microbial communities

We further expected that *M. brunneum* application and establishment of an endophytic fungus in plant tissues creates a long-lasting effect on bacterial and fungal communities. Fungal diversity and richness in wheat rhizosphere and leaves were affected by *M. brunneum* application and fungi showed a decrease in diversity (KW-test, rz, x^2=4.81, df=1, p=0.028; lv, x^2=9.42, p=0.002) and richness (KW-test, rz, x^2=7.41, df=1, p=0.006; lv, x^2=9.42, p=0.002) in inoculated wheat plants for harvest 2 (Table 9). In contrast, bacterial diversity (KW-test, rz, x^2=8.59, df=1, p=0.003; lv, x^2=4.48, p=0.03) and richness (KW-test, rz, x^2=4.5, df=1, p=0.034; lv, x^2= 4.65, p=0.031) in wheat rhizosphere and leaves were significantly higher or lower in inoculated plants specific for harvest 2 (Table 8). Previously, Rabiey et al. (2017) reported that soil inoculation with the fungus *Piriformospora indica* increased the fungal diversity in soil and root endosphere of wheat in a pot experiment which was in contrast to our results; however, they also reported an increase of bacterial diversity which was in line with our results. We speculate that these observations are related to specific responses of bacteria and fungi to our applied strain.

To identify the influence of *M. brunneum* application on microbial community composition, we performed a NMDS (non-metric multidimensional scaling) analysis based on Bray-Curtis dissimilarities. NMDS showed distinct clustering according to application which was mainly observed in bacterial leaf endosphere (Figure 4). Permutational multivariate analysis of variance (PERMANOVA) also showed that *M. brunneum* application significantly influenced bacterial (H1, R^2=11.7%, p=0.004, H2, R^2=4.1, p=0.028) and fungal (H1, R^2=4.6%, p=0.028, H2, R^2=7.4, p=0.001) community composition in the rhizosphere, whereas root community was unaffected (Table 7). In accordance to this, Ardanov and coworkers (2012) showed compartment specific effects of the substrate inoculation of *Methylobacterium* spp. on bacteria. Application only changed bacterial composition in potato shoots, whereas root endophytes were not influenced. In contrast to our study, Zimmermann et al. (2016) observed that seed coating with *Fusarium oxysporum* did not alter indigenous fungal community structure in the rhizosphere of maize, whereas soil type and plant growth stage showed the strongest effect on fungi.

In the present study, we still found significant effects of *M. brunneum* application on bacterial and fungal communities in the plant endosphere and rhizosphere after seven weeks of plant growing. As we investigated only in one plant growth stage (vegetative phase), we do not know how long effects of *M. brunneum* application last on plant and its plant microbiome which is also important to know in case of biological control options. In contrast to our findings, several studies reported that application of an entomopathogenic fungus showed no or transient changes on microbial communities (Hirsch et al., 2013; Hong et al., 2017; Mayerhofer et al., 2017). We speculate that these discrepancies to our results might be related to different inoculated agents, crop species or even growth stage of plant (Aguilar-Trigueros and Rillig 2016; Gadhave et al. 2018; Liu et al. 2018). Recently, Gadhave et al. (2018) showed that impacts of application on bacterial community in the broccoli root endosphere were dependent on *Bacillus* species. Similarly, Aguilar-Trigueros and Rillig (2016) observed that each plant species responded differently to application such as improved plant growth which was dependent on the identity of the inoculated fungus. Further reasons for contradictory results might be the usage of different inoculation methods or different investigated compartments. It is well-known that these factors can influence inoculation or colonization efficiency (Tefera and Vidal, 2009; Akutse et al., 2013; Greenfield et al., 2016; Jaber and Enkerli, 2016) and as a result may also affect plant microbiome.

3.6 Cropping system influenced effect of *M. brunneum* application on plant microbiome

We further evaluated whether cropping system had an influence on microbial communities. We found that cropping system exhibited significantly higher microbial diversity and richness in XW compared to MW (Table 8, 9). However, this was only observed for fungi in the root endosphere (KW-test, shannon, x^2= 4.51, df=1, p=0.03, richness, x^2=4.17, p=0.04) and for bacteria in the rhizosphere (KW-test, shannon, x^2=5.81, df=1, p=0.015; richness, x^2= 7.82, p=0.005). In faba bean rhizosphere, we showed the opposite trend, thus bacterial diversity (KW-test, x^2=5.22, df=1, p=0.022) and richness (KW-test, x^2=4.17, p=0.04) was significantly higher in MB compared to XB. Our observations were in line with previous studies which found that intercropping increased microbial diversity in the rhizosphere and bulk soil (Yang et al., 2016; Li and Wu, 2018). Yang et al. (2016) investigated in ten different spring crops grown in monoculture and intercropping system and found that intercropping increased bacterial diversity in the rhizosphere, however, responses were also crop species dependent. In line with this, Li and Wu (2018) showed that from seven intercropping systems only the combination of cucumber/mustard and cucumber/trifolium increased bacterial and fungal diversity in bulk soil compared to cucumber monoculture.

In the present study, we showed that cropping system significantly affected bacterial community composition in plant endosphere whereas fungi were only influenced in rhizosphere soil (Table 7). Previous studies also reported significant effects of cropping system on bacterial and fungal community composition in different compartments (Zhang et al., 2010; Wang et al., 2012; Granzow et al., 2017). Recently, Granzow et al. (2017)

showed that cropping system affected bacterial and fungal communities in bulk and rhizosphere soil as well as fungal composition in roots, whereas leaf endophytes were unaffected which is in contrast to our results. In contrast to our study, they used commercially available soil and different crop genotypes which might influence the response of the microbiome in a different way (Wagner et al., 2016).

Consistent with our third hypothesis, we observed that cropping system influenced the effect of *M. brunneum* application on bacterial and fungal communities. For example, we found significantly lower bacterial diversity and richness in XBF rhizosphere compared to MB, whereas XWF rhizosphere had significantly higher diversity and richness compared to MWF at harvest 1 (Table 8). In the root endosphere, bacteria showed highest diversity and richness under MBF compared to MB and/or XBF at harvest 2. In contrast, effects of cropping regimes on fungal alpha-diversity were only found in harvest 2 (Table 9). For example, fungi had significantly higher diversity and richness in the root endopshere of XWF compared to MW or MWF. Partly in line with this observation, previous studies showed that the combination of cropping system and *rhizobium* inoculation can influence bacterial diversity or abundance in the root endosphere and rhizosphere soil of faba bean or maize (Zhang et al., 2011; Zhang et al., 2015). Zhang et al. (2011) reported that intercropping and rhizobial seed inoculation increased bacterial diversity compared to monoculture in the root endosphere of faba bean, but they also indicated that plant growth stage was the main factor influencing bacterial community and its response towards rhizobial inoculation and cropping system.

Evaluation of microbial community composition with NMDS showed similar clustering of fungal root community towards cropping regimes in harvest 2 (Figure 5). Pairwise Adonis analysis confirmed that fungal root community was significantly altered between the cropping regimes XW and XWF, MW and XWF as well as between MWF and XWF (Table S9). Here, we demonstrated that plant compartment but also sampling time exhibited strong effects on bacterial and fungal communities and their response towards cropping regimes which is consistent with our last hypothesis. Although we could not confirm the occurrence of the applied *Metarhizium* species with our approach, we speculate that *M. brunneum* application played an important role in the observed results for alpha- and beta-diversity; however, we cannot clearly say whether inoculation influenced bacterial or fungal response towards cropping systems or that cropping system changed response of microbial communities towards inoculation. We further hypothesize that *M. brunneum* application might have changed nutrient accumulation or indirectly changed plant root exudation pattern of plants (Gu et al., 2016). Previous studies reported that entomopathogenic fungi produce siderophores and/or organic acids that can alter availability of certain nutrients (Krasnoff et al., 2014; Sánchez-Rodriguez et al., 2016; Krell et al., 2018). It is well-known that nitrogen and carbon can strongly influence microbial communities (Wan et al., 2015; Kaiser et al., 2016). In agreement with our assumption, we observed that especially nutrient availability (nitrogen and carbon) within the different plant compartments changed according to inoculation and cropping system. For example, C:N ratio under XWF in the rhizosphere increased similar to bacterial diversity and richness. Carbon in roots significantly decreased in XBF compared to MBF similar to bacterial diversity.

3.7 Predominant and associated bacterial taxa differ between *M. brunneum* application and control treatment

Fungal inoculum as invading agent might antagonistic interact or compete for resources and niches with other microorganisms which in turn results in an increase or depletion of specific microorganisms (de Boer, 2017). Results of the relative abundance might indicate this assumption, as we observed that inoculation induced the proliferation of specific bacteria. For example, *Halomonas* (17.23 %) and *Shewanella* (11.21 %) were frequently found in inoculated rhizosphere samples, but were almost absent (< 0.01 % abundance) in rhizosphere soil of non-inoculated treatments (Figure 2). The opposite effect was observed in the leaf endosphere. Here, *Halomonas* (62.25 %) and *Shewanella* (28.46 %) were mainly found in non-inoculated faba bean plants. In addition, both bacterial genera were completely absent in the soil for harvest 2. Interestingly, these bacteria were mainly reported in plant growth promoting, biological control or in bioremediation (Tiwari et al., 2011; Jha, Gontia and Hartmann, 2012; Gong et al., 2015). Previous studies also showed that inoculation of bacteria or fungi may change abundances of specific (beneficial) bacterial and fungal taxa in the rhizosphere, endosphere or episphere (Hong et al., 2017; Gadhave et al., 2018; Liu et al., 2018).

To identify bacterial and fungal taxa responsible for the observed differences among application, we performed a multipattern analysis to investigate which microorganisms are significantly associated with those treatments (Table S8). In general, soil communities harbored more associated zOTUs than endophyte communities which is most probably related with higher sequence numbers in soil compared to endosphere samples. Although we did not observe a significant difference in microbial community composition between control and inoculated unplanted soil samples, we found 199 (2.47 % of all bacterial taxa included in the analysis) and 65 (2.31 %) significantly associated bacterial taxa for fungal application and control treatment in harvest 1. Several associated bacterial taxa for fungal application included zOTUs belonging to *Haliangium*. In contrast, 48 (1.70 %) fungal taxa were significantly associated with *M. brunneum* application, whereas only one taxon with control samples for unplanted soil. Members of *Haliangium* spp. are known to produce haliangicin, an antifungal metabolite that can inhibit growth of several fungi (Fudou et al., 2001). Moreover, few zOTUs belonging to *Halomonas* and *Shewanella* were significantly and uniquely associated with fungal application in the rhizosphere of harvest 1. In plant endosphere, we found that both zOTUs were significantly associated with control plants for harvest 1. Furthermore, we found several zOTUs of *Sphingomonas* significantly and uniquely associated with *M. brunneum* application in leaf and root endosphere as well as in the rhizosphere soil for harvest 2. Plant-associated *Sphingomonas* members have been shown to be effective agents against the plant pathogen *Pseudomonas syringae* (Innerebner et al., 2011). In addition, Hong et al., (2017) reported an increase of *Sphingomonas* abundance after *M. anisopliae* leaf application and, similar to Innerebner et al. (2011), they indicated that these bacteria might contribute to facilitate plant disease resistance and/or stress response. Because of this complexity of interactions between microorganisms and between microorganisms and plant, the effects of fungal application on plant and microbiome are difficult to predict.

4 Conclusion

To date, the combined effect of cropping system and fungal inoculation on fungal and bacterial communities in endosphere and rhizosphere soil of two important crop species has not been studied using large-scale metabarcoding. In line with our first hypothesis we found that crop species and compartment exhibited a strong influence on the plant microbiome. *M. brunneum* application changed bacterial and fungal community composition at both sampling times, whereas response of alpha-diversity showed some variance between sampling time. Furthermore, cropping system, crop species and plant compartment were important in changing the effect of fungal application on bacterial and fungal communities. The present findings increase our understanding of how the application of an entomopathogenic fungus affects microbial community which might gain further importance for biological control strategies in the future. However, we need long-term studies to properly quantify observed effects under field conditions.

5 Acknowledgment

This study is part of the project IMPAC^3 and was funded by the Federal Ministry of Education and Research (FKZ 031A351A). The authors thank Prof. Dr. Wolfgang Link from the Working Group "Breeding Research Faba Bean" (Division of Plant Breeding at the University of Göttingen) for providing the seed material.

Tables

Table 1. Sampling numbers for each container and harvest.

Treatments/ Compartments	Unplanted soil	Rhizosphere	Roots	Leaves	Plants/ treatment
CS	1 (3/3)	-	-	-	-
		Harvest 1			
C	1 (5/5)	-	-	-	-
CF	1 (5/5)	-	-	-	-
MB	-	1 (5/5)	1 (5/5)	1 (3/4)	5
XB	-	1 (5/5)	1 (5/4)	1 (4/2)	5
MW	-	2 (5/5)	2 (5/5)	2 (4/5)	10
XW	-	2 (5/5)	2 (5/5)	2 (5/5)	10
MBF	-	1 (5/5)	1 (5/4)	1 (3/4)	5
XBF	-	1 (5/5)	1 (5/5)	1 (2/4)	5
MWF	-	2 (5/5)	2 (5/4)	2 (4/5)	10
XWF	-	2 (5/5)	2 (5/5)	2 (5/5)	10
		Harvest 2			
MB	-	2 (5/5)	2 (5/4)	2 (0/3)	10
XB	-	2 (5/5)	2 (5/4)	2 (1/2)	10
MW	-	3 (5/5)	3 (5/5)	3 (5/5)	15
XW	-	3 (5/5)	3 (5/5)	3 (5/5)	15
MBF	-	2 (4/5)	2 (5/3)	2 (4/5)	10
XBF	-	2 (5/5)	2 (5/5)	2 (2/2)	10
MWF	-	3 (4/5)	3 (5/5)	3 (5/5)	15
XWF	-	3 (5/5)	3 (5/5)	3 (5/4)	15
Total	13	80	160	160	60(FB), 100(W)

Each container was replicated five times. Abbreviations: W, wheat; FB, faba bean; F, inoculated samples; MW, wheat in monoculture; MB, faba bean in monoculture, XB, faba bean samples in intercropping; XW, wheat samples in intercropping; SC, starting soil; C unplanted control soil. Number in brackets refers to the number of samples left after removal of samples with too low sequencing numbers. The first number refers to bacteria, the second to fungi.

Table 2. Edaphic properties (mean ± SE).

	Harvest 1			Harvest 2		
Treatment	**C_{total} (%)**	**N_{total} (%)**	**C:N ratio**	**C_{total} (%)**	**N_{total} (%)**	**C:N ratio**
C	2.01±0.01	0.21±0.00	9.69±0.16	1.95±0.01	0.20±0.00	9.60±0.03
CF	1.98±0.02	0.21±0.00	9.48±0.14	1.93±0.02	0.21±0.00	9.45±0.09
B	2.03±0.02	0.22±0.00	9.53±0.10	2.01±0.03	**0.20±0.00a**	**10.06±0.12a**
B_F	2.01±0.01	0.21±0.00	9.49±0.06	1.93±0.05	**0.18±0.01b**	**10.73±0.30b**
W	2.06±0.02	**0.21±0.00a**	9.61±0.05	2.00±0.02	**0.20±0.00b**	10.15±0.05
W_F	2.05±0.03	**0.21±0.00b**	9.79±0.15	1.88±0.07	**0.18±0.01a**	10.74±0.15
XB	2.04±0.01	0.21±0.00	**9.63±0.09a**	1.91±0.05	0.18±0.01	10.51±0.32
MB	2.00±0.02	0.21±0.00	**9.39±0.05b**	2.02±0.03	0.20±0.00	10.32±0.19
XW	2.08±0.03	0.21±0.00	9.84±0.14	1.87±0.07	0.18±0.01	10.43±0.30
MW	2.03±0.02	0.21±0.00	9.56±0.06	2.00±0.03	0.19±0.00	10.46±0.15
XB	2.04±0.03	0.21±0.00	9.66±0.18	1.98±0.04	**0.20±0.00a**	9.99±0.12
XBF	2.03±0.01	0.21±0.00	9.59±0.05	1.84±0.12	**0.17±0.00b**	11.04±0.57
MB	2.02±0.04	0.22±0.00	9.39±0.06	2.04±0.05	**0.20±0.00a**	10.13±0.21
MBF	2.00±0.02	0.21±0.00	9.37±0.08	2.01±0.04	**0.19±0.00ab**	10.48±0.3
XW	2.05±0.02	0.21±0.00	**9.57±0.09b**	1.97±0.02	**0.20±0.00a**	9.97±0.1
XWF	2.11±0.05	0.21±0.00	**10.11±0.2a**	1.77±0.12	**0.17±0.01b**	10.88±0.54
MW	2.08±0.02	0.21±0.00	**9.65±0.03ab**	2.03±0.03	**0.20±0.00a**	10.33±0.15
MWF	1.98±0.02	0.21±0.00	**9.47±0.1b**	1.98±0.04	**0.19±0.00ab**	10.59±0.28

Different letters in columns indicate statistically significant differences between treatments ($p \leq 0.05$, means ± SE). Abbreviations: MB/MW, faba bean/wheat grown in monoculture; XB/XW, faba bean/wheat intercropped; C, control unplanted soil; F, inoculated samples.

Table 3. Biomass [g] and pH value (mean ± SE).

	Roots		Aerial parts		Shoot/Root ratio		pH-bidest	pH-KCl
Treat-ment	**H1**	**H2**	**H1**	**H2**	**H1**	**H2**	**H1**	**H1**
XB	3.72 ±0.35	8.75 ±0.50	13.27 ±0.95	36.70 ±2.55	3.73 ±0.31	4.19 ±0.16	7.46 ±0.02	6.88 ±0.04
MB	4.40 ±0.47	8.04 ±0.58	16.21 ±1.35	37.94 ±1.32	3.91 ±0.37	4.91 ±0.32	7.41 ±0.03	6.89 ±0.02
XW	1.53 ±0.28	4.61 ±0.58	3.50 ±0.34	**10.61 ±0.71a**	2.93 ±0.52	2.44 ±0.20	**7.45 ±0.01a**	**6.92 ±0.02a**
MW	1.82 ±0.41	4.56 ±0.74	3.73 ±0.49	**7.50 ±0.33b**	2.45 ±0.27	1.94 ±0.25	**7.50 ±0.01b**	**6.96 ±0.01b**
B	3.98 ±0.29	8.84 ±0.58	13.96 ±1.32	37.29 ±2.48	3.54 ±0.27	4.31 ±0.32	7.45 ±0.01	6.86 ±0.03
B_F	4.14 ±0.53	7.95 ±0.49	15.52 ±1.15	37.35 ±1.48	4.10 ±0.38	4.79 ±0.20	7.42 ±0.03	6.90 ±0.02
W	1.70 ±0.26	4.98 ±0.75	3.54 ±0.36	9.34 ±0.86	2.36 ±0.26	2.17 ±0.28	7.48 ±0.01	6.94 ±0.01
W_F	1.65 ±0.43	4.09 ±0.40	3.69 ±0.47	8.70 ±0.60	3.02 ±0.52	2.22 ±0.17	7.47 ±0.02	6.94 ±0.02

Different letters in columns indicate statistically significant differences between treatments ($p \leq 0.05$, means ± SE). Abbreviations: B, faba bean; W, wheat; MB/MW, faba bean/wheat grown in monoculture; XB/XW, faba bean/wheat intercropped; F, inoculated samples.

Table 4. Plant height [cm] and thermal images [°C] from 4th to 7th week of the experiment.

	4 weeks	5 weeks (H1)	6 weeks	7 weeks (H2)
		Height		
XB	23.1±0.9	29.4±0.9	37.7±0.9	44.4±1.2
MB	22.8±0.6	31.3±0.8	38.3±0.8	46.1±0.9
XW	33.9±1.3	**37.1±1.2a**	**37.9±1.1a**	**38.5±0.8a**
MW	33.5±0.4	**34.9±0.6b**	**35.0±0.6b**	**37.0±1.3b**
B	**21.8±0.7b**	29.6±1.0	37.6±1.0	43.7±1.1
B_F	**24.1±0.6a**	31.1±0.7	38.4±0.7	46.8±0.9
W	34.6±0.5	36.3±0.9	36.8±0.8	37.4±0.6
W_F	32.9±1.2	35.6±1.1	36.1±1.1	38.0±1.4
		Thermal Images		
X	17.04±0.32	17.15±0.33	**17.40±0.18a**	19.37±0.32
M	17.64±0.21	17.78±0.24	**17.92±0.18b**	19.82±0.31
Non-inoculated	17.39±0.27	17.88±0.29	17.93±0.16	19.95±0.31
Inoculated	17.48±0.25	17.25±0.25	17.56±0.21	19.39±0.34

Different letters in columns indicate statistically significant differences between treatments ($p \leq 0.05$, means ± SE). Abbreviations: B, faba bean; W, wheat; MB/MW, faba bean/wheat grown in monoculture; XB/XW, faba bean/wheat intercropped; F, inoculated samples; M, includes MB and MW; X, intercropping.

Table 5. Water content, total organic carbon and nitrogen in roots and leaves for harvest 2 (mean ± SE).

	C_{total} (%)	N_{total} (%)	C:N ratio	Water content [%]
		Roots		
B	11.06±1.01	0.91±0.08	**12.14±0.26a**	64.67±3.37
B_F	14.24±2.44	0.87±0.08	**16.31±2.67b**	54.42±10.21
W	4.12±0.68	0.28±0.03	14.37±0.88	32.33±4.91
W_F	3.23±0.23	0.24±0.01	13.36±0.70	54.27±7.78
XB	**9.98±0.85a**	0.79±0.07	12.70±0.24	60.29±5.34
MB	**15.32±2.28b**	1.00±0.08	15.74±2.76	58.79±9.63
XW	3.24±0.16	0.25±0.00	12.90±0.46	46.64±7.86
MW	4.21±0.77	0.27±0.03	14.94±0.96	37.52±6.32
XB	**10.55±1.07ab**	0.85±0.09	12.38±0.42	60.79±6.00
XBF	**9.40±1.40a**	0.72±0.10	13.01±0.19	59.79±9.61
MB	**11.56±1.82ab**	0.96±0.13	11.88±0.33	68.55±2.76
MBF	**19.07±3.63b**	1.02±0.09	19.60±5.16	49.04±19.03
XW	3.23±0.16	0.25±0.00	12.84±0.52	34.46±10.02
XWF	3.24±0.34	0.25±0.01	12.97±0.92	61.87±8.04
MW	5.01±1.30	0.30±0.06	15.89±1.43	30.20±2.40
MWF	3.20±0.37	0.23±0.01	13.74±1.15	46.68±13.41
		Leaves		
B	45.22±1.23	3.53±0.11	12.85±0.20	88.61±0.20
B_F	44.12±0.16	3.39±0.11	13.15±0.49	88.95±0.36
W	42.42±0.09	1.94±0.12	22.66±1.52	82.12±0.68
W_F	42.73±0.17	2.20±0.21	20.96±2.04	82.21±1.40
XB	44.17±0.23	3.44±0.04	12.83±0.13	88.89±0.23
MB	45.17±1.22	3.48±0.15	13.16±0.51	88.67±0.34
XW	42.43±0.08	**2.40±0.15a**	**18.29±1.16a**	**84.18±0.55a**
MW	42.69±0.16	**1.77±0.12b**	**25.06±1.52b**	**80.14±0.89b**
XB	44.01±0.38	3.40±0.05	12.96±0.12	88.65±0.29
XBF	44.32±0.29	3.49±0.06	12.71±0.23	89.13±0.37
MB	46.44±2.43	3.66±0.22	12.74±0.39	88.57±0.31
MBF	43.91±0.11	3.29±0.21	13.58±0.96	88.77±0.65
XW	42.40±0.10	**2.21±0.11ab**	**19.37±1.00ab**	**83.73±0.53ab**
XWF	42.47±0.15	**2.63±0.30a**	**16.94±2.31a**	**84.74±1.06a**
MW	42.44±0.15	**1.67±0.12b**	**25.95±2.01b**	**80.52±0.73ab**
MWF	42.93±0.25	**1.87±0.21ab**	**24.17±2.46ab**	**79.67±1.94b**

Different letters in columns indicate statistically significant differences between treatments ($p \leq 0.05$, means ± SE). Abbreviations: MB/MW, faba bean/wheat grown in monoculture; XB/XW, faba bean/wheat intercropped; F, inoculated samples. Note: water content refers here to the absolute water content.

Table 6. Spatial and temporal distribution of *Metarhizium* species in soil and plants.

	inoculated				non-inoculated			
	Lv	Ro	Rz	Soil	Lv	Ro	Rz	Soil
	Harvest 1							
*M. anisopliae*_SH200393.07FU	0.285	0.522	0.0048	0	0	0	0.0004	0
M. carneum	0	0	0.0016	0.003	0	0	0.0024	0.0026
*M. flavoviride*_SH214395.07FU	0	0	0	0	0	0	0.0001	0.042
*M. marquandii*_SH217934.07FU	0	0	0.0028	0.0026	0	0	0.0014	0.003
Total	**0.285**	**0.522**	**0.009**	**0.0055**	**0**	**0**	**0.0044**	**0.0477**
	Harvest 2							
*M. anisopliae*_SH200393.07FU	0	0.1316	0.0394	NA	0	0.0002	0	NA
M. carneum	0	0.0047	0.0017	NA	0	0	0.0123	NA
*M. flavoviride*_SH214395.07FU	0	0	0.0006	NA	0	0	0.0006	NA
*M. marquandii*_SH217934.07FU	0	0.0216	0.0003	NA	0	0	0.0026	NA
Total	**0**	**0.1579**	**0.042**		**0**	**0.0002**	**0.0155**	

Numbers refer to the relative abundance in the dataset. Abbreviations: Lv, leaves; Ro, roots; Rz, rhizosphere soil; Soil, unplanted soil.

Table 7. Effect of cropping system, inoculation and crop species on bacterial and fungal community compositions.

	Bacteria				Fungi			
	H1		H2		H1		H2	
	R^2 (%)	*p*	R^2 (%)	*p*	R^2 (%)	*p*	R^2 (%)	*p*
	Rhizosphere soil							
Inoculation	**11.7**	**0.004**	**4.1**	**0.028**	**4.6**	**0.028**	**7.4**	**0.001**
Cropping System	2.4	0.324	3.6	0.095	2.5	0.475	**4.5**	**0.021**
Crop species	3.4	0.221	**8.9**	**0.001**	3.8	0.064	**8.8**	**0.001**
	Roots							
Inoculation	1.6	0.833	1.4	0.668	2.8	0.366	3.2	0.273
Cropping System	**3.6**	**0.001**	0.9	0.899	2.2	0.633	5.4	0.076
Crop species	1.6	0.649	**6.1**	**0.001**	**16.9**	**0.001**	**24.7**	**0.001**
	Leaves							
Inoculation	**12.7**	**0.006**	**9.8**	**0.038**	2.4	0.713	5.2	0.096
Cropping System	2.5	0.56	**9.4**	**0.032**	2.8	0.532	3.3	0.367
Crop species	**12.9**	**0.004**	7.9	0.06	**9.1**	**0.001**	**20.1**	**0.001**

Results of the permutational multivariate analysis of variance (PERMANOVA) with Bray-Curtis distances testing for the different treatments. Cropping system compares monoculture versus intercropping. Inoculation compares inoculated versus non-inoculated samples. Statistically significant differences ($p \leq 0.05$) between the treatments for each plant compartment are written in bold.

Table 8. Bacterial richness and diversity with regard to plant compartment, harvest and cropping regime.

		Bacteria			
		Richness		Diversity	
	Treatment	H1	H2	H1	H2
		Unplanted soil			
	C	**2688.74±386.25ab**	NA	6.82±0.35	NA
	CF	**2974.66±106.16a**	NA	7.00±0.15	NA
	CS	**2658.17±110.06b**	NA	6.85±0.05	NA
		Rhizosphere soil			
Inocula-tion	Faba bean control	**1245.54±76.63a**	1225.68±110.67	6.30±0.19	6.22±0.34
	Faba bean inoculated	**790.00±590.35b**	1220.20±171.63	4.65±2.34	6.27±0.17
	Wheat control	1282.82±114.47	**1209.49±152.08a**	6.38±0.21	**6.30±0.22a**
	Wheat inoculated	**1046.07±574.83A**	**1449.94±100.94Bb**	5.42±2.05	**6.64±0.14b**
Cropping System	XB	**933.89±490.80a**	1209.72±156.12	**5.20±1.95a**	6.25±0.18
	MB	**1136.28±427.48b**	1243.00±102.62	**5.87±1.60b**	6.23±0.39
	XW	**1375.46±137.84a**	1313.05±178.68	**6.51±0.24a**	6.46±0.24
	MW	**953.43±506.66b**	1320.49±188.54	**5.29±1.97b**	6.44±0.27
Cropping Regime	XB	**1213.38±80.60ab**	1229.08±119.21	**6.20±0.21ab**	6.25±0.20
	XBF	**654.40±538.27a**	1190.36±199.28	**4.20±2.46a**	6.25±0.18
	MB	**1277.70±64.27b**	1222.28±115.50	**6.40±0.13b**	6.20±0.47
	MBF	**959.50±637.81ab**	1294.80±48.08	**5.20±2.40ab**	6.31±0.17
	XW	**1298.36±152.90ab**	1194.12±149.56	**6.36±0.27ab**	6.31±0.23
	XWF	**1452.56±67.16a**	1431.98±118.78	**6.66±0.09a**	6.61±0.16
	MW	**1267.28±74.16ab**	1224.86±170.53	**6.39±0.15ab**	6.30±0.24
	MWF	**639.58±570.82b**	1479.87±72.76	**4.19±2.38b**	6.68±0.10
		Roots			
Inocula-tion	Faba bean control	39.69±108.20	9.18±4.44	1.03±1.62	**0.55±0.12a**
	Faba bean inoculated	16.76±14.99	13.13±9.59	0.76±0.28	**0.70±0.18b**
	Wheat control	**87.49±43.34A**	**170.35±67.61B**	**2.86±0.99A**	**4.29±0.76B**
	Wheat inoculated	**80.86±58.82A**	**173.49±56.66B**	**2.42±1.71A**	**4.13±0.88B**
Cropping System	XB	9.77±12.15	8.23±4.93	0.59±0.22	0.59±0.11
	MB	46.68±106.40	14.08±8.78	1.20±1.58	0.67±0.21
	XW	98.21±47.95	162.00±72.34	3.17±1.13	4.05±0.95
	MW	70.14±51.24	181.84±48.33	2.11±1.46	4.37±0.63
Cropping Regime	XB	6.22±2.20	**11.62±5.07ab**	0.54±0.11	**0.61±0.14ab**
	XBF	13.32±17.20	**4.84±0.55a**	0.64±0.29	**0.58±0.08ab**
	MB	73.16±153.42	**6.74±1.95a**	1.53±2.30	**0.50±0.08a**
	MBF	20.20±13.43	**21.42±5.91b**	0.87±0.24	**0.83±0.16b**
	XW	90.28±37.01	184.02±87.05	2.99±0.67	4.45±0.77
	XWF	106.14±60.38	139.98±54.64	3.35±1.54	3.65±1.01
	MW	84.70±53.26	156.68±47.34	2.74±1.31	4.13±0.79
	MWF	55.58±50.41	207.00±37.83	1.49±1.44	4.62±0.35

Table continued on the following page

		Leaves			
Inoculation	Faba bean control	5.36±0.89	4.10±NA	1.44±0.15	1.04±NA
	Faba bean inoculated	6.82±3.17	9.26±2.08	1.38±0.47	1.87±0.32
	Wheat control	**4.59±3.35A**	**7.97±1.87Ba**	0.88±0.72	**1.51±0.37a**
	Wheat inoculated	6.70±5.36	**4.87±3.10b**	1.18±0.98	**0.88±0.72b**
Cropping System	XB	6.35±2.88	8.90±1.56	1.50±0.40	1.82±0.26
	MB	5.58±1.25	8.52±3.06	1.33±0.17	1.75±0.48
	XW	6.11±4.34	7.26±2.22	1.18±0.84	1.40±0.45
	MW	5.06±4.86	5.58±3.46	0.83±0.88	0.99±0.76
Cropping Regime	XB	5.70±0.82	4.10±NA	1.50±0.13	1.04±NA
	XBF	7.65±5.87	8.90±1.56	1.52±0.87	1.82±0.26
	MB	4.90±0.92	NA	1.37±0.16	NA
	MBF	6.27±1.30	9.40±2.41	1.29±0.19	1.89±0.37
	XW	5.40±4.37	8.00±2.59	1.11±0.90	1.53±0.52
	XWF	6.82±4.70	6.52±1.73	1.25±0.87	1.26±0.36
	MW	3.58±1.40	7.94±1.08	0.58±0.33	1.48±0.18
	MWF	6.55±6.87	3.22±3.44	1.08±1.24	0.50±0.82

Diversity is expressed as Shannon values and richness is based on the number of unique OTUs. Small and large letters in columns and rows indicate statistically significant differences between the treatments in each compartment ($p \leq 0.05$, means ± SD). Abbreviations: MB/MW, faba bean/wheat grown in monoculture; XB/XW, faba bean/wheat intercropping, C, control unplanted soil; F, inoculated samples, CS, control starting soil.

Table 9. Fungal richness and diversity with regard to plant compartment, harvest and cropping regime.

		Fungi			
	Treatment	**Richness**		**Diversity**	
		H1	**H2**	**H1**	**H2**
		Unplanted soil			
	C	380.00±71.74	NA	4.14±0.51	NA
	CF	399.02±47.94	NA	4.37±0.14	NA
	CS	410.77±26.57	NA	4.15±0.11	NA
		Rhizosphere soil			
Inoculation	Faba bean control	131.02±21.72	113.63±22.03	3.81±0.36	3.58±0.54
	Faba bean inoculated	**139.29±17.50A**	**95.39±34.79B**	**3.90±0.31A**	**3.36±0.70B**
	Wheat control	**127.11±10.17A**	**143.02±12.11Ba**	**3.76±0.24A**	**4.01±0.15Ba**
	Wheat inoculated	129.62±16.02	**111.42±35.58b**	3.89±0.26	**3.30±1.13b**
Cropping System	XB	135.71±21.83	117.40±18.24	3.84±0.38	3.73±0.37
	MB	134.60±18.41	91.62±34.33	3.88±0.28	3.21±0.73
	XW	131.46±11.72	135.12±18.26	3.90±0.19	**3.93±0.33a**
	MW	125.27±14.31	119.32±38.58	3.76±0.30	**3.39±1.15b**

Table continued on the following page

Cropping Regime	XB	134.48±29.40	121.60±19.28	3.85±0.46	3.83±0.33
	XBF	136.94±14.28	113.20±18.24	3.83±0.35	3.62±0.41
	MB	127.56±12.93	105.66±23.69	3.77±0.28	3.33±0.63
	MBF	141.64±21.71	77.58±39.97	3.98±0.28	3.10±0.87
	XW	129.18±11.49	**143.30±11.20a**	3.82±0.22	**4.06±0.16a**
	XWF	133.74±12.81	**126.94±21.39ab**	3.97±0.14	**3.80±0.42ab**
	MW	125.04±9.49	**142.74±14.30a**	3.70±0.28	**3.97±0.14ab**
	MWF	125.50±19.25	**95.90±42.11b**	3.81±0.34	**2.80±1.44b**
Roots					
Inoculation	Faba bean control	14.40±2.69	16.64±5.56	1.90±0.24	1.97±0.46
	Faba bean inoculated	18.16±7.16	22.39±6.81	2.20±0.59	2.51±0.50
	Wheat control	13.12±5.65	14.01±6.40	1.67±0.63	1.65±0.83
	Wheat inoculated	12.01±3.40	18.18±7.71	1.70±0.54	2.14±0.78
Cropping System	XB	16.22±5.84	21.95±6.47	1.98±0.49	2.38±0.50
	MB	17.14±6.48	16.27±5.92	2.23±0.47	2.05±0.57
	XW	**14.66±5.32a**	**19.80±7.07a**	**1.95±0.58a**	**2.26±0.79a**
	MW	**10.30±2.27b**	**12.39±5.45b**	**1.39±0.42b**	**1.52±0.70b**
Cropping Regime	XB	14.60±2.25	19.80±5.51	1.87±0.27	2.15±0.31
	XBF	17.52±7.73	23.24±7.25	2.07±0.64	2.52±0.57
	MB	14.00±4.53	14.28±4.91	1.97±0.24	1.83±0.54
	MBF	19.23±7.56	20.25±7.42	2.40±0.55	2.47±0.44
	XW	16.72±6.20	**14.38±5.85a**	2.06±0.68	**1.69±0.76a**
	XWF	12.60±3.82	**25.22±2.20b**	1.84±0.51	**2.84±0.16b**
	MW	9.52±1.01	**13.64±7.60a**	1.28±0.22	**1.60±0.98a**
	MWF	11.28±3.18	**11.14±2.27a**	1.53±0.60	**1.44±0.36a**
Leaves					
Inoculation	Faba bean control	8.95±2.89	7.54±3.34	1.96±0.42	1.64±0.71
	Faba bean inoculated	6.44±1.76	5.66±2.78	1.57±0.37	1.24±0.61
	Wheat control	4.47±2.32	**5.52±1.03a**	1.02±0.68	**1.27±0.26a**
	Wheat inoculated	4.95±2.72	**4.13±0.73b**	1.10±0.70	**0.90±0.24b**
Cropping System	XB	6.78±1.33	6.10±0.85	1.68±0.24	1.49±0.05
	MB	8.06±3.18	6.51±3.34	1.78±0.55	1.39±0.72
	XW	4.73±2.20	4.98±1.37	1.07±0.62	1.11±0.37
	MW	4.69±2.84	4.76±0.92	1.05±0.75	1.08±0.26
Cropping Regime	XB	6.15±1.06	8.75±2.47	1.54±0.22	1.99±0.34
	XBF	7.10±1.48	6.10±0.85	1.75±0.25	1.49±0.05
	MB	10.35±2.40	6.73±4.10	2.17±0.32	1.41±0.86
	MBF	5.78±1.96	5.48±3.36	1.39±0.42	1.15±0.71
	XW	4.72±2.46	5.78±1.33	1.06±0.67	1.34±0.35
	XWF	4.74±2.21	3.98±0.46	1.08±0.65	0.82±0.11
	MW	4.22±2.44	5.26±0.65	0.98±0.77	1.20±0.16
	MWF	5.16±3.41	4.26±0.93	1.12±0.83	0.96±0.30

Diversity is expressed as Shannon values and richness is based on the number of unique OTUs. Small and large letters in columns and rows indicate statistically significant differences between the treatments in each compartment ($p \leq 0.05$, means ± SD). Abbreviations: MB/MW, faba bean/wheat grown in monoculture; XB/XW, faba bean/wheat intercropping, C, control unplanted soil; F, inoculated samples; CS, control starting soil.

Figures

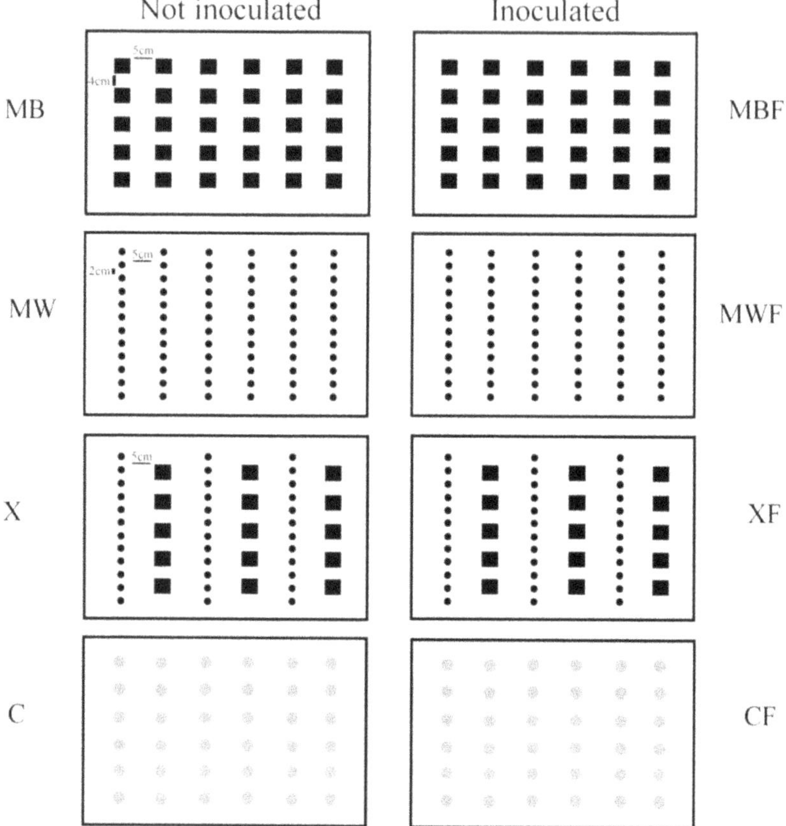

Figure 1. Experimental design of the study. MB/MW, faba bean/wheat grown in monoculture; X, intercropping; C, unplanted soil which containd glassbeads; F, inoculated samples.

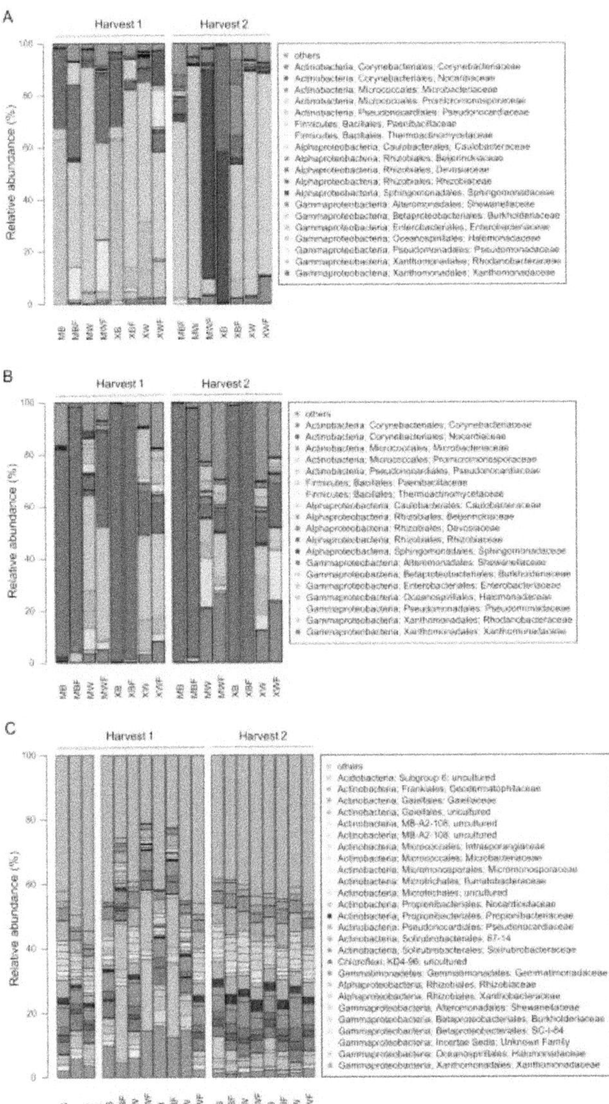

Figure 2. Abundant bacterial families with regard to plant compartment and cropping regime. Only treatments with an average abundance > 0.05 % are shown. Mean relative abundances of each taxa were calculated for each sample. MB/MW, faba bean/wheat grown in monoculture; XB/XW, faba bean/wheat intercropped; C, control unplanted soil; F, inoculated samples; CS, control starting soil. **A**, leaf; **B**, root; **C**, unplanted and rhizosphere soil.

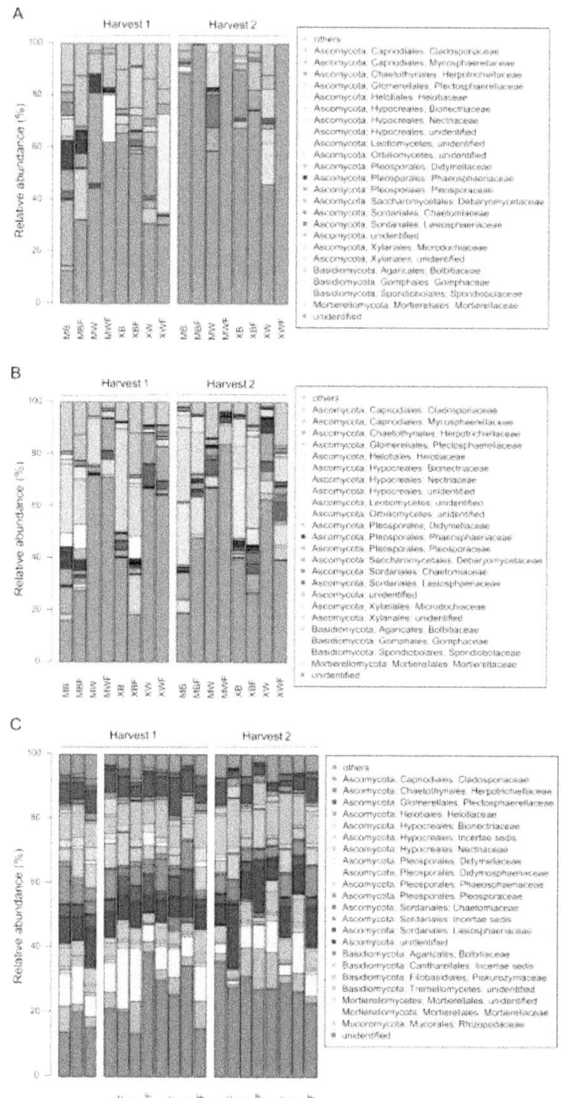

Figure 3. Abundant fungal families with regard to plant compartment and cropping regime. Only treatments with an average abundance > 0.05 % are shown. Mean relative abundances of each taxa were calculated for each sample. MB/MW, faba bean/wheat grown in monoculture; XB/XW, faba bean/wheat intercropped; C, control unplanted soil; F, inoculated samples; CS, control starting soil. **A**, leaf; **B**, root; **C**, unplanted and rhizosphere soil.

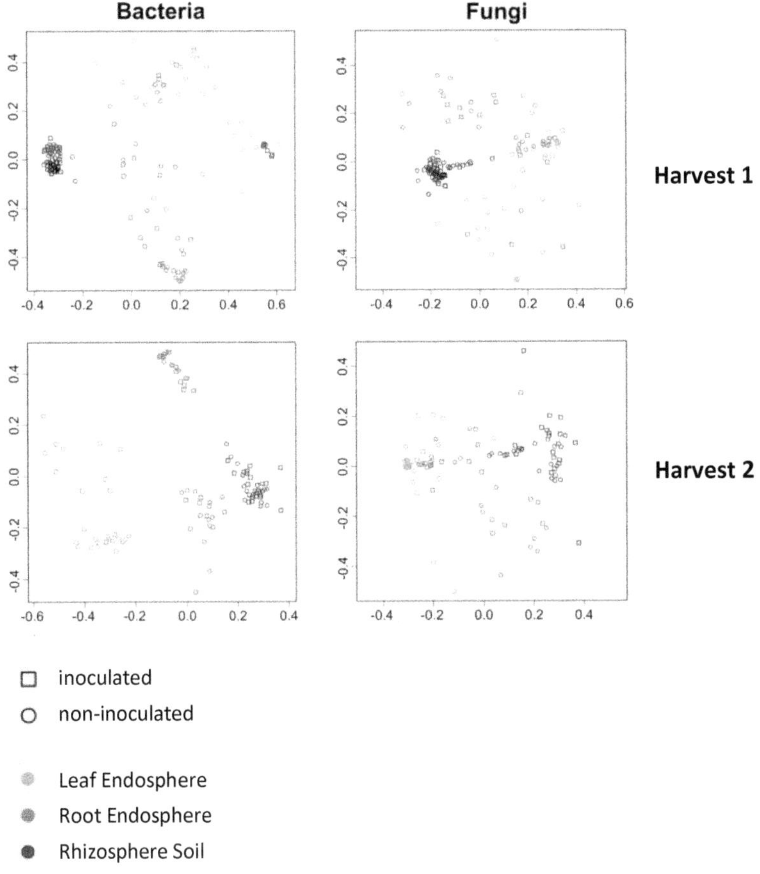

Figure 4. Response of bacterial and fungal communities in different compartments towards inoculation. NMDS ordination of microbial communities is color-coded by the respective compartment and different symbols indicate inoculation. Ordination is based on Bray-Curtis dissimilarities.

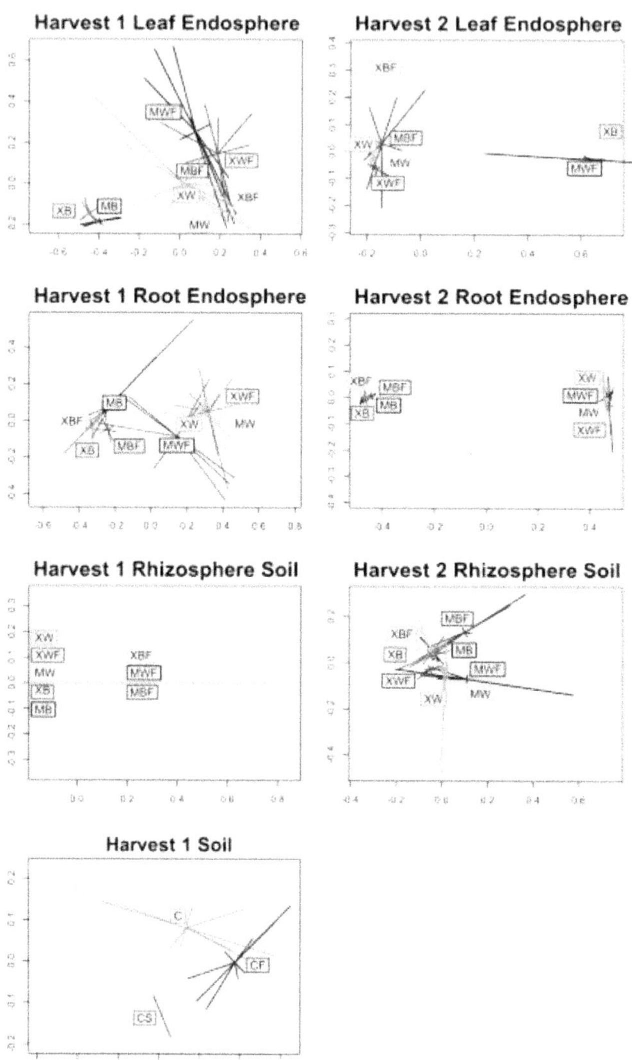

Figure continued on the following page

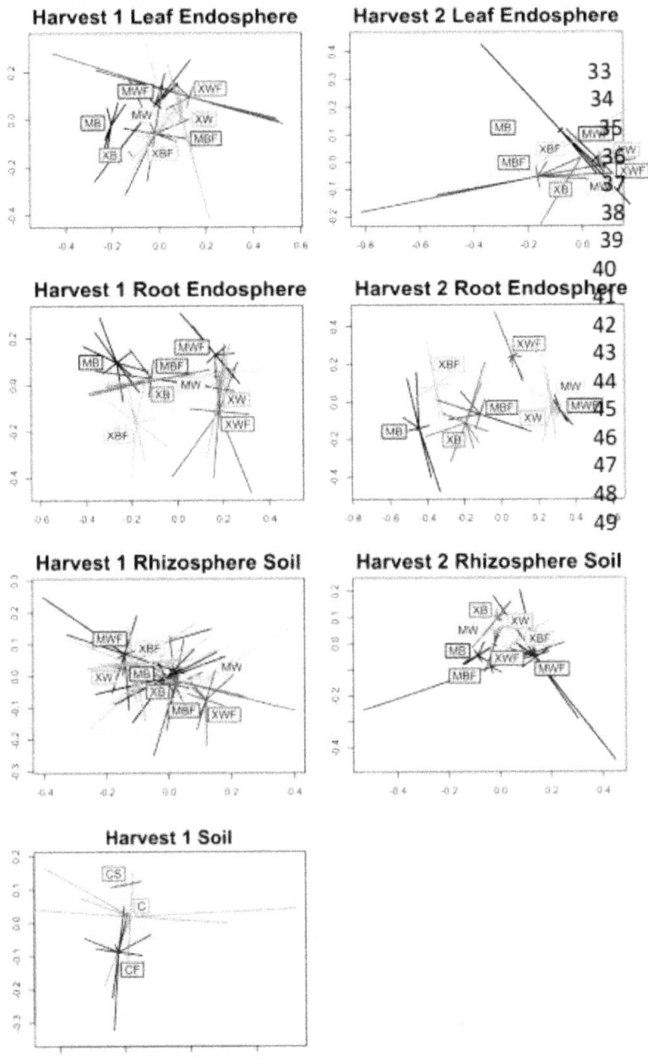

Figure 5. Response of bacterial and fungal communities in different compartments towards cropping regimes. NMDS ordination of microbial communities is color-coded by the respective cropping regime. Ordination is based on Bray-Curtis dissimilarities between samples. MB/MW, faba bean/wheat grown in monoculture; XB/XW, faba bean/wheat intercropped; C, control unplanted soil; F, inoculated samples; CS, control starting soil.

6 References

Aguilar-Trigueros CA, Rillig MC. 2016. Effect of different root endophytic fungi on plant community structure in experimental microcosms. Ecology and Evolution. Nov;6:8149-8158.

Akutse KS, Maniania NK, Fiaboe KKM, Van den Berg J, Ekesi S. 2013. Endophytic colonization of Vicia faba and Phaseolus vulgaris (Fabaceae) by fungal pathogens and their effects on the life-history parameters of Liriomyza huidobrensis (Diptera: Agromyzidae). Fungal Ecology. Aug;6:293-301.

Anderson MJ. 2001. A new method for non-parametric multivariate analysis of variance. Austral Ecology. Feb;26:32-46.

Anderson MJ. 2006. Distance-based tests for homogeneity of multivariate dispersions. Biometrics. Mar;62:245-253.

Andreote FD, da Rocha UN, Araujo WL, Azevedo JL, van Overbeek LS. 2010. Effect of bacterial inoculation, plant genotype and developmental stage on root-associated and endophytic bacterial communities in potato (Solanum tuberosum). Antonie Van Leeuwenhoek International Journal of General and Molecular Microbiology. May;97:389-399.

Ardanov P, Sessitsch A, Haggman H, Kozyrovska N, Pirttila AM. 2012. Methylobacterium-Induced Endophyte Community Changes Correspond with Protection of Plants against Pathogen Attack. Plos One. Oct;7.

Barrs HD, Weatherley PE. 1962. A Re-Examination of the Relative Turgidity Technique for Estimating Water Deficits in Leaves. Australian Journal of Biological Sciences 15 (3):413–428.

Batta YA. 2013. Efficacy of endophytic and applied Metarhizium anisopliae (Metch.) Sorokin (Ascomycota: Hypocreales) against larvae of Plutella xylostella L. (Yponomeutidae: Lepidoptera) infesting Brassica napus plants. Crop Protection. Feb;44:128-134.

Behie SW, Bidochka MJ. 2014. Nutrient transfer in plant-fungal symbioses. Trends in Plant Science. Nov;19:734-740.

Bing LA, Lewis LC. 1991. SUPPRESSION OF OSTRINIA-NUBILALIS (HUBNER) (LEPIDOPTERA, PYRALIDAE) BY ENDOPHYTIC BEAUVERIA-BASSIANA (BALSAMO) VUILLEMIN. Environmental Entomology. Aug;20:1207-1211.

Bolger AM, Lohse M, Usadel B. 2014. Trimmomatic: a flexible trimmer for Illumina sequence data. Bioinformatics. Aug;30:2114-2120.

Chelius MK, Triplett EW. 2001. The diversity of archaea and bacteria in association with the roots of Zea mays L. Microbial Ecology. Apr;41:252-263.

Clifton EH, Jaronski ST, Coates BS, Hodgson EW, Gassmann AJ. 2018. Effects of endophytic entomopathogenic fungi on soybean aphid and identification of Metarhizium isolates from agricultural fields. Plos One. Mar;13.

Clifton EH, Jaronski ST, Hodgson EW, Gassmann AJ. 2015. Abundance of Soil-Borne Entomopathogenic Fungi in Organic and Conventional Fields in the Midwestern USA with an Emphasis on the Effect of Herbicides and Fungicides on Fungal Persistence. Plos One. Jul;10.

de Boer W. 2017. Upscaling of fungal-bacterial interactions: from the lab to the field. Current Opinion in Microbiology. Jun;37:35-41.

de Caceres M, Legendre P. 2009. Associations between species and groups of sites: indices and statistical inference. Ecology. Dec;90:3566-3574.

de Faria MR, Wraight SP. 2007. Mycoinsecticides and Mycoacaricides: A comprehensive list with worldwide coverage and international classification of formulation types. Biological Control. Dec;43:237-256.

de Mendiburu F. 2014. Statistical Procedures for Agricultural Research. Package "Agricolae" Version 1.4-4. Comprehensive R Archive Network, Institute for Statistics and Mathematics, Vienna.

Edgar RC. 2010. Search and clustering orders of magnitude faster than BLAST. Bioinformatics. Oct;26:2460-2461.

Edgar RC, Haas BJ, Clemente JC, Quince C, Knight R. 2011. UCHIME improves sensitivity and speed of chimera detection. Bioinformatics. Aug;27:2194-2200.

Fox J, Weisberg S. 2011. An R companion to applied regression. Second Edition. Sage.

Fudou R, Iizuka T, Sato S, Ando T, Shimba N, Yamanaka S. 2001. Haliangicin, a novel antifungal metabolite produced by a marine myxobacterium 2. Isolation and structural elucidation. Journal of Antibiotics. Feb;54:153-156.

Gadhave KR, Devlin PF, Ebertz A, Ross A, Gange AC. 2018. Soil inoculation with Bacillus spp. modifies root endophytic bacterial diversity, evenness and community composition in a context-specific manner. Microbiology Ecology. Mar; 1-10.

Gdanetz K, Trail F. 2017. The wheat microbiome under four management strategies, and potential for endophytes in disease protection. Phytobiomes. Oct;158-168.

Gong AD, Li HP, Shen L, Zhang JB, Wu AB, He WJ, Yuan QS, He JD, Liao YC. 2015. The Shewanella algae strain YM8 produces volatiles with strong inhibition activity against Aspergillus pathogens and aflatoxins. Frontiers in Microbiology. Oct;6.

Granzow S, Kaiser K, Wemheuer B, Pfeiffer B, Daniel R, Vidal S, Wemheuer F. 2017. The Effects of Cropping Regimes on Fungal and Bacterial Communities of Wheat and Faba Bean in a Greenhouse Pot Experiment Differ between Plant Species and Compartment. Frontiers in Microbiology. May;8.

Greenfield M, Jimenez MIG, Ortiz V, Vega FE, Kramer M, Parsa S. 2016. Beauveria bassiana and Metarhizium anisopliae endophytically colonize cassava roots following soil drench inoculation. Biological Control. Apr;95:40-48.

Gu Y, Wei Z, Wang XQ, Friman VP, Huang JF, Wang XF, Mei XL, Xu YC, Shen QR, Jousset A. 2016. Pathogen invasion indirectly changes the composition of soil microbiome via shifts in root exudation profile. Biology and Fertility of Soils. Oct;52:997-1005.

Hess M, Barralis G, Bleiholder H, Buhr L, Eggers T, Hack H, Stauss R. 1997. Use of the extended BBCH scale - general for the descriptions of the growth stages of mono- and dicotyledonous weed species. Weed Research. Dec;37:433-441.

Hirsch J, Galidevara S, Strohmeier S, Devi KU, Reineke A. 2013. Effects on Diversity of Soil Fungal Community and Fate of an Artificially Applied Beauveria bassiana Strain Assessed Through 454 Pyrosequencing. Microbial Ecology. Oct;66:608-620.

Hodge A. 2003. Plant nitrogen capture from organic matter as affected by spatial dispersion, interspecific competition and mycorrhizal colonization. New Phytologist. Jan; 157:303-314.

Hong MS, Peng GX, Keyhani NO, Xia YX. 2017. Application of the entomogenous fungus, Metarhizium anisopliae, for leafroller (Cnaphalocrocis medinalis) control and its effect on rice phyllosphere microbial diversity. Applied Microbiology and Biotechnology. Sep;101:6793-6807.

Hummel RL, Walgenbach JF, Barbercheck ME, Kennedy GG, Hoyt GD, Arellano C. 2002. Effects of production practices on soil-borne entomopathogens in western North Carolina vegetable systems. Environmental Entomology. Feb;31:84-91.

Innerebner G, Knief C, Vorholt JA. 2011. Protection of Arabidopsis thaliana against Leaf-Pathogenic Pseudomonas syringae by Sphingomonas Strains in a Controlled Model System. Applied and Environmental Microbiology. May;77:3202-3210.

Jaber LR, Enkerli J. 2016. Effect of seed treatment duration on growth and colonization of Vicia faba by endophytic Beauveria bassiana and Metarhizium brunneum. Biological Control. Dec;103:187-195.

Jha B, Gontia I, Hartmann A. 2012. The roots of the halophyte Salicornia brachiata are a source of new halotolerant diazotrophic bacteria with plant growth-promoting potential. Plant and Soil. Jul;356:265-277.

Kaiser K, Wemheuer B, Korolkow V, Wemheuer F, Nacke H, Schoning I, Schrumpf M, Daniel R. 2016. Driving forces of soil bacterial community structure, diversity, and function in temperate grasslands and forests. Scientific Reports. Sep;6.

Kembel SW, Cowan PD, Helmus MR, Cornwell WK, Morlon H, Ackerly DD, Blomberg SP, Webb CO. 2010. Picante: R tools for integrating phylogenies and ecology. Bioinformatics. Jun;26:1463-1464.

Kepler RM, Ugine TA, Maul JE, Cavigelli MA, Rehner SA. 2015. Community composition and population genetics of insect pathogenic fungi in the genus Metarhizium from soils of a long-term agricultural research system. Environmental Microbiology. Aug;17:2791-2804.

Kõljalg U, Nilsson RH, Abarenkov K, Tedersoo L, Taylor AFS, Bahram M, Bates ST, Bruns TD, Bengtsson-Palme J, Callaghan TM, et al. 2013. Towards a unified paradigm for sequence-based identification of fungi. Molecular Ecology. Nov;22:5271-5277.

Krasnoff SB, Keresztes I, Donzelli BGG, Gibson DM. 2014. Metachelins, Mannosylated and N-Oxidized Coprogen-Type Siderophores from Metarhizium robertsii. Journal of Natural Products. Jul;77:1685-1692.

Krell V, Unger S, Jakobs-Schoenwandt D, Patel AV. 2018. Endophytic Metarhizium brunneum mitigates nutrient deficits in potato and improves plant productivity and vitality. Fungal Ecology. Aug;34:43-49.

Lacey LA, Grzywacz D, Shapiro-Ilan DI, Frutos R, Brownbridge M, Goettel MS. 2015. Insect pathogens as biological control agents: Back to the future. Journal of Invertebrate Pathology. Nov;132:1-41.

Lane DJ. 1991. “16S/23S rRNA sequencing,” in Nucleic Acids Techniques in Bacterial Systematics, eds E. Stackebrandt andM. Goodfellow (Chichester: John Wiley & Sons), 115–147.

Le Cocq K, Gurr SJ, Hirsch PR, Mauchline TH. 2017. Exploitation of endophytes for sustainable agricultural intensification. Molecular Plant Pathology. Apr;18:469-473.

Li P, Ye SF, Liu H, Pen AH, Ming F, Tang XM. 2018. Cultivation of Drought-Tolerant and Insect-Resistant Rice Affects Soil Bacterial, but Not Fungal, Abundances and Community Structures. Frontiers in Microbiology. Jun;9.

Li S, Wu FZ. 2018. Diversity and Co-occurrence Patterns of Soil Bacterial and Fungal Communities in Seven Intercropping Systems. Frontiers in Microbiology. Jul;9.

Liu CM, Yang ZF, He PF, Munir S, Wu YX, Ho HH, He YQ. 2018. Deciphering the bacterial and fungal communities in clubroot-affected cabbage rhizosphere treated with Bacillus Subtilis XF-1. Agriculture Ecosystems & Environment. Mar;256:12-22.

Liu YC, Qin XM, Xiao JX, Tang L, Wei CZ, Wei JJ, Zheng Y. 2017. Intercropping influences component and content change of flavonoids in root exudates and nodulation of Faba bean. Journal of Plant Interactions.12:187-192.

Martinez Arbizu P. 2017. Pairwiseadonis: Pairwise multilevel comparison using adonis. R Package Version 0.0.1.

Mayerhofer J, Eckard S, Hartmann M, Grabenweger G, Widmer F, Leuchtmann A, Enkerli J. 2017. Assessing effects of the entomopathogenic fungus Metarhizium brunneum on soil microbial communities in Agriotes spp. biological pest control. Fems Microbiology Ecology. Oct;93.

McKinnon AC, Glare TR, Ridgway HJ, Mendoza-Mendoza A, Holyoake A, Godsoe WK, Bufford JL. 2018. Detection of the Entomopathogenic Fungus Beauveria bassiana in the Rhizosphere of Wound-Stressed Zea mays Plants. Frontiers in Microbiology. Jun;9.

McMurdie PJ, Holmes S. 2014. Waste Not, Want Not: Why Rarefying Microbiome Data Is Inadmissible. Plos Computational Biology. Apr;10.

Moora M, Zobel M. 1996. Effect of arbuscular mycorrhiza on inter- and intraspecific competition of two grassland species. Oecologia. Oct;108:79-84.

Nübel U, Engelen B, Felske A, Snaidr J, Wieshuber A, Amann RI, Ludwig W, Backhaus H. 1996. Sequence heterogeneities of genes encoding 16S rRNAs in Paenibacillus polymyxa detected by temperature gradient gel electrophoresis. Journal of Bacteriology. Oct;178:5636-5643.

Ogle DH. 2016. Introductory fisheries analyses with R. Chapman & Hall/CRC.

Oksanen O, Blanchet FG, Kindt R, Legendre P, Minchin PR, O'Hara RB, Simpson GL, Solymos P, Stevens MHH, Wagner H. 2016. Vegan: Community Ecology Package. R Package Version 2.3-5.

Ortiz-Urquiza A, Luo ZB, Keyhani NO. 2015. Improving mycoinsecticides for insect biological control. Applied Microbiology and Biotechnology. Feb;99:1057-1068.

Quast C, Pruesse E, Yilmaz P, Gerken J, Schweer T, Yarza P, Peplies J, Glockner FO. 2013. The SILVA ribosomal RNA gene database project: improved data processing and web-based tools. Nucleic Acids Research. Jan;41:D590-D596.

R Core Team. 2016. R: A Language and Environment for Statistical Computing. Vienna: R Foundation for Statistical Computing. Available online at: http://www.R-project.org

Rabiey M, Ullah I, Shaw LJ, Shaw MW. 2017. Potential ecological effects of Piriformospora indica, a possible biocontrol agent, in UK agricultural systems. Biological Control. Jan;104:1-9.

Raya-Diaz S, Sánchez-Rodriguez AR, Segura-Fernandez JM, del Campillo MD, Quesada-Moraga E. 2017. Entomopathogenic fungi-based mechanisms for improved Fe nutrition in sorghum plants grown on calcareous substrates. Plos One. Oct;12.

Ritz C, Streibig JC. 2016. Package 'drc' Analysis of dose-response curves. Version 3.0-1.

Roumet C, Lafont F, Sari M, Warembourg F, Garnier E. 2008. Root traits and taxonomic affiliation of nine herbaceous species grown in glasshouse conditions. Plant and Soil. Nov;312:69-83.

Sánchez-Rodriguez AR, Barron V, Del Campillo MC, Quesada-Moraga E. 2016. The entomopathogenic fungus Metarhizium brunneum: a tool for alleviating Fe chlorosis. Plant and Soil. Sep;406:295-310.

Sánchez-Rodriguez AR, Raya-Diaz S, Zamarreno AM, Garcia-Mina JM, del Campillo MC, Quesada-Moraga E. 2018. An endophytic Beauveria bassiana strain increases spike production in bread and durum wheat plants and effectively controls cotton leafworm (Spodoptera littoralis) larvae. Biological Control. Jan;116:90-102.

Sasan RK, Bidochka MJ. 2012. The insect-pathogenic fungus *Metarhizium robertsii* (Clavicipitaceae) is also an endophyte that stimulates plant root development. American Journal of Botany. Jan; 99:101-107.

Schneider S, Widmer F, Jacot K, Kolliker R, Enkerli J. 2012. Spatial distribution of Metarhizium clade 1 in agricultural landscapes with arable land and different semi-natural habitats. Applied Soil Ecology. Jan;52:20-28.

Schwarzenbach K, Enkerli J, Widmer F. 2009. Effects of biological and chemical insect control agents on fungal community structures in soil microcosms. Applied Soil Ecology. May;42:54-62.

Siczek A, Frac M, Kalembasa S, Kalembasa D. 2018. Soil microbial activity of faba bean (Vicia faba L.) and wheat (Triticum aestivum L.) rhizosphere during growing season. Applied Soil Ecology. Sep;130:34-39.

Steinwender BM, Enkerli J, Widmer F, Eilenberg J, Thorup-Kristensen K, Meyling NV. 2014. Molecular diversity of the entomopathogenic fungal Metarhizium community within an agroecosystem. Journal of Invertebrate Pathology. Nov;123:6-12.

Tefera T, Vidal S. 2009. Effect of inoculation method and plant growth medium on endophytic colonization of sorghum by the entomopathogenic fungus Beauveria bassiana. Biocontrol. Oct;54:663-669.

Tichy L, Chytry M. 2006. Statistical determination of diagnostic species for site groups of unequal size. Journal of Vegetation Science. Dec;17:809-818.

Tiwari S, Singh P, Tiwari R, Meena KK, Yandigeri M, Singh DP, Arora DK. 2011. Salt-tolerant rhizobacteria-mediated induced tolerance in wheat (Triticum aestivum) and chemical diversity in rhizosphere enhance plant growth. Biology and Fertility of Soils. Nov;47:907-916.

Toju H, Tanabe AS, Yamamoto S, Sato H. 2012. High-Coverage ITS Primers for the DNA-Based Identification of Ascomycetes and Basidiomycetes in Environmental Samples. Plos One. Jul;7.

Vidal S, Jaber LR. 2015. Entomopathogenic fungi as endophytes: plant-endophyte-herbivore interactions and prospects for use in biological control. Current Science. Jul;109:46-54.

Wagner MR, Lundberg DS, del Rio TG, Tringe SG, Dangl JL, Mitchell-Olds T. 2016. Host genotype and age shape the leaf and root microbiomes of a wild perennial plant. Nature Communications. Jul;7.

Wan XH, Huang ZQ, He ZM, Yu ZP, Wang MH, Davis MR, Yang YS. 2015. Soil C:N ratio is the major determinant of soil microbial community structure in subtropical coniferous and broadleaf forest plantations. Plant and Soil. Feb;387:103-116.

Wang Y, Marschner P, Zhang FS. 2012. Phosphorus pools and other soil properties in the rhizosphere of wheat and legumes growing in three soils in monoculture or as a mixture of wheat and legume. Plant and Soil. May;354:283-298.

Wallace J, Laforest-Lapointe I, Kembel SW. 2018. Variation in the leaf and root microbiome of sugar maple (Acer saccharum) at an elevational range limit. Peerj. Aug;6.

Wemheuer F, Kaiser K, Karlovsky P, Daniel R, Vidal S, Wemheuer B. 2017. Bacterial endophyte communities of three agricultural important grass species differ in their response towards management regimes. Scientific Reports. Jan;7.

Wemheuer F, Wemheuer B, Kretzschmar D, Pfeiffer B, Herzog S, Daniel R, Vidal S. 2016. Impact of grassland management regimes on bacterial endophyte diversity differs with grass species. Letters in Applied Microbiology. Apr;62:323-329.

Wemheuer B, and Wemheuer F. 2017. "Assessing bacterial and fungal diversity in the plants endosphere," in Metagenomics - Methods and Protocols, Vol. 1539, edsW. Streit, and R. Daniel (New York, NY: Humana Press), 75–84.

White TJ, Bruns T, Lee S, and Taylor J. 1990. "Amplification and direct sequencing of fungal ribosomal RNA genes for phylogenetics," in PCR Protocols: a Guide to Methods and Applications, eds M. A. Innis, D. H. Gelfand, J. J. Sninsky, and T. J.White (New York, NY: Academic Press). 18, 315–322.

Xiao YB, Li L, Zhang FS. 2004. Effect of root contact on interspecific competition and N transfer between wheat and fababean using direct and indirect N-15 techniques. Plant and Soil. May;262:45-54.

Yang ZP, Yang WP, Li SC, Hao JM, Su ZF, Sun M, Gao ZQ, Zhang CL. 2016. Variation of Bacterial Community Diversity in Rhizosphere Soil of Sole-Cropped versus Intercropped Wheat Field after Harvest. Plos One. Mar;11.

Yao Z, He D, Lei Y. 2018. Thermal imaging for early nondestructive detection of wheat stripe rust. ASABE Annual International Meeting;1801728.

Zhang NN, Sun YM, Li L, Wang ET, Chen WX, Yuan HL. 2010. Effects of intercropping and Rhizobium inoculation on yield and rhizosphere bacterial community of faba bean (Vicia faba L.). Biology and Fertility of Soils. Aug;46:625-639.

Zhang YZ, Wang ET, Li M, Li QQ, Zhang YM, Zhao SJ, Jia XL, Zhang LH, Chen WF, Chen WX. 2011. Effects of rhizobial inoculation, cropping systems and growth stages on endophytic bacterial community of soybean roots. Plant and Soil. Oct;347:147-161.

Zhang NN, Sun YM, Wang ET, Yang JS, Yuan HL, Scow KM. 2015. Effects of intercropping and Rhizobial inoculation on the ammonia-oxidizing microorganisms in rhizospheres of maize and faba bean plants. Applied Soil Ecology. Jan;85:76-85.

Zhou Y, Zhu HH, Fu SL, Yao Q. 2017. Variation in Soil Microbial Community Structure Associated with Different Legume Species Is Greater than that Associated with Different Grass Species. Frontiers in Microbiology. May;8.

Zimmermann J, Musyoki MK, Cadisch G, Rasche F. 2016. Biocontrol agent Fusarium oxysporum f.sp strigae has no adverse effect on indigenous total fungal communities and specific AMF taxa in contrasting maize rhizospheres. Fungal Ecology. Oct;23:1-10.

Chapter 3:
Intercropping with winter faba bean in arable land and white clover in grassland improves water use efficiency of the cropping system

Annika Meißner, Merle Tränkner, Klaus Dittert

In preparation for submission

Intercropping with winter faba bean in arable land and white clover in grassland improves water use efficiency of the cropping system

Meißner, Annika[1,2*†], Tränkner, Merle[1], Dittert, Klaus[3]

[1] Institute of Applied Plant Nutrition, University of Goettingen, Carl-Sprengel-Weg 1, D-37075 Goettingen, Germany

[2] Center of Biodiversity and Sustainable Land Use, University of Goettingen, Grisebachstr. 6, D-37077 Goettingen, Germany

[3] Section of Plant Nutrition and Crop Physiology, Department of Crop Sciences, University of Goettingen, Carl-Sprengel-Weg 1, D-37075 Goettingen, Germany

*** Correspondence:**
Annika Meißner, M.Sc.
annika.meissner@mein.gmx

†Present Address:

KWS SAAT SE & Co. KGaA, Grimsehlstrasse 31, D-37574 Einbeck, Germany

Abstract

Intercropping of legumes with non-legumes is well known to increase the productivity of the systems by mutual facilitation in arable and grassland crops. One aspect is the complementary use of water resources. This complementary water use increasingly gains importance with respect to the predicted changes in precipitation patterns towards extreme weather events. Under these conditions, the performance of cropping systems also depends on the genotypic performance of species and varieties.

The aim of this study was to investigate the water use and productivity of pure and intercropped stands of the legumes winter faba bean (eight genotypes) and white clover (four genotypes) as well as the non-legumes winter wheat and perennial ryegrass and chicory. Plants were grown at two sites with contrasting growth conditions: the fertile site was a deep silty loam with high water availability, while the marginal site was a shallow loamy soil with generally lower water availability. In additional non-legume plots, effects of nitrogen fertilizer were studied. We investigated carbon stable isotope discrimination ($\delta^{13}C$), transpiration by thermal imaging as well as instantaneous water use efficiency (WUE) by gas exchange.

Results show that in both land-use systems, there were no differences between pure legume stands and intercropped stands with respect to transpiration, $\delta^{13}C$ and WUE. In contrast, non-legumes had significantly lower WUE. Application of nitrogen fertilizer in non-legumes did not increase the WUE to similar levels as intercropping with N_2-fixing species. Apart from yield increments frequently reported in other studies, intercropping with legumes had significant advantages in the use efficiency of water resources compared to pure winter wheat or perennial ryegrass.

The intercropping effect was more pronounced than the site effect. Plant stands at the fertile or at the marginal site showed only small differences. Only $\delta^{13}C$ of winter faba bean was higher at the deep loamy site compared to the shallow soil, being indicative of higher water use efficiency. Overall, genotypic variation of the tested lines was low. Winter faba bean lines showed some genetic variation in $\delta^{13}C$ while there were no differences among white clover genotypes. In grassland, cropping of chicory as mixture component had substantial influence on the water acquisition. In conclusion, the mixture and identity of species were found to be important drivers of water use in field crops, and the study shows that intercropping with legumes considerably improved crop stand WUE.

1 Introduction

Annual or perennial intercropping systems have the potential to reduce the need for cropping management practices such as fertilization and tillage, while promoting soil fertility and biomass production (Weißhuhn et al., 2017). Similar effects are generally accredited to legumes, which usually are rare in conventional European crop rotations where cereals dominate (Cernay et al., 2015; Reckling et al., 2016). Adding legumes to non-legumes generally reduces the need for inputs and simultaneously support productivity and biodiversity of the ecosystem (Altieri, 1999; Lüscher et al., 2014). Intercropping systems including legumes grown in arable land and grassland have been reported to significantly increase the crop productivity and yield stability as compared to crops grown in pure stands (Bedoussac et al., 2015; Raseduzzaman & Jensen, 2017; Sturludóttir et al., 2014).

Additionally, enhanced species richness and diversity of cropping systems decrease vulnerability to abiotic stresses due to different plant architecture and flowering characteristics resulting in complementary use of resources such as light, nutrients and water (Frison et al. 2011; Hauggaard-Nielsen et al., 2008; Picasso et al., 2011). This is particularly beneficial considering the abiotic stress factor drought which has major impacts on yield stability worldwide (Zampieri et al., 2017). An early plant response to drought is stomatal closure in order to reduce transpiration and thus restrict water loss (Boyer & Westgate, 2004). However, stomatal closure also reduces the influx of CO_2 into the leaf resulting in decreased carbon assimilation rates which can in turn impair biomass production (Benešová et al., 2012). The ratio of biomass production or CO_2 assimilation rates to water loss is generally defining the water use efficiency (Tambussi et al., 2007).

In a crop canopy, the net ecosystem exchange of CO_2 describes the canopy CO_2 fluxes consisting of plant CO_2 uptake as well as plant and soil respiration, hence reflecting the instantaneous net canopy productivity. Complemented with evapotranspiration, the instantaneous WUE of a specific crop stand ecosystem can be obtained and carbon-water balances quantified. A time-integrated index for WUE is the carbon stable isotope discrimination ($\delta^{13}C$) which reflects the relation between photosynthetic CO_2 demand and diffusive supply of CO_2 for carboxylation reactions (Farquhar et al., 1989). Environmental conditions such as evaporative demand, soil water status and air temperature, as well as plant characteristics such as leaf N content influence the CO_2 supply and the plants demand for CO_2. All these parameters affect the stomatal conductance and the intercellular CO_2 concentration, ultimately changing the carbon stable isotope signature (Seibt et al., 2008).

In terms of plant characteristics, WUE has become an increasingly important target in breeding of cereals (Richards et al., 2002). In contrast to cereals, yield improvements by breeding of most legumes were comparatively small during the last decades (Foyer et al., 2016; Sharma et al., 2013). Moreover, breeding is traditionally performed in pure stands so that the pool of suitable cultivars for intercropping systems is scarce and needs to be validated (Finney et al., 2016; Reckling et al., 2016). Additionally, productivity under water limitation is highly dependent on the species (Schilling et al., 2016) and is therefore difficult to predict in mixtures. Intercropping ecosystems include complex environmental dynamics that influence the carbon-water balance. Hence, assessing the potential of intercropping for enhancing the WUE is highly interesting, but only rarely performed up to date.

A crucial aspect with regard to productivity is the nitrogen availability which is quantitatively one of the most important nutrients for biomass production. Nitrogen deficiency is compensated by increased transpiration for increasing the nutrient acquisition from the soil (Matimati et al., 2014). In turn, reduced transpiration as described above may induce or aggravate nitrogen deficiency (Kunrath et al., 2018). Application of nitrogen fertilizer is therefore well known to enhance WUE (e.g. Toft, Anderson, and Nowak 1989; Brueck and Senbayram 2009). Sufficient nitrogen supply either by biological N_2 fixation or by N fertilizer application is thus indispensable in environments prone to temporal water scarcity. Hence, including N_2-fixing species into crop stands provides beneficial effects in terms of water use and productivity.

Positive mixture effects are, however, affected by the characteristics of the sites regarding soil and climatic conditions (Suter et al., 2015). It is well established that selection for differences should be done under diverse conditions to identify the genetic potentials of different lines in multiple unfavorable environments (Boyer, 1982; Muktadir et al., 2020). Considering this aspect, WUE of suitable genotypes for intercropping of legumes and non-legumes may differ when they are grown under contrasting environmental conditions. In this context, studies on WUE of intercropping systems at the field scale are rare.

In order to deepen our understanding of WUE within the concept of multi-species crop stands, two field experiments were carried out with diverse genotypes of arable (*Vicia faba*

L., *Triticum aestivum* L.) and grassland species (*Trifolium repens* L., *Lolium perenne* L., *Cichorium intybus* L.). The experiments were run in parallel at two study sites with contrasting growth conditions regarding fertility and water availability. Water use and WUE were evaluated throughout the growing seasons. We hypothesize that (1) intercropping with legumes as well as the application of nitrogen fertilizer can improve WUE; (2) different genotypes of the species differ in their responses and (3) the WUE at the marginal site is higher than at the fertile site.

2 Material and methods

Two field experiments with identical set up were conducted in the years 2015, 2016 and 2017 at the fertile site Reinshof and the marginal site Deppoldshausen, both located close to Goettingen in central Germany. The fertile site had high nutrient and water availability whereas the marginal site had a shallow soil with lower water availability.

The fertile site Reinshof (51.484 °N, 9.923 °E, 157 m asl.) is characterized by its deep Gleyic Fluvisol with land values between 81 and 89 (an index of up to 100 for site characteristics and yield potential (Rust, 2006)), an average pH of 7.0, a high silt content of 68 %, 21 % clay, 11 % sand and 2.81 % humus. The marginal site Deppoldshausen (51.581 °N, 9.967 °E, 342 m asl.) has a shallow Calcaric Leptosol with land values from 38 to 48, an average pH of 7.3, a comparably high clay content of 43 %, 55 % silt, 2 % sand and 3.29 % humus.

Weather conditions were recorded by Deutscher Wetterdienst (DWD) at the nearby station 'Göttingen' (51.500 °N, 9.951 °E, 167 m asl.) and at both field sites there were custom-made weather stations (Adolf Thies GmbH & Co. KG, Göttingen, Germany). Overall, the climate is temperate with a mean annual temperature of 9.4 °C and there is significant rainfall throughout the year (632 mm). At the marginal site, monthly air temperature was by 0.65 °C lower than at the fertile site, leading to a lower yield potential (Fig.1A). In 2015, rainfall in May and June was lower than the long-term average, while in June 2016 there were higher-than-average rainfall and temperatures in contrast to a following dry summer (Fig.1B). The year 2017 was characterized by a comparatively warm spring with average rainfall and very high rainfall in summer (up to 243.8 mm in July).

Experimental design

The factorial field experiments were part of the collaborative research project '*Novel genotypes for mixed cropping allow for improved sustainable land use across arable land, grassland and woodland*' (IMPAC³). At each of the two sites, two different land-use systems were established: i) arable land, and ii) grassland. In both land-use systems arable land and grassland, the tested crop stands were grown with four replicates in a split-plot design. While grassland was established permanently, the arable systems were rotated with rye in a four-year crop rotation.

In the arable system, pure stands consisted of either winter faba bean (*Vicia faba* L.; 8 genotypes) or winter wheat (*Triticum aestivum* L.; variety Genius) (Tab. 1). The pure stands of winter wheat were fertilized either with 0 kg nitrogen (N) ha^{-1} or 120 kg N ha^{-1} in two dressings. Within the arable system, intercropping was done by growing each winter faba bean genotype with winter wheat as substitutive mixtures. Seeds were provided by the Institute of Plant Breeding at the Department of Crop Sciences, University of Goettingen, and Saaten-Union (Isernhagen, Germany).

In the grassland system, pure stands were either cropped with white clover (*Trifolium repens* L.; 4 genotypes), with perennial ryegrass (*Lolium perenne* L.; genotype Elp 060687) or with chicory (*Cichorium intybus* L.; variety Puna II) (Tab. 1). The pure stands of perennial ryegrass and chicory were fertilized either with 0 kg N ha^{-1} or 240 kg N ha^{-1} split into four dressings. Within the grassland system, intercropping was done by growing each white clover genotype with either perennial ryegrass or chicory. Seeds were provided by Deutsche Saatveredelung (DSV; Lippstadt, Germany) and British Seed Houses (Lincoln, Great Britain).

Tab.1: Tested genotypes and seeding densities of legumes (winter faba bean Vf; white clover Tr) and non-legumes (winter wheat Ta; perennial ryegrass Lp; chicory Ci) in crop stands of arable land and grassland.

Land-use system	Species	Genotype	Seeding density pure stands	Seeding density intercropping
Arable land	Winter faba bean	Vf1: S_004 Vf 2: S_062 Vf 3: S_069 Vf 4: S_265 Vf 5: Hiverna/2 Vf 6: Côte d'Or/1 Vf 7: WAB-Fam157 Vf 8: WAB-EP98-267	40 seeds/m^2	20 seeds/m^2
	Winter wheat	Ta: Genius	320 seeds/m^2	160 seeds/m^2
Grassland	White clover	Tr1: EGB PX 90305 Tr2: EGB PX 90312 Tr3: EGB PX 90914 Tr4: EGB PX 90915	1000 seeds/m^2	400 seeds/m^2
	Perennial Ryegrass	Lp: Elp 060687	1000 seeds/m^2	600 seeds/m^2
	Chicory	Ci: Puna II	1000 seeds/m^2	600 seeds/m^2

Management

In arable systems, soil cultivation comprised plowing and harrowing before seeding. Plot size was 3 m * 9 m including 12 rows of plants. Intercropped plots were arranged in alternating rows with half of the seeding density per species in comparison to the respective pure stands (Tab.1). When needed, herbicides were applied in combination with hand weeding to remove emerging weeds and insecticides were used against aphid pests. N fertilizer was applied in two rates of 60 kg N/ha each at tillering and stem elongation of winter wheat. N source was calcium ammonium nitrate.

In grassland, plots of 3 m * 5 m were established in summer 2014. Seeding density in intercropped plots was reduced in order to arrange a substitutive intercropping system. Pest control was regularly undertaken against mice and slugs. Biomass harvests were performed four times a year with an interval of approximately six weeks to simulate the common grassland cutting. The first cut was done approximately two months after the start of the vegetation period (middle of May) and the last approximately one month before the end of the vegetation period (beginning of October) in each year. N fertilizer was used in four dressings: 80 kg N/ha in March, 60 kg N/ha after the first and the second cut each, 40 kg N/ha after the third cut. N source was also calcium ammonium nitrate.

Carbon stable isotope analysis

The carbon stable isotope signature ($\delta^{13}C$) was measured as the ratio of ^{13}C to ^{12}C in aboveground dry matter. In arable crops, whole-plant samples for analysis of $\delta^{13}C$ were taken at the developmental stage of full flowering of faba bean (BBCH 65). Faba bean and wheat were harvested and analyzed separately in intercropped plots. In grassland crops, samples were taken as aliquots from the second biomass harvest. Accordingly, plant samples were a mixture of biomass of the respective species, representing the current composition of species. Those sampling dates in arable and grassland plots were chosen to cover a part of the vegetation period with high biomass production.

After drying at 60 °C for 48 h, the samples were ground to powder and homogenized in a ball mill (Retsch MM400, Haan, Germany). Aliquots were weighed into tin capsules. Samples were analyzed with a MAT Delta Plus IRMS (Finnigan MAT GmbH, Bremen, Germany) at the stable isotope analysis center of the University of Goettingen KOSI. Isotope ratios were related to PeeDee Belemnite (PDB) according to Smith and Epstein (1971).

For carbon isotope analysis, all pure stands and all possible mixtures of the abovementioned species and genotypes were investigated. Though ANOVA results showed significant year effects (Supplement Table S.1), values of arable and grassland crops were calculated as multi-annual means over the investigated years 2015 until 2017. As $\delta^{13}C$ is not as affected by daily weather fluctuations as gas exchange measurements, this was done in order to obtain the year-independent potential of the crop stands and genotypes in terms of intrinsic water use efficiency.

Gas exchange measurements and canopy surface temperature

Carbon assimilation (net ecosystem exchange of CO_2; NEE) as well as evapotranspiration (ET) were estimated under field conditions by measuring gas flux gradients of the canopy (Lindner et al., 2015). At the fertile site, measurements were performed during the vegetation periods of 2015 and 2016 and at the marginal site in 2016 on a weekly basis during the vegetation periods.

Canopies were covered with a transparent chamber (6 mm Plexiglas-XT, custom-made by Hecker Kunststofftechnik, Dortmund, Germany). This chamber had a base area of 0.36 m^2

and there was an option to expand it in height depending on plant growth. The air inside the chamber was circulated by battery driven fans to prevent concentration gradients. In order to avoid influx or efflux of air, the chamber was equipped with a sharp aluminum frame at the bottom that was pressed into the ground. Additionally, the chamber was mounted with sensors for photosynthetically active radiation (PAR) (Apogee original, Apogee instruments, Logan, UT, USA) as well as sensors for air temperature and relative humidity (funky_climate, ESYS GmbH, Berlin, Germany) to record the conditions inside and outside the chamber.

Within the arable field plots, measurements were conducted in the northern 2 m, whereas in grassland, gas flux measurements took place in the western half of the plots. Measurements were taken on clear and sunny days. Before each measurement, the chamber was flushed with ambient air. When covering the canopy with the chamber, changes in CO_2 and H_2O concentrations were recorded for a period of 2 minutes. In 2015, the gas exchange measurements were done using a GFS-3000 (Heinz Walz GmbH, Effeltrich, Germany) while in 2016, an infrared gas analyzer EGM-5 (PP Systems, Amesbury, MA, USA) was used.

Net ecosystem exchange of CO_2 (NEE) and evapotranspiration (ET) were calculated according to Jákli et al. (2018)

$$\mathrm{NEE} = \frac{\mathrm{dCO_2}}{\mathrm{V_m(T)}} * \frac{\mathrm{V}}{\mathrm{A}} * (-1)$$

$$\mathrm{ET} = \frac{\mathrm{dH_2O}}{\mathrm{V_m(T)}} * \frac{\mathrm{V}}{\mathrm{A}}$$

where dCO_2 and dH_2O are the changes in CO_2 and H_2O concentration as linear regression; $V_m(T)$ is the molar volume of the air, adjusted for the specific temperature at the measurement; V and A are the volume and the ground area of the chamber. Relating NEE (Supplement Table S.2 and S.4) to ET (Supplement Table S.3 and S.5), the instantaneous water use efficiency (WUE) of the canopy was calculated according to Tallec et al. (2013)

$$\mathrm{WUE} = \frac{\mathrm{NEE}}{\mathrm{ET}}$$

An indicator for transpiration of the crop stands was assessed as canopy surface temperature by thermography. Thermal images were taken with a T640 infrared camera (FLIR Systems, OR, USA) at approximately 2 m above ground of the same spot which was used for gas exchange measurements. Image analysis was performed with the software FLIR ResearchIR, generating mean values from the central area within the image. In order to even out differences between dates by varying weather conditions, the average air temperature of the measurement day was subtracted from the canopy surface temperature according to Jackson et al. (1981)

$$\mathrm{dT} = \mathrm{T(c)} - \mathrm{T(a)}$$

where T(c) is the surface temperature of the canopy and T(a) is the air temperature.

For gas exchange measurements and thermal images, one genotype of faba bean (i.e. A6: Côte d'Or/1) and white clover (i.e. A2: EGB PX 90312) was chosen for evaluation in pure stands and in intercropped stands with the non-legumes.

Yield data

Grain yield of the arable crops was harvested as described in (Siebrecht-Schöll, 2019). In brief, a core area of the plots (7 m * 1.5 m) was harvested with a parcel combine harvester (Hege 160, Haldrup GmbH, Ilshofen, Germany) at seed maturity and afterwards dried with cold air. Intercropping plots of faba bean and wheat were harvested in one passage and separated by sieving.

Biomass of the grassland crops was harvested as described in (Heshmati et al., 2020). In brief, core parts of the plots (5 m * 1.4 m) were harvested with a combine forage harvester (Wintersteiger hd 1500, Wintersteiger AG, Ried im Innkreis, Austria) to a stubble height of 5 cm. During the vegetation period, four cuts were carried out in intervals of six weeks. Aliquots were dried at 60 °C until weight constancy to determine and calculate the dry matter yield.

The yield data of both domains were previously published as condensed forms in publications and dissertations within the framework of the IMPAC³ project (Heshmati, 2019; Heshmati et al., 2020; Nelson et al., 2021; Siebrecht-Schöll, 2019).

Statistics

Statistical analyses were performed with R version 3.4.1 (R Core Team, 2017). For data analysis and statistical evaluation, linear mixed effects models were calculated using the R package *nlme* (Pinheiro et al., 2018) with block and year as random effects and treatment and site as fixed effects. Tests for normal distribution with Shapiro-Wilk-Test (Shapiro & Wilk, 1965) and for homogeneity of variance with Levene-Test (Levene, 1960) proved that the requirements were complied.

Analysis of Variance (ANOVA) was performed to determine differences between treatments. Those differences were assessed by the Tukey test at $p < 0.05$ using the package *emmeans* (Lenth, 2018). Additionally, the R packages *plyr* (Wickham, 2011) and *muMin* (Barton, 2018) were used for data analysis. In order to compare $\delta^{13}C$ results from intercropping with the respective pure stands, *a priori* pairwise t-tests were conducted at $p < 0.05$ (wheat at the fertile site) and $p < 0.1$ (faba bean; wheat at the marginal site).

3 Results

In terms of carbon stable isotope discrimination ($\delta^{13}C$) of the arable crops, there were no significant differences for winter faba bean between pure and intercropped stands (Fig.2, Supplement Table S.1). However, site and winter faba bean genotype had a strong influence on this parameter. At the fertile site, genotypes Vf3, Vf5, Vf6 and Vf7 had higher $\delta^{13}C$ values (between -27.2 and -27.5 ppm) than the four other faba bean genotypes. At the marginal site and independent of whether grown in pure or intercropped stands, most faba bean genotypes had significantly higher $\delta^{13}C$ that at the fertile site, whereby Vf5 and Vf7 had highest values (between -26.5 and -26.7 ppm). In contrast, pure stands of winter wheat at the fertile site exhibited significantly highest $\delta^{13}C$ (-26.4 ppm), higher than at the marginal site (-26.8 ppm) and higher than wheat grown in legume mixtures. Intercropping with different faba bean genotypes did not affect wheat $\delta^{13}C$.

In grassland, pure stands of white clover had highest $\delta^{13}C$ at both sites (between -27.4 and -27.6 ppm) (Fig.3, Supplement Table S.1). Intercropping with perennial ryegrass led to decreased $\delta^{13}C$ to a range of -27.8 to -28.3 ppm, while intercropping with chicory further decreased $\delta^{13}C$ (-29.1 to -29.4 ppm at the fertile site and -28.5 to -28.7 ppm at the marginal site). There were no significant differences between white clover genotypes. In pure stands of non-legumes, ryegrass showed higher $\delta^{13}C$ than chicory, whereby chicory showed significant differences between both experimental sites. Nitrogen fertilizer had no effect on ryegrass or chicory $\delta^{13}C$.

Generally, in both land-use systems the water use efficiency (WUE), calculated as ratio of NEE and ET, was highest in 2015, when only the fertile site was measured (Fig.4, Supplement Table S.6). At the beginning of ripening and senescence of arable crops in July 2015, WUE decreased by reduced NEE (Supplement Table S.2). Additionally, WUE of all grassland crops decreased drastically after the harvests due to negative NEE (Supplement Table S.4). The WUE of intercropped and pure stands of all crops did not generally differ within each year or each site. Yet, depending on the date, pure legumes and intercropped stands had a significantly higher WUE than the non-legumes, while this effect was more pronounced at the marginal site for both land-use systems. Nitrogen fertilization had no significant effect on the WUE of wheat, whereas WUE of ryegrass was significantly enhanced at the marginal site by application of nitrogen. Here, WUE of fertilized ryegrass was on equal levels compared to crop stands including white clover.

The analysis of transpiration, assessed by canopy surface temperatures, revealed that the crop stand affected the temperature difference between canopy surface and air (dT) (Fig.5, Supplement Table S.7). At both sites and in both years, lowest dT were observed for pure legume stands as well as for intercropping, which was similar in arable crops and slightly increased for intercropping of grassland crops. Non-legumes without nitrogen fertilizer had overall highest dT, while fertilized non-legumes also showed increased dT. In arable land, these differences were significant only on few dates throughout the vegetation period, whereas in grassland biggest and significant differences in dT appeared during May and June in each year.

The average grain yield of arable crops at the fertile site was at 43.2 dt/ha (2015) and 36.0 dt/ha (2016), while the crops at the marginal site had an average grain yield of 35.6 dt/ha (2015) and 36.0 dt/ha (2016) (Fig.6, Supplement Table S.8). Variations in the grain yield of pure faba bean and intercropped treatments within one site were little in 2015, whereas in 2016 genotypes Vf2 and Vf7 had greater grain yield in pure as well as in intercropped stands at both sites. Pure wheat had lower yields at the marginal site in both years (20.9 dt/ha in 2015 and 22.0 dt/ha in 2016) compared to the fertile site. Nitrogen fertilization in 2016 significantly increased grain yield of wheat by 20 dt/ha at both sites.

Similar to faba bean, the variation in dry matter yield of the white clover genotypes was low (Fig.7, Supplement Table S.8). Furthermore, pure white clover and intercropping of white clover and ryegrass reached similar levels of dry matter yield at both sites and in both years (average 60.7 dt/ha). Pure ryegrass and pure chicory produced significantly more dry matter at the fertile site than at the marginal site in both years. Chicory in non-fertilized pure stands (63.7 dt/ha) also generated significantly more dry matter than ryegrass (27.9 dt/ha). In both non-legumes, N fertilizer led to significant increases in dry matter production.

4 Discussion

In the present study, the effect of species intercropping with legumes on the system's water use efficiency was investigated. We tested a number of factors influencing the water use of the crop stands, i.e. contrasting growth conditions at two study sites, various leguminous genotypes as well as the application of nitrogen fertilizer to pure non-leguminous species. It is discussed to what extent intercropping can influence water use efficiency and in which ways these factors interfere.

Sites and growth conditions

In grassland crops, possible site effects were masked by the strong impact of defoliations by the biomass harvests. The courses of moisture and temperature at the study sites also had great effects on the crop stand performance as high air temperature was a strong driver of the evapotranspiration. Very low precipitation at the marginal site in July 2016 further led to higher respiration of the grassland crops compared to CO_2-assimilation, which resulted in negative WUE and lower yields.

Nevertheless, certain differences in the crop stand performance at the two study sites were observed that indicate responses to the contrast in soil and climate conditions. This was expected because the dominating crop in mixtures and thus the overall performance have been reported to differ in dependence of the soil type (Knudsen et al., 2004).

At the marginal site, $\delta^{13}C$ of most faba bean genotypes was significantly higher than at the fertile site, indicating greater WUE at the marginal site. This might be due to the lower water availability at the marginal site with poorer soil characteristics (Nelson et al., n.d.). Lower water availability might have caused small stomata conductance in relation to CO_2 fixation capacity, which induces low internal CO_2 concentrations in the leaves, and thus less discrimination against ^{13}C.

Interestingly, wheat in pure stands had lower $\delta^{13}C$ at the marginal site, indicating lower WUE compared to the fertile site. This might be caused by effects resulting from different canopy architecture. At the fertile site (higher WUE and greater grain yield), plant canopy might have been denser in response to a deeper soil profile and higher nutrient availability and thus favoring a microclimate in the canopy with higher humidity and decreased vapor pressure deficit. As a result, stomata conductance was low leading to less discrimination against ^{13}C.

Significant differences in WUE and surface temperature between pure non-legumes and other crop stands in grassland occurred at the beginning of the vegetation period in April. In arable land, significant differences between the crop stands occurred widely during the months May and June, when arable crops were flowering. This is in line with a compilation of Sadras and Angus (2006) on wheat, who showed that variability in WUE is mostly determined by evapotranspiration at flowering. Afterwards different microclimatic conditions lead to reduced CO_2 fluxes (Jákli et al., 2016). The use of light resources is also impaired in dense vegetation. A study on intercropping with soybean showed drastic reductions in photosynthetic capacity due to shading (Yao et al., 2017). Following that, measurements at full flowering of faba bean and before the first and second harvest of grassland plots can be considered representative for the crop stands.

Genotypes and species

The carbon isotope discrimination showed clear differences between faba bean and wheat resulting from differences in plant anatomy. For faba bean, there was variability in genotype specific characteristics in WUE, which was also observed by several research groups (Ali et al., 2016; Belachew et al., 2018; Khazaei et al., 2013). These characteristics determined by $\delta^{13}C$ were independent of the crop stand, thus pointing to general advantages of specific genotypes with respect to WUE. At both sites, $\delta^{13}C$ of genotypes Vf1 (S_004) and Vf2 (S_062) indicated lowest WUE, whereas $\delta^{13}C$ of Vf5 (Hiverna/2) and Vf7 (WAB-Fam157) suggested highest WUE. In addition, Vf5 and Vf7 showed relatively higher $\delta^{13}C$ at the marginal site being indicative of the ability to increase WUE in comparison to the fertile site. These genotypes are therefore promising candidates for drought-prone environments. The results confirm hypotheses (2), that there is genotypic variability and (3), that WUE is higher at the marginal site.

Including chicory in grassland crop stands increased the dry matter yield and lowered WUE (decreased $\delta^{13}C$ values). Apart from general species-specific effects derived from different plant characteristics within the mixed samples, this can most likely be explained by looking at a study of Skinner (2008). The author found that the deep rooting chicory increases water availability for the whole system due to its ability of water uptake from deeper soil layers and even transfers it to some extent to companion species. This could be the reason that white clover/chicory intercropping significantly differs from pure white clover but not from pure chicory.

Beside this significant effect, among the genotypes of white clover there were no differences in pure or intercropped stands. Here, the genotypic variability seems to be too low to detect differences. This is in line with a related study of Heshmati et al. (2020), who found that the chosen white clover genotypes may not reflect the overall yield potential of this species. The authors further point to a higher relevance of the chosen species grown in intercropping rather than the white clover genotype. This finding is supported by the above-mentioned results of the present study.

Nitrogen

In our experiment, increased N availability by N_2-fixation or mineral fertilizer in grassland crops led to improved instantaneous WUE at the marginal site compared to pure ryegrass or chicory as also shown by Kunrath et al. (2018). This improved WUE by higher N availability under water limitations can be related to the spatial effect of increased biomass production as documented in a related study (Heshmati, 2019) and the observed significant increase in dry matter yields. In denser crop stands, evaporation from the soil, i.e. unproductive water loss, is reduced leading to a greater proportion of transpiration, the productive loss of water (Caviglia & Sadras, 2001). These attributes probably contributed to the high WUE of the investigated intercropped grassland species.

In contrast to the N-driven enhancement of gas exchange in both grassland and arable crops, the results of $\delta^{13}C$ and canopy temperature were not positively affected by N application. Lower dT of grassland crops even indicated higher transpiration with N application. Nitrogen fertilizer has been reported to lead to decreased $\delta^{13}C$ and intrinsic WUE (Hussain et al., 2015; Shangguan et al., 2000). However, our results on $\delta^{13}C$ and dT are in line with a study on tobacco, where it was shown that the nitrogen nutritional status did not alter carbon isotope discrimination, while transpiration was significantly increased by N application (Senbayram et al., 2015). A reason for our findings might be that the nutrient uptake is accompanied by increased water extraction from the soil (Angus et al., 2001) leading to a parallel increase in transpiration and assimilation. This water use for nutrient acquisition uncouples photosynthesis and transpiration leading to reduced correlation between $\delta^{13}C$ and WUE (Hobbie & Colpaert, 2004).

Intercropping

In our study, measurements of gas exchange as well as carbon isotope discrimination displayed no absolute enhancement of WUE in intercropped stands. In contrast to hypothesis (1), WUE was not highest in intercropping but in pure stands of the legumes winter faba bean (arable system) and white clover (grassland system). It is very likely that this is due to the nutrient status of the low-input system which led to advantages for pure legumes over mixtures and non-legumes. With higher N supply, the non-legumes were able to develop a greater root system and access water from greater soil depths.

The $\delta^{13}C$ results indicate similar stomatal adjustment for faba bean in the different stands. It is therefore crucial to compare the behavior of pure wheat stands and intercropped stands. Although $\delta^{13}C$ values and therefore the WUE are lower for wheat in intercropping, the overall yield production of the crop stand was comparable to pure faba bean. Additionally, intercropping of faba bean and wheat realized its potential of improved WUE measured by gas exchange at later stages of the vegetation period. That can be related to different plant architecture and thus complementary leaf area index and light use efficiency (Lindner et al., 2015). This way, intercropping systems were able to maintain productivity at temporarily improved WUE compared to pure non-legumes.

This interpretation also accounts for the grassland species. From dT it can be concluded that stomata opening and thus transpiration in non-legumes without fertilizer was generally low. The main reason is low yield production of non-fertilized chicory and ryegrass. A mixture of species in grassland significantly improved dry matter yields as well as temporarily enhancements of WUE. This was facilitated by synergistic effects in soil water usage as white clover improved water storage ability of the soil as well as hydraulic conductivity compared to perennial ryegrass (Marshall et al., 2016). Therefore, addition of legumes to non-legumes in grassland systems also resulted in a better use of resources.

5 Conclusion

The results obtained in this study indicate higher productivity of legume-based intercropping. Although water use is enhanced for supporting yield production, it was shown that intercropping positively affected stomatal control which was driven by complementary plant architecture. However, WUE of intercropping was not generally superior compared to both leguminous and non-leguminous pure stands. Comparing the intercropped stands with pure legume stands, there was no improvement in WUE, which was most likely due to the low-input system benefitting from N_2-fixation. Nevertheless, in comparison to non-legumes in both land-use systems, intercropping significantly improved WUE. With regard to common cereal-based crop rotations, intercropping is thus a considerable option to improve WUE of agricultural systems without yield losses.

6 Acknowledgments

We thank Jonas Lotze and Phillip Walter, who had substantial contribution in conducting the gas exchange and thermography measurements. Furthermore, we thank Regina Martsch for harvesting and organizing the samples for the stable isotope analysis. Funded by the Federal Ministry of Education and Research (FKZ 031A351A). IMPAC³ is a project of the Center of Biodiversity and sustainable Land Use at the University of Goettingen.

Figures

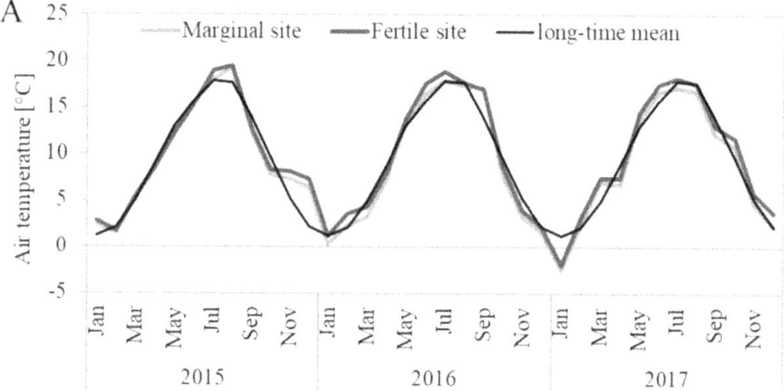

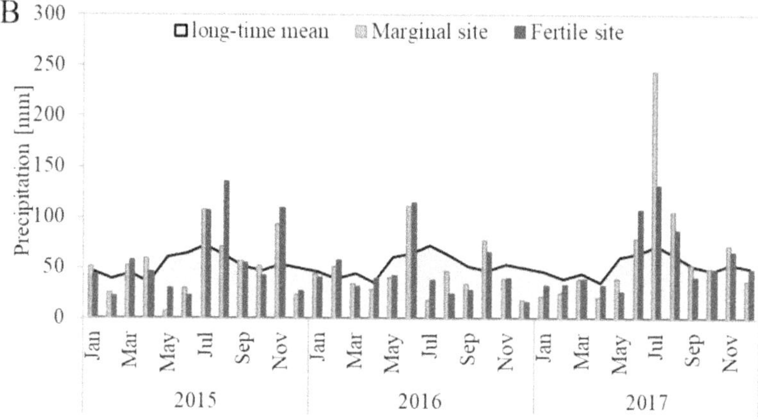

Fig.1: Monthly air temperature (A) and monthly precipitation (B) at the marginal and fertile site in contrast to the long-time mean of the years 1987 to 2017.

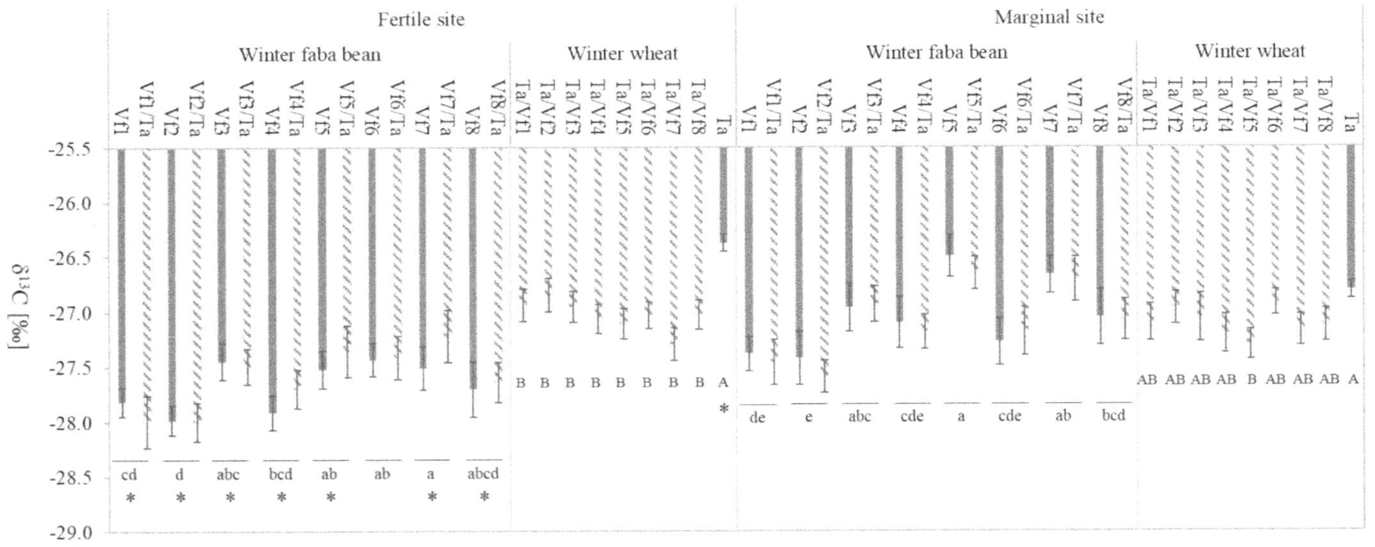

Fig.2: Carbon stable isotope discrimination in different genotypes of winter faba bean (Vf1 – Vf8) and winter wheat (Ta) in pure stands (full bars) and intercropped stands (hatched bars) at the fertile and marginal site. Means were calculated as averages over all three investigated years (2015-2017). Error bars indicate standard error; different letters indicate significant differences within one site. Stars indicate significant differences of each treatment comparing the two sites. Tukey-test at $p < 0.05$.

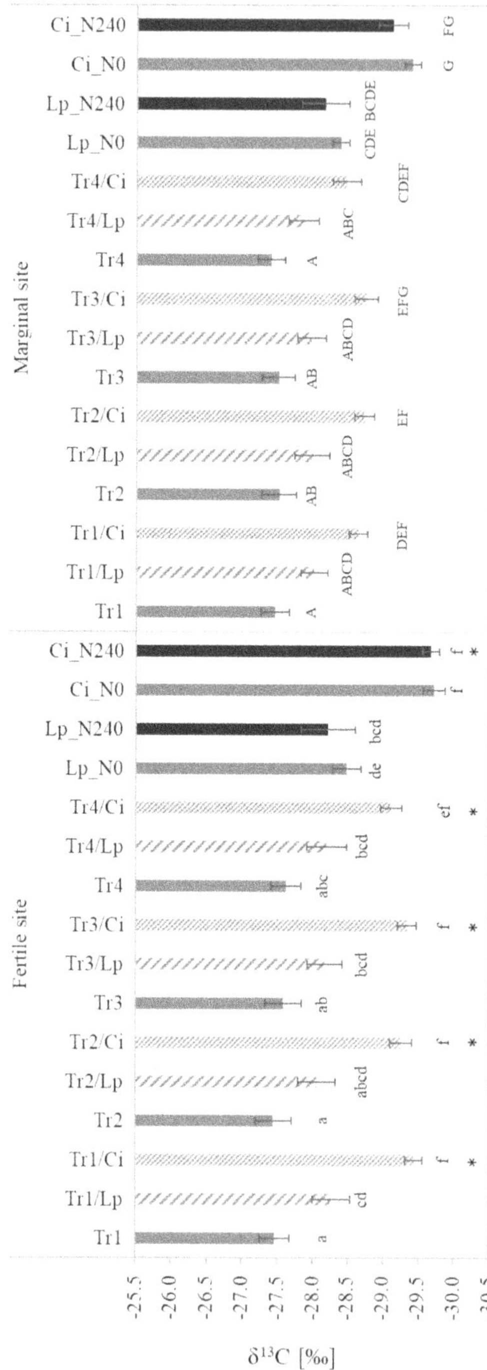

Fig.3: Carbon stable isotope discrimination in different genotypes of white clover (Tr1 – Tr4), perennial ryegrass (Lp) and chicory (Ci) in pure stands (full bars) and intercropped stands (interrupted bars) at the fertile and marginal site. Intercropping data were assessed as mixed sample from the biomass harvests. N0: no N-fertilizer, N240: fertilized with 240 kg N/ha. Means were calculated as average over all three investigated years (2015-2017). Error bars indicate standard error; different letters indicate significant differences within one site. Stars indicate significant differences of treatments comparing the two sites. Tukey-test at $p < 0.05$.

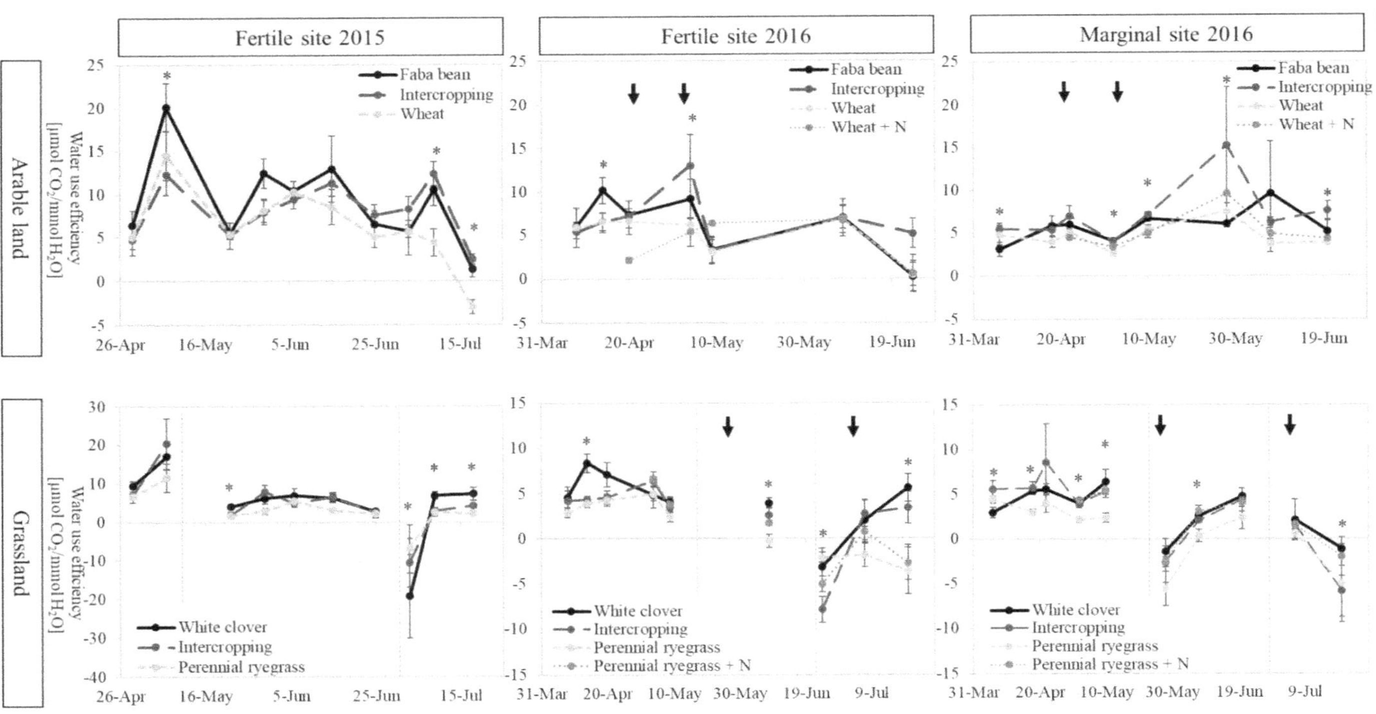

Fig.4: Water use efficiency of arable and grassland crops at the fertile and marginal site in the years 2015 and 2016. Interruptions in the lines of grassland stands indicate defoliations, while dashed lines mark the dates of biomass harvests. N: nitrogen fertilizer; split applications are indicated by arrows. Error bars indicate standard error; stars indicate significant differences among treatments within one date. Tukey-test at $p < 0.05$.

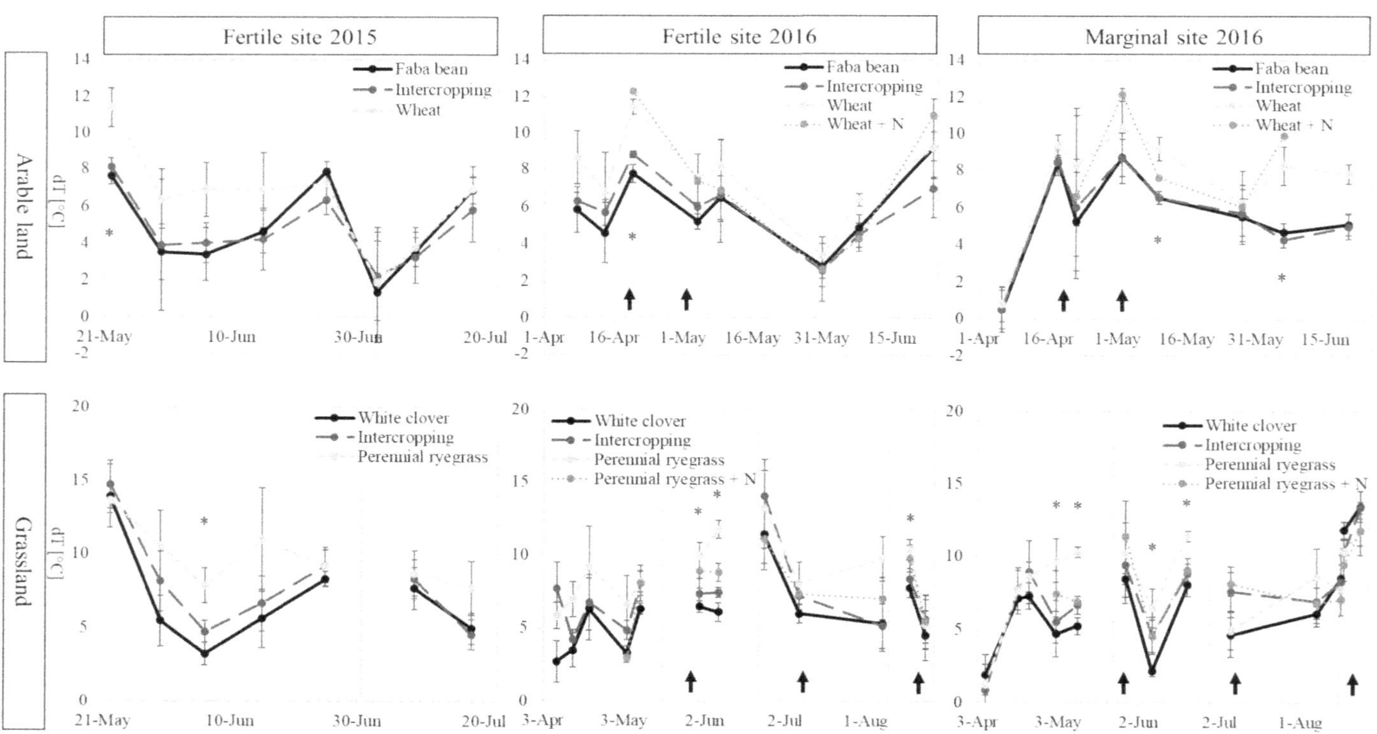

Fig.5: Temperature difference between canopy and air (dT) of arable and grassland crops at the fertile and marginal site in the years 2015 and 2016. Interruptions in the lines of grassland stands indicate defoliations, while dashed lines mark the dates of biomass harvests. N: nitrogen fertilizer; split applications are indicated by arrows. Error bars indicate standard error; stars indicate significant differences among treatments within one date. Tukey-test at $p < 0.05$.

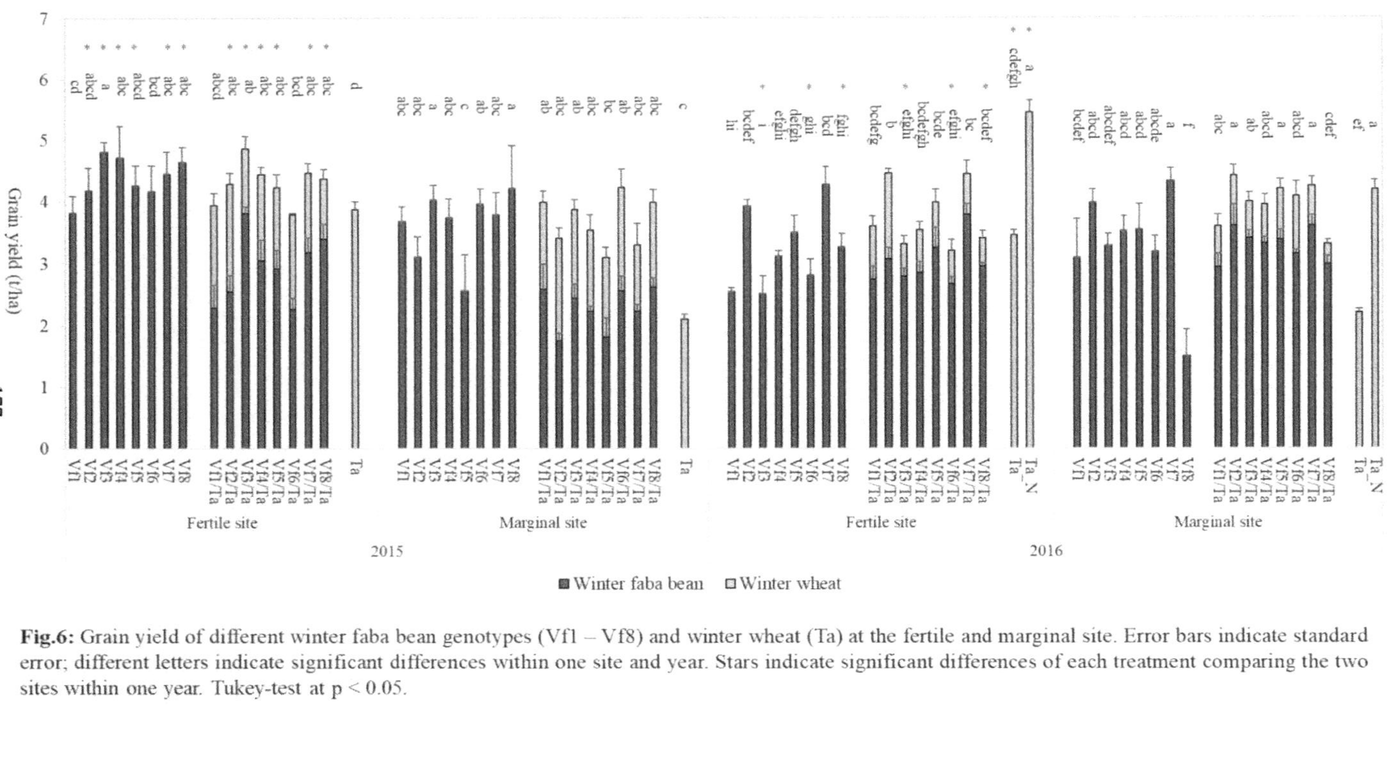

Fig.6: Grain yield of different winter faba bean genotypes (Vf1 – Vf8) and winter wheat (Ta) at the fertile and marginal site. Error bars indicate standard error; different letters indicate significant differences within one site and year. Stars indicate significant differences of each treatment comparing the two sites within one year. Tukey-test at $p < 0.05$.

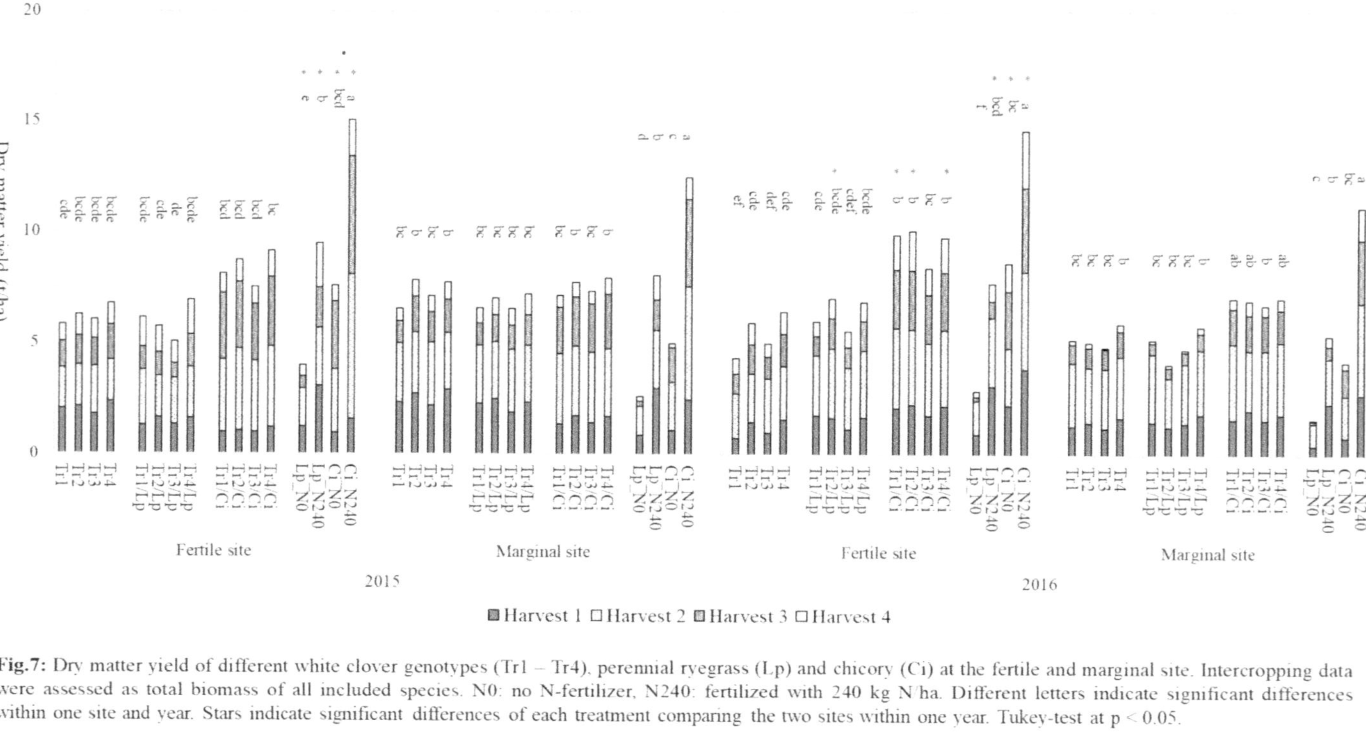

Fig.7: Dry matter yield of different white clover genotypes (Tr1 – Tr4), perennial ryegrass (Lp) and chicory (Ci) at the fertile and marginal site. Intercropping data were assessed as total biomass of all included species. N0: no N-fertilizer, N240: fertilized with 240 kg N/ha. Different letters indicate significant differences within one site and year. Stars indicate significant differences of each treatment comparing the two sites within one year. Tukey-test at $p < 0.05$.

7 References

Ali, M. B. M., Welna, G. C., Sallam, A., Martsch, R., Balko, C., Gebser, B., Sass, O., & Link, W. (2016). Association Analyses to Genetically Improve Drought and Freezing Tolerance of Faba Bean (Vicia faba L.). *Crop Science*, *56*(3), 1036–1048. https://doi.org/10.2135/cropsci2015.08.0503

Altieri, M. A. (1999). The ecological role of biodiversity in agroecosystems. In M. G. Paoletti (Ed.), *Invertebrate Biodiversity as Bioindicators of Sustainable Landscapes* (pp. 19–31). Elsevier. https://doi.org/10.1016/B978-0-444-50019-9.50005-4

Angus, J. F., Gault, R. R., Peoples, M. B., Stapper, M., & Herwaarden, A. F. van. (2001). Soil water extraction by dryland crops, annual pastures, and lucerne in south-eastern Australia. *Australian Journal of Agricultural Research*, *52*(2), 183–192. https://doi.org/10.1071/ar00103

Barton, K. (2018). *MuMIn: Multi-model inference.* R package version 1.40.4. https://CRAN.R-project.org/package=MuMIn

Bedoussac, L., Journet, E.-P., Hauggaard-Nielsen, H., Naudin, C., Corre-Hellou, G., Jensen, E. S., Prieur, L., & Justes, E. (2015). Ecological principles underlying the increase of productivity achieved by cereal-grain legume intercrops in organic farming. A review. *Agronomy for Sustainable Development*, *35*(3), 911–935. https://doi.org/10.1007/s13593-014-0277-7

Belachew, K. Y., Nagel, K. A., Fiorani, F., & Stoddard, F. L. (2018). Diversity in root growth responses to moisture deficit in young faba bean (Vicia faba L.) plants. *PeerJ*, *6*, e4401. https://doi.org/10.7717/peerj.4401

Benešová, M., Holá, D., Fischer, L., Jedelský, P. L., Hnilička, F., Wilhelmová, N., Rothová, O., Kočová, M., Procházková, D., Honnerová, J., Fridrichová, L., & Hniličková, H. (2012). The Physiology and Proteomics of Drought Tolerance in Maize: Early Stomatal Closure as a Cause of Lower Tolerance to Short-Term Dehydration? *PLoS ONE*, *7*(6), e38017. https://doi.org/10.1371/journal.pone.0038017

Boyer, J. S. (1982). Plant Productivity and Environment. *Science*, *218*(4571), 443–448. https://doi.org/10.1126/science.218.4571.443

Boyer, J. S., & Westgate, M. E. (2004). Grain yields with limited water. *Journal of Experimental Botany*, *55*(407), 2385–2394. https://doi.org/10.1093/jxb/erh219

Brueck, H., & Senbayram, M. (2009). Low nitrogen supply decreases water-use efficiency of oriental tobacco. *Journal of Plant Nutrition and Soil Science*, *172*(2), 216–223. https://doi.org/10.1002/jpln.200800097

Caviglia, O. P., & Sadras, V. O. (2001). Effect of nitrogen supply on crop conductance, water- and radiation-use efficiency of wheat. *Field Crops Research*, *69*(3), 259–266. https://doi.org/10.1016/S0378-4290(00)00149-0

Cernay, C., Ben-Ari, T., Pelzer, E., Meynard, J.-M., & Makowski, D. (2015). Estimating variability in grain legume yields across Europe and the Americas. *Scientific Reports*, *5*, 11171. https://doi.org/10.1038/srep11171

Ehrmann, J., & Ritz, K. (2014). Plant: Soil interactions in temperate multi-cropping production systems. *Plant and Soil*, *376*(1–2), 1–29. https://doi.org/10.1007/s11104-013-1921-8

Farquhar, G. D., Hubick, K. T., Condon, A. G., & Richards, R. A. (1989). Carbon Isotope Fractionation and Plant Water-Use Efficiency. In P. W. Rundel, J. R. Ehleringer, & K. A. Nagy (Eds.), *Stable Isotopes in Ecological Research* (Vol. 68, pp. 21–40). Springer New York. https://doi.org/10.1007/978-1-4612-3498-2_2

Finney, D. M., White, C. M., & Kaye, J. P. (2016). Biomass Production and Carbon/Nitrogen Ratio Influence Ecosystem Services from Cover Crop Mixtures. *Agronomy Journal*, *108*(1), 39–52. https://doi.org/10.2134/agronj15.0182

Foyer, C. H., Lam, H.-M., Nguyen, H. T., Siddique, K. H. M., Varshney, R. K., Colmer, T. D., Cowling, W., Bramley, H., Mori, T. A., Hodgson, J. M., Cooper, J. W., Miller, A. J., Kunert, K., Vorster, J., Cullis, C., Ozga, J. A., Wahlqvist, M. L., Liang, Y., Shou, H., ... Considine, M. J. (2016). Neglecting legumes has compromised human health and sustainable food production. *Nature Plants*, *2*(8), 16112. https://doi.org/10.1038/nplants.2016.112

Frison, E. A., Cherfas, J., & Hodgkin, T. (2011). Agricultural Biodiversity Is Essential for a Sustainable Improvement in Food and Nutrition Security. *Sustainability*, *3*(1), 238–253. https://doi.org/10.3390/su3010238

Hauggaard-Nielsen, H., Jørnsgaard, B., Julia Kinane, & Jensen, E. S. (2008). Grain legume–cereal intercropping: The practical application of diversity, competition and facilitation in arable and organic cropping systems. *Renewable Agriculture and Food Systems*, *23*(1), 3–12. https://doi.org/10.1017/S1742170507002025

Heshmati, S. (2019). *Effect of white clover and perennial ryegrass genotype on yield and forage quality of grass-clover and grass-clover-forb mixtures* [Georg-August-Universität Göttingen]. http://hdl.handle.net/11858/00-1735-0000-002E-E63A-2

Heshmati, S., Tonn, B., & Isselstein, J. (2020). White clover population effects on the productivity and yield stability of mixtures with perennial ryegrass and chicory. *Field Crops Research*, *252*, 107802. https://doi.org/10.1016/j.fcr.2020.107802

Hobbie, E. A., & Colpaert, J. V. (2004). Nitrogen availability and mycorrhizal colonization influence water use efficiency and carbon isotope patterns in Pinus sylvestris. *New Phytologist*, *164*(3), 515–525. https://doi.org/10.1111/j.1469-8137.2004.01187.x

Hussain, K., Wongleecharoen, C., Hilger, T., Vanderborght, J., Garré, S., Onsamrarn, W., Sparke, M.-A., Diels, J., Kongkaew, T., & Cadisch, G. (2015). Combining δ13C measurements and ERT imaging: Improving our understanding of competition at the crop-soil-hedge interface. *Plant and Soil*, *393*(1–2), 1–20. https://doi.org/10.1007/s11104-015-2455-z

Jackson, R. D., Idso, S. B., Reginato, R. J., & Pinter, P. J. (1981). Canopy temperature as a crop water stress indicator. *Water Resources Research*, *17*(4), 1133–1138. https://doi.org/10.1029/WR017i004p01133

Jákli, B., Hauer-Jákli, M., Böttcher, F., Meyer zur Müdehorst, J., Senbayram, M., & Dittert, K. (2018). Leaf, canopy and agronomic water-use efficiency of field-grown sugar beet in response to potassium fertilization. *Journal of Agronomy and Crop Science*, *204*(1), 99–110. https://doi.org/10.1111/jac.12239

Jákli, Bálint, Tränkner, M., Senbayram, M., & Dittert, K. (2016). Adequate supply of potassium improves plant water-use efficiency but not leaf water-use efficiency of spring wheat. *Journal of Plant Nutrition and Soil Science*, *179*(6), 733–745. https://doi.org/10.1002/jpln.201600340

Khazaei, H., Street, K., Santanen, A., Bari, A., & Stoddard, F. L. (2013). Do faba bean (Vicia faba L.) accessions from environments with contrasting seasonal moisture availabilities differ in stomatal characteristics and related traits? *Genetic Resources and Crop Evolution*, *60*(8), 2343–2357. https://doi.org/10.1007/s10722-013-0002-4

Knudsen, M. T., Hauggaard-Nielsen, H., Jørnsgård, B., & Jensen, E. S. (2004). Comparison of interspecific competition and N use in pea–barley, faba bean–barley and lupin–barley intercrops grown at two temperate locations. *The Journal of Agricultural Science*, *142*(6), 617–627. https://doi.org/10.1017/S0021859604004745

Kunrath, T. R., Lemaire, G., Sadras, V. O., & Gastal, F. (2018). Water use efficiency in perennial forage species: Interactions between nitrogen nutrition and water deficit. *Field Crops Research*, *222*, 1–11. https://doi.org/10.1016/j.fcr.2018.02.031

Lenth, R. V. (2018). *emmeans: Estimated Marginal Means, aka Least-Squares Means*. R package version 1.1.

Levene, H. (1960). Robust tests for equality of variances. *Contributions to Probability and Statistics*, *1*, 278–292.

Lindner, S., Otieno, D., Lee, B., Xue, W., Arnhold, S., Kwon, H., Huwe, B., & Tenhunen, J. (2015). Carbon dioxide exchange and its regulation in the main agro-ecosystems of Haean catchment in South Korea. *Agriculture, Ecosystems & Environment*, *199*, 132–145. https://doi.org/10.1016/j.agee.2014.09.005

Lüscher, A., Mueller-Harvey, I., Soussana, J. F., Rees, R. M., & Peyraud, J. L. (2014). Potential of legume-based grassland-livestock systems in Europe: A review. *Grass and Forage Science*, *69*(2), 206–228. https://doi.org/10.1111/gfs.12124

Marshall, A. H., Collins, R. P., Humphreys, M. W., & Scullion, J. (2016). A new emphasis on root traits for perennial grass and legume varieties with environmental and ecological benefits. *Food and Energy Security*, *5*(1), 26–39. https://doi.org/10.1002/fes3.78

Matimati, I., Verboom, G. A., & Cramer, M. D. (2014). Nitrogen regulation of transpiration controls mass-flow acquisition of nutrients. *Journal of Experimental Botany*, *65*(1), 159–168. https://doi.org/10.1093/jxb/ert367

Muktadir, M. A., Adhikari, K. N., Merchant, A., Belachew, K. Y., Vandenberg, A., Stoddard, F. L., & Khazaei, H. (2020). Physiological and Biochemical Basis of Faba Bean Breeding for Drought Adaptation—A Review. *Agronomy*, *10*(9), 1345. https://doi.org/10.3390/agronomy10091345

Nelson, W. (2020). *Investigating resource competition in cereal-legume intercropping systems* [Georg-August-Universität Göttingen]. http://hdl.handle.net/21.11130/00-1735-0000-0005-14AA-5

Nelson, W.C.D., Siebrecht-Schöll, D. J., Hoffmann, M. P., Rötter, R. P., Whitbread, A. M., & Link, W. (2021). What determines a productive winter bean-wheat genotype combination for intercropping in central Germany? *European Journal of Agronomy*, *128*, 126294. https://doi.org/10.1016/j.eja.2021.126294

Picasso, V. D., Brummer, E. C., Liebman, M., Dixon, P. M., & Wilsey, B. J. (2011). Diverse perennial crop mixtures sustain higher productivity over time based on ecological complementarity. *Renewable Agriculture and Food Systems*, *26*(4), 317–327. https://doi.org/10.1017/S1742170511000135

Pinheiro, J., Bates, D., DebRoy, S., Sarkar, D., & R Core Team. (2018). *nlme: Linear and Nonlinear Mixed Effects Models*. R package version 3.1-131.1. https://CRAN.R-project.org/package=nlme

R Core Team. (2017). *R: A language and environment for statistical computing*. R Foundation for Statistical Computing, Vienna, Austria. http://www.R-project.org/

Raseduzzaman, Md., & Jensen, E. S. (2017). Does intercropping enhance yield stability in arable crop production? A meta-analysis. *European Journal of Agronomy*, *91*, 25–33. https://doi.org/10.1016/j.eja.2017.09.009

Reckling, M., Bergkvist, G., Watson, C. A., Stoddard, F. L., Zander, P. M., Walker, R. L., Pristeri, A., Toncea, I., & Bachinger, J. (2016). Trade-Offs between Economic and Environmental Impacts of Introducing Legumes into Cropping Systems. *Frontiers in Plant Science*, *7*, 669. https://doi.org/10.3389/fpls.2016.00669

Richards, R. A., Rebetzke, G. J., Condon, A. G., & van Herwaarden, A. F. (2002). Breeding Opportunities for Increasing the Efficiency of Water Use and Crop Yield in Temperate Cereals. *Crop Science*, *42*(1), 111–121. https://doi.org/10.2135/cropsci2002.1110

Rust, I. (2006). *Amendment of the German land appraisal (Reichsbodenschätzung) in consideration of climatic conditions* [University of Goettingen]. https://ediss.uni-goettingen.de/handle/11858/00-1735-0000-0006-AB70-D

Sadras, V. O., & Angus, J. F. (2006). Benchmarking water-use efficiency of rainfed wheat in dry environments. *Australian Journal of Agricultural Research*, *57*(8), 847–856. https://doi.org/10.1071/AR05359

Schilling, G., Eißner, H., Schmidt, L., & Peiter, E. (2016). Yield formation of five crop species under water shortage and differential potassium supply. *Journal of Plant Nutrition and Soil Science*, *179*(2), 234–243. https://doi.org/10.1002/jpln.201500407

Senbayram, M., Tränkner, M., Dittert, K., & Brück, H. (2015). Daytime leaf water use efficiency does not explain the relationship between plant N status and biomass water-use efficiency of tobacco under non-limiting water supply. *Journal of Plant Nutrition and Soil Science*, *178*(4), 682–692. https://doi.org/10.1002/jpln.201400608

Shangguan, Z. P., Shao, M. A., & Dyckmans, J. (2000). Nitrogen nutrition and water stress effects on leaf photosynthetic gas exchange and water use efficiency in winter wheat. *Environmental and Experimental Botany*, *44*(2), 141–149. https://doi.org/10.1016/S0098-8472(00)00064-2

Shapiro, S. S., & Wilk, M. B. (1965). An Analysis of Variance Test for Normality (Complete Samples). *Biometrika*, *52*(3/4), 591–611. https://doi.org/10.2307/2333709

Sharma, S., Upadhyaya, H. D., Varshney, R. K., & Gowda, C. L. L. (2013). Pre-breeding for diversification of primary gene pool and genetic enhancement of grain legumes. *Frontiers in Plant Science*, *4*, 309. https://doi.org/10.3389/fpls.2013.00309

Siebrecht-Schöll, D. J. (2019). *Breeding analysis of eight winter faba bean genotypes for mixed cropping with winter wheat* [Georg-August-Universität Göttingen]. http://hdl.handle.net/21.11130/00-1735-0000-0005-128E-7

Skinner, R. H. (2008). Yield, Root Growth, and Soil Water Content in Drought-Stressed Pasture Mixtures Containing Chicory. *Crop Science*, *48*(1), 380–388. https://doi.org/10.2135/cropsci2007.04.0201

Smith, B. N., & Epstein, S. (1971). Two Categories of 13C/12C Ratios for Higher Plants. *Plant Physiology*, *47*(3), 380–384. https://doi.org/10.1104/pp.47.3.380

Streit, J., Meinen, C., Nelson, W. C. D., Siebrecht-Schöll, D. J., & Rauber, R. (2019). Above- and belowground biomass in a mixed cropping system with eight novel winter faba bean genotypes and winter wheat using FTIR spectroscopy for root species discrimination. *Plant and Soil*, *436*(1–2), 141–158. https://doi.org/10.1007/s11104-018-03904-y

Sturludóttir, E., Brophy, C., Bélanger, G., Gustavsson, A.-M., Jørgensen, M., Lunnan, T., & Helgadóttir, Á. (2014). Benefits of mixing grasses and legumes for herbage yield and nutritive value in Northern Europe and Canada. *Grass and Forage Science*, *69*(2), 229–240. https://doi.org/10.1111/gfs.12037

Suter, M., Connolly, J., Finn, J. A., Loges, R., Kirwan, L., Sebastià, M.-T., & Lüscher, A. (2015). Nitrogen yield advantage from grass–legume mixtures is robust over a wide range of legume proportions and environmental conditions. *Global Change Biology*, *21*(6), 2424–2438. https://doi.org/10.1111/gcb.12880

Tallec, T., Béziat, P., Jarosz, N., Rivalland, V., & Ceschia, E. (2013). Crops' water use efficiencies in temperate climate: Comparison of stand, ecosystem and agronomical approaches.

Agricultural and Forest Meteorology, *168*, 69–81. https://doi.org/10.1016/j.agrformet.2012.07.008

Tambussi, E. A., Bort, J., & Araus, J. L. (2007). Water use efficiency in C3 cereals under Mediterranean conditions: A review of physiological aspects. *Annals of Applied Biology*, *150*(3), 307–321. https://doi.org/10.1111/j.1744-7348.2007.00143.x

Toft, N. L., Anderson, J. E., & Nowak, R. S. (1989). Water use efficiency and carbon isotope composition of plants in a cold desert environment. *Oecologia*, *80*(1), 11–18. https://doi.org/10.1007/BF00789925

Waha, K., Müller, C., Bondeau, A., Dietrich, J. P., Kurukulasuriya, P., Heinke, J., & Lotze-Campen, H. (2013). Adaptation to climate change through the choice of cropping system and sowing date in sub-Saharan Africa. *Global Environmental Change*, *23*(1), 130–143. https://doi.org/10.1016/j.gloenvcha.2012.11.001

Weißhuhn, P., Reckling, M., Stachow, U., & Wiggering, H. (2017). Supporting Agricultural Ecosystem Services through the Integration of Perennial Polycultures into Crop Rotations. *Sustainability*, *9*(12), 2267. https://doi.org/10.3390/su9122267

Wickham, H. (2011). The Split-Apply-Combine Strategy for Data Analysis. *Journal of Statistical Software*, *40*(1). https://doi.org/10.18637/jss.v040.i01

Yao, X., Zhou, H., Zhu, Q., Li, C., Zhang, H., Wu, J.-J., & Xie, F. (2017). Photosynthetic Response of Soybean Leaf to Wide Light-Fluctuation in Maize-Soybean Intercropping System. *Frontiers in Plant Science*, *8*, 1695. https://doi.org/10.3389/fpls.2017.01695

Zampieri, M., Ceglar, A., Dentener, F., & Toreti, A. (2017). Wheat yield loss attributable to heat waves, drought and water excess at the global, national and subnational scales. *Environmental Research Letters*, *12*(6), 064008. https://doi.org/10.1088/1748-9326/aa723b

Chapter 4:
Legume-based mixed intercropping systems may lower agricultural born N_2O emissions

Mehmet Senbayram, Christian Wenthe, Annika Lingner, Johannes Isselstein, Horst Steinmann, Cengiz Kaya, Sarah Köbke

Published in Energy, Sustainability and Society (2016) 6:2
DOI: 10.1186/s13705-015-0067-3

Senbayram *et al. Energy, Sustainability and Society* (2016) 6:2
DOI 10.1186/s13705-015-0067-3

Energy, Sustainability and Society
a SpringerOpen Journal

ORIGINAL ARTICLE Open Access

Legume-based mixed intercropping systems may lower agricultural born N_2O emissions

Mehmet Senbayram[1,2*], Christian Wenthe[1], Annika Lingner[1], Johannes Isselstein[3], Horst Steinmann[4], Cengiz Kaya[2] and Sarah Köbke[5]

Abstract

Background: The area used for bioenergy crops (annual row crops (e.g., wheat, maize), herbaceous perennial grasses, and short-rotation woody crops (e.g., poplar)) is increasing because the substitution of fossil fuels by bioenergy is promoted as an option to reduce greenhouse gas (GHG) emissions. However, biomass used for bioenergy production is not per se environmentally benign, since bioenergy crop production is associated with negative side effects such as GHG emissions from soil (dominated by N_2O). N_2O emissions vary greatly in space and time; thus, direct comparison of soil N_2O fluxes from various agro-ecosystems is certainly crucial for the assessment of the GHG reduction potential from energy crops.

Methods: Therefore, our study aimed to evaluate the two different agro-ecosystems (cropland and agro-forestry) cultivated in central Germany for their environmental impact. In a 1-year field experiment, we compared N_2O fluxes from cropland (non-fertilized wheat, N-fertilized wheat, non-fertilized faba bean, and wheat mixed intercropping with faba bean) and agro-forestry (non-fertilized poplar, N-fertilized poplar, non-fertilized Robinia, and poplar mixed intercropping with Robinia) as a randomized split-block design.

Results: Rainfall at the field site was slightly over average during the period from 1 April to 1 July in 2014 (201 mm rain) and considerably below average during the same period in 2015 (100 mm rain). Cumulative mean N_2O fluxes were up to five fold higher in agro-forestry than in arable crop treatments during 2014 growing period. We hypothesized that the difference in N_2O emissions when comparing arable land and agro-forestry was mainly due to the limited water and nutrient uptake of plantations during the first year. Among the arable crops (wheat, N-fertilized wheat, wheat mixed intercropped with bean, and bean), seasonal and annual N_2O emissions were highest in soils when faba bean was grown as a mono-crop. On the other hand, cumulative mean N_2O fluxes were 31 % lower ($p < 0.05$) when faba bean mixed with wheat than in soils planted with N-fertilized wheat.

Conclusions: The latter clearly suggests that using legume crops as intercrop or mixed crop in wheat may significantly mitigate fertilizer-derived N_2O fluxes and may be an effective proxy for increasing GHG emission savings for energy crops.

* Correspondence: mehmetsenbayram6@yahoo.co.uk
[1]Institute of Applied Plant Nutrition, University of Gottingen, Carl-Sprengel-Weg 1, Gottingen 37075, Germany
[2]Present Address: Department of Plant Nutrition and Soil Science, University of Harran, SanliUrfa TR-63000, Turkey
Full list of author information is available at the end of the article

Background

The observed increase in global average temperatures over the last decades is very likely due to the observed increase in anthropogenic greenhouse gas concentrations in the atmosphere. Nitrous oxide (N_2O) is a potent greenhouse gas as it absorbs long-wave radiation and contributes to the reduction of the ozone layer in the stratosphere [1]. Data from the ice-core analysis show that for thousands of years, mean atmospheric N_2O concentrations were close to 270 ppbv; however, the latter increased about 20 % in recent years [2, 3]. The Intergovernmental Panel on Climate Change (IPCC) Third Assessment Report identified microbial production of N_2O in expanding and fertilized agricultural lands as a primary driver of this increase [4]. However, large uncertainties remain on the estimates of N_2O fluxes from the biosphere [2, 5] due to complex interactions between the related processes and controls on production, consumption, and transport through the soil and on the release into the atmosphere [6, 7].

High-yielding agricultural systems have specifically high nitrogen (N) demand which cannot be supplied by soil N reserves. Therefore, additional N input is needed; however, in most agricultural land-use systems, the application of organic and inorganic nitrogen fertilizers triggers the emissions of anthropogenic N_2O emissions. The present IPCC default factor for direct N_2O emissions arising from the nitrogenous fertilizer application to managed soils is 1 %. However, soil N_2O emissions vary significantly depending on soil type, plant species, climate, crop rotation, tillage method, and fertilizer application rates [4]. At high N_2O-emitting sites, most N_2O release is characterized by short peak emissions (up to 90 % of the annual emissions) connected mainly to (i) precipitation events and change in soil moisture [8–10], (ii) N fertilization [11–13], (ii) freeze-thaw cycle [14], and (iv) soil tillage [15]. Such peak N_2O emissions from soils are highly variable in space and time; thus, measuring and quantifying variance in N_2O emissions are rather difficult, and there are only a few field experiments available allowing long-term comparison of various crops and other factors [16]. It is absolutely crucial to simulate such N_2O peak events and to identify the driving factors in order to be able to develop mitigation options.

The main microbial reactions involved in the production of N_2O are nitrification (oxidation of NH_4^+ to NO_3^-) and denitrification (reduction of NO_3^-, via NO, N_2O, to N_2). There is now growing evidence that denitrification (bacterial or fungal) is the dominating process responsible for N_2O losses from agricultural soils [8, 10, 17]. Therefore, in many agricultural field studies, N_2O emission events are found in periods when high mineral nitrogen concentrations in soils coincide with high soil moisture (increasing the rate of denitrification) [7, 18]. The relevance of soil moisture can also be seen in field studies with similar climatic, soil, and substrate conditions, where N_2O emissions nevertheless show significant inter-annual variability caused by the differences in soil water content [19]. These differences can usually be deduced from the weather records as they are related to the differences in the input of water through rainfall [18]. But the soil-water balance is also strongly influenced by the offtake of water through evapotranspiration. Crop plants play an important role, as they may transpire 500 to 600 mm of water per growth cycle which is 30 to 90 % of the precipitation input in that period. In this respect, crops differ substantially, not only in the total amount of water that is transpired but also in the growth and transpiration pattern in the course of the year [19, 20].

Biofuels are often called "CO_2 neutral" in the sense that CO_2 which is emitted in the course of their combustion has previously been fixed from the atmosphere via photosynthesis during plant growth. Many industrialized countries have established ambitious policy targets and often offer financial incentives to stimulate the production or use of bioenergy. The main reasons for the promotion of biofuel production are that it is made from renewable resources (organic manures, plant materials, food waste), that it is expected to have no or even positive effect on the atmospheric greenhouse gas balance, and that it may reduce the dependency on fossil fuel [21]. However, biomass used for bioenergy generation is not per se environmentally benign, since its production is inevitably associated with negative side effects such as GHG emissions or N leaching. Soil N_2O emissions are likely to be the dominating greenhouse gas emissions associated with bioenergy crop production [22]. It has been reported that the production and use of biofuels compared with the use of conventional fossil fuels may lead to a reduction or even increase in the total greenhouse gas emissions (72 to 107 %), depending on the type of bioenergy crop used and combustion technology chosen [23]. Here, the authors showed that N_2O would typically make up 10 to 80 % of the total greenhouse gas emissions in the biofuel production chain.

Intercropping, defined as any system of multiple cropping within the same space can be used as an alternate bioenergy cropping system [24]. The intercropping of cereals with legumes is particularly common, and introducing N_2-fixing legumes into cereal-based crop rotations may reduce synthetic mineral N-fertilizer use and thought to mitigate N_2O fluxes. A reduction of N_2O in tree-based intercropping systems has been reported [25]. In contrast, in a review study, Rochette and Janzen [26] concluded that legumes can produce substantial N_2O emissions. The cultivation of N_2-fixing legume species (e.g., faba bean (*Vicia faba L.*) as arable crop or Robinia

(*Robinia pseudoacacia*) as woody plant) could stimulate N_2O emissions simply by increasing N input to soils, thus providing additional substrate for nitrification and denitrification [27]. Authors reported that faba bean could release 13 % of their fixed N as rhizodeposition [28]. Furthermore, denitrification by N-fixing bacteria can be another source of N_2O in legume domains [29].

As discussed above, N_2O emissions vary greatly in space and time; thus, the direct comparison of soil N_2O fluxes from various agro-ecosystems is certainly necessary and only very few studies are available on this subject. Therefore, our study aimed to evaluate two different agro-ecosystems (cropland and agro-forestry) cultivated in central Germany for their environmental impact. In a 1-year field experiment, we compared N_2O fluxes from cropland (non-fertilized wheat, N-fertilized wheat, non-fertilized faba bean, and wheat mixed intercropping with faba bean) and agro-forestry (non-fertilized poplar, N-fertilized poplar, non-fertilized Robinia, and poplar mixed intercropping with Robinia) as a randomized split-block design. Therefore, the objectives of the present study were (i) to obtain year-round N_2O emission data in various agro-ecosystems (cropland and agro-forestry) in central Germany and (ii) to provide more knowledge on how a legume-based crop production system may affect N_2O budget.

Methods

Field experiment

A field experiment was conducted at the research farm Reinshof (51.49° N, 9.93° E, 150 m asl) of the Georg-August-University Göttingen, Germany. The continental climate leads to an average precipitation of 651 mm and a mean temperature of 9.2 °C. The rainfall and temperature at field site during the investigation period can be found in Fig. 1. Here, annual mean temperature and cumulative rainfall were 10.1 °C and 677 mm in 2014, respectively. The soil was classified as Haplic Luvisol according to the FAO classification system. At 0–30 cm soil section, the soil contained 15 % clay, 73 % silt, and 12 % sand with a pH of 6.7, and 0.1 % total N and 1.0 % total organic carbon content [30].

The 1-year field experiment (part of a large field trial) was set up in April 2014 with the aim of comparing N_2O fluxes from the cropland (non-fertilized wheat (WT), N-fertilized wheat (NWT), non-fertilized faba bean (FB), wheat mixed intercropping with faba bean (WFB)), agro-forestry (non-fertilized poplar (PL), N-fertilized poplar (NPL), non-fertilized Robinia (RB), and poplar mixed intercropping with Robinia (PRB)) as a randomized split-block design. In the first year of the study, spring wheat (*Triticum aestivum L.*) cultivar Tybald and faba bean (*V. faba L.*) cultivar Fuego were sown at the same time at the optimum planting date for

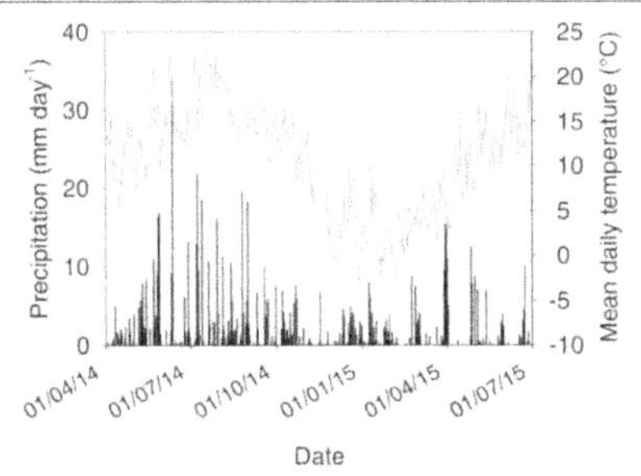

Fig. 1 Mean daily air temperature [°C] at canopy level (*gray lines*) and daily precipitation [mm day^{-1}] (*black bars*) at experimental site Reinshof, Göttingen, in central Germany during the investigation period (April 2014–July 2015)

spring wheat (see Table 1 for details). Agro-forestry treatments consisted of poplar (*Populus* "Hybride 275") (PL); poplar mixed with Robinia (*R. pseudoacacia* "HKG 81901") (PLR); Robinia (RB); and N-fertilized poplar (PLN). In the agro-forestry treatments, all trees were planted with 1 tree m^{-2} on 30 April 2014. Plot size was 5 × 5 m for forest and 3 × 9 m for arable treatments with three replications for each treatment. After the harvest of crop land in 2014, winter wheat (*T. aestivum* "genus"), wheat intercropped with winter bean (*V. faba* "S_004"), mono-winter bean, and N-fertilized wheat were seeded on 28 October 2014. Seeding density in the mono-cropped winter wheat plots was 320 plants m^{-2}, inter-cropped plots were alternately seeded with wheat density of 160 plants m^{-2} and a bean density of 20 plants m^{-2}, and mono-bean plots had a density of 40 plants m^{-2}. Granular N fertilizer in the form of calcium-ammonium-nitrate was applied at a rate of 80 kg N ha^{-1} as a single dressing to the soil surface on 15 May 2014 and 25 March 2015 in the respective treatments.

Soil mineral N

For the analysis of soil mineral N, soil from 0–15 cm depth was sampled extracted with a 0.0125 M $CaCl_2$ solution (1:5 *w*/*v*) and shaked for 1 h. The extracts were then filtered with Whatman 602 filter paper and stored at −20 °C until analysis. The extracts were analyzed colorimetrically for the concentrations of NO_3^- and NH_4^+ using the San++ continuous flow analyzer (Skalar Analytical B.V., Breda, The Netherlands).

Trace gas flux measurement

After the N application, gas samples were taken daily for a period of 1 week, followed by intervals of 2–3 days

Table 1 Sowing dates and fertilizer application rate of each treatment

Abbreviation	Crop	Fertilizer (kg CAN-N ha^{-1})	Stand	Date of sowing spring and winter crops
WT	Spring wheat/winter wheat	None	Mono	25 March 2014/28 October 2014
NWT	Spring wheat + N/winter wheat + N	80	Mono	25 March 2014/28 October 2014
WFB	Mixed intercropping (wheat and faba bean)[a]	None	Mixed intercrop	25 March 2014/28 October 2014
FB	Field beans[a]	None	Mono	25 March 2014/28 October 2014
PL	Poplar	None	Mono	30 April 2014
NPL	Poplar	80	Mono	30 April 2014
PRB	Mixed planting Poplar/Robinia	None	Mixed planting	30 April 2014
RB	Robinia	None	Mono	30 April 2014

[a]Information was given for both spring and winter crops

until the end of the vegetation period and once per week during the winter period using the closed chamber method described by Hutchinson and Mosier [31]. On each of 24 plots, basal rings made of polyvinyl chloride (PVC) (height 10 cm, diameter 60 cm) were pressed 5 cm into the soil. For the measurements, PVC chambers with an inner diameter of 60 cm and a height of 30 cm were put onto the basal rings, and tightened by a butyl rubber band all around the junction. The chambers were closed for 40 min within the period of 10.00–14.00 h at each sampling day. Gas samples were taken from the chamber atmosphere at 0–20–40 min after closing the chamber using pre-evacuated 12-ml glass vials (Labco, Crewe, UK). The top of the PVC chambers were covered with white polystyrene to reflect solar radiation and have temperature stability within the chamber.

Concentrations of N_2O were analyzed by a Bruker gas chromatography system (456-GC, Bruker, Billerica, USA) by deploying an electron capture detector (ECD) for N_2O. Operating conditions for the GC were as follows: injector temperature 95 °C, column temperature 85 °C, and detector temperature 320 °C. Samples were introduced using a Gilson auto-sampler (GX-281) (Gilson Inc., Middleton, WI, USA). Data processing was performed using the CompassCDS (vers. 3.0) software.

Statistics

Daily N_2O flux rates for dates between sampling dates were calculated using linear interpolation, and annual cumulative N_2O emissions were calculated as the sum of all daily flux rates for the vegetation period of each crop and the entire investigation periods during March 2014–August 2015. We used general linear model and Tukey's test for pairwise comparison of cumulative N_2O flux between the treatments within each year. Statistical analyses were done using SPSS version 13.0.

Results and discussion

Rainfall and crop production

Rainfall at the site Reinshof was slightly over average during the period from 1 April to 1 July in 2014 (201 mm) and considerably below average during the period from 1 April to 1 July in 2015 (100 mm). Therefore, the 2014 spring period was much wetter than spring 2015. Overall, 2014–2015 winter period was quite mild with no significant freeze-thaw event.

Soil mineral N

The time course of the soil mineral-N concentrations are shown in Fig. 2. In soils planted with arable crops, soil NH_4^+ concentrations during the investigation period remained rather low (below 10 kg N ha^{-1} at 30 cm layer) in non-fertilized treatments. The application of mineral-N fertilizer in NWT treatment caused a slight increase in topsoil NH_4^+ concentrations for a short time period which decreased rapidly to the background concentrations within a week. The low concentration of soil NH_4^+ even in N-fertilized treatments (in the form of calcium-ammonium-nitrate) can be attributed to a rapid nitrification as soil texture and pH (6.7) serve ideal conditions for nitrification. In all treatments, there was a significant increase in topsoil NH_4^+ concentrations in spring 2015 regardless from the N application. Early spring period in 2015 was reasonably dry compared to the same period in 2014. Therefore, the latter can be attributed to the processes related to the soil wetting after a long dry period (such as in spring 2015) which can accelerate N release from the mineralization of soil organic matter immediately after rewetting [32].

Overall treatments, soil NO_3^- concentrations in 0–15 cm soil segment varied between 10 and 85 kg NO_3^--N ha^{-1}. Here, NO_3^- was generally the dominant soil N form and highly variable when sampled soon after additions of fertilizer N. In all soils, concentrations of NO_3^- in the 0–15-cm layer decreased over time with the largest

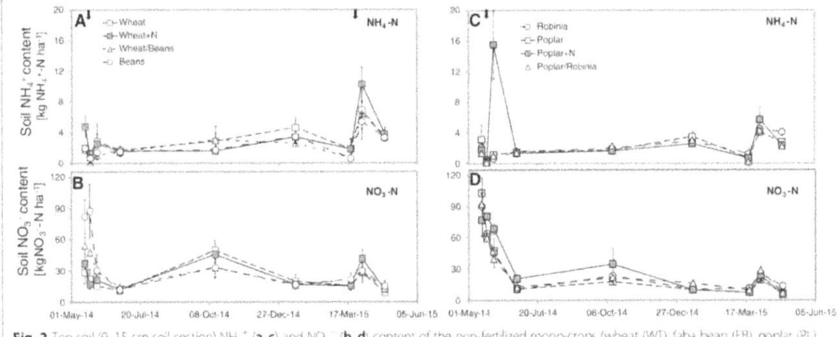

Fig. 2 Top soil (0–15 cm soil section) NH_4^+ (**a**, **c**) and NO_3^- (**b**, **d**) content of the non-fertilized mono-crops (wheat (WT), faba bean (FB), poplar (PL), and Robinia (RB)), N-fertilized mono-crops (80 kg N ha^{-1} in the form of calcium-ammonium-nitrate; wheat (NWT) and poplar (NPL)) and non-fertilized mixed crops (wheat with faba bean (WFB) and poplar with Robinia (PRB)). *Vertical arrows* above the charts indicate application dates of mineral fertilizers. *Error bars* show the standard error of the mean of each treatment ($n = 3$)

decrease found in the arable crops specifically in WT and NWT treatments. During the vegetation period in 2014, soil NO_3^- content was generally higher in agro-forestry soils than in arable soils. Plant nutrient and water uptake was expected to be higher in cropland compared to the young agro-forestry treatments in 2014 due to small size and low growth rate of young trees. Thus, more rapid depletion of soil mineral N in arable crops than agro-forestry treatments can mainly be attributed to the differences in plant N uptake.

In arable land stand, soil NO_3^- concentrations were clearly higher (significant in 2014, $p < 0.01$) in FB than in other non-fertilized treatments during the vegetation period (Fig. 2a, b). For legume crops, inputs of biologically fixed N largely supplement to the uptake of soil mineral N to meet crop N demand. Thus, the legume species also take up mineral N from soils for growth before fixing additional N. The preferential use of soil mineral N helps explain why there is also significant depletion of soil NO_3^- in FB treatment [33]. In a review study, authors reported that average 41 % (for chickpea), 65 % (for faba bean), and 66 % (for field peas) of N that were present in legumes were derived from soil N [33]. However, slightly higher soil NO_3^- concentration in FB treatments than non-legume soils suggests that there were still reasonably more NO_3^- available for potential denitrification losses during the legume-growing season.

Seasonal N_2O emissions

In both years, flux data indicate that N_2O emissions were dominated by specific event periods (Fig. 3). Overall, maximum daily emissions of N_2O in the early summer period in 2014 were 0.16 ± 0.07 and 0.04 ± 0.01 kg N_2O–N ha^{-1} day^{-1} in agro-forestry and arable land treatments, respectively. In a 2-year field study, Lebender et al. [12] observed similar flux rates over a nearby site with similar soil conditions and agricultural practices (wheat and spring barley). Maximum N_2O emissions measured in agro-forestry treatments (0.16 ± 0.07 kg N_2O–N ha^{-1} day^{-1}) have been usually observed in young agro-forest ecosystems [7, 27]. Almost all significant N_2O fluxes occurred as daily peak N_2O emissions and were measured only during the early summer period in 2014. The importance of these peak emissions in early summer period on the annual budget of N_2O emissions highlights the necessity of continuous flux monitoring to accurately determine the N loss from agro-ecosystems specifically in spring and early summer seasons [7, 20].

In 2014, N_2O emissions gradually decreased to the background levels (below 10 g N_2O–N ha^{-1} day^{-1}) from the months of May to July and remained in background levels until March 2015 (Fig. 3). Interestingly, the latter was less than 0.01 kg N_2O–N ha^{-1} day^{-1} during the early summer period in 2015 for all treatments (including N-fertilized treatments). As seen in Fig. 1, early summer period in 2014 was relatively wet (April–June, 201 mm rainfall) compared to the same period in 2015 (April–June, only 100 mm rainfall). Therefore, we may attribute higher daily N_2O fluxes in 2014 than in 2015 (in early summer period) to the differences in mineral N and moisture content of the soil. Mineral N content of all soils in June 2014 was almost similar as compared to the same period in 2015, whereas N_2O fluxes were still about 10-fold higher in June 2014 than in June 2015. In this context, we may conclude that soil moisture seems to be the major

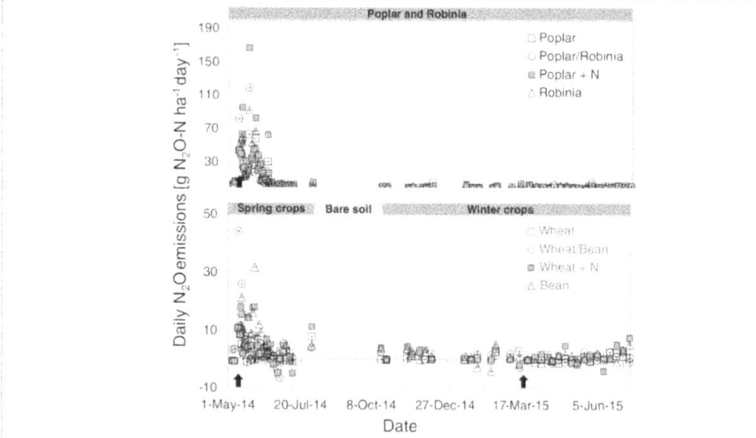

Fig. 3 Daily N_2O fluxes of the non-fertilized mono-crops (wheat (WT), faba bean (FB), poplar (PL), and Robinia (RB)), N-fertilized mono-crops (80 kg N ha^{-1} in the form of calcium ammonium nitrate, wheat (NWT) and poplar (NPL)), and non-fertilized mixed crops (wheat with faba bean (WFB) and poplar with Robinia (PRB)). *Vertical arrows* under the charts indicate application dates of mineral fertilizers. *Horizontal bars* above the charts indicate growing periods of crops. *Error bars* show the standard error of the mean of each treatment ($n = 3$)

driving factor of higher N_2O emissions in the 2014 summer period than in 2015.

All the abovementioned factors (e.g., high moisture, high soil temperature in early summer, and moderate or high NO_3^- content of the soil) are known to trigger specifically the denitrification rate in soils. Thus, we may speculate that denitrification (fungal or bacterial) was the potential key source of measured large N_2O fluxes in the 2014 early summer period. Our earlier report supports this hypothesis in which sandy loam soil was incubated under laboratory conditions, and similar to the field experiment, large N_2O peak events were observed immediately after rewetting of the soil. Here, a stable isotope-labeling study clearly showed that denitrification was the major source (over 90 % of emitted N_2O) of large N_2O peaks that occurred in wet seasons [8, 19]. Furthermore, there is now a growing evidence that fungal denitrification may be the key process producing N_2O in such situations rather than bacterial denitrification [17]. Surely, more research is needed to reveal (i) the dominant processes and (ii) key microbial or fungal strains producing N_2O under these specific conditions (especially during early summer period).

Effect of plant species on N_2O emissions

Mean cumulative N_2O fluxes during the vegetation period in 2014 were 152 ± 58, 217 ± 29, and 441 ± 10 g N_2O–N ha^{-1} in WT, WFB, and FB treatments, respectively. Among the non-fertilized arable crops (wheat, wheat mixed intercropped with bean, and bean), N_2O emission over the 2014 growing seasons is highest in soils when faba bean (FB) was grown (Fig. 4). Introducing N-fixing legumes into cereal-based crop rotations may reduce synthetic mineral-N fertilizer use and thought to mitigate N_2O fluxes. However, the present study clearly showed that when faba bean was grown as a mono-crop, N_2O fluxes were about threefold higher compared to WT treatment. In contrast to the present study, authors reported that growing season N_2O emissions from N_2-fixing legumes are significantly lower than from non-legumes and are often comparable to unfertilized background emissions [26, 34]. In line with the present study, Rochette and Janzen [26] (in a review study) concluded that legumes can produce substantial N_2O emissions. They speculated that the main source of N_2O emissions from soils planted with N_2-fixing legumes during the vegetation period may be attributable to the N release from root exudates and/or from the decomposition of dead root residues. An alternative process that may contribute to the latter would be the N_2O emission during the N_2-fixation process in the nodules where N_2 is fixed. Authors reported that several Rhizobium species in the free-living forms or in legume roots can denitrify NO_3^- and release N_2O from active

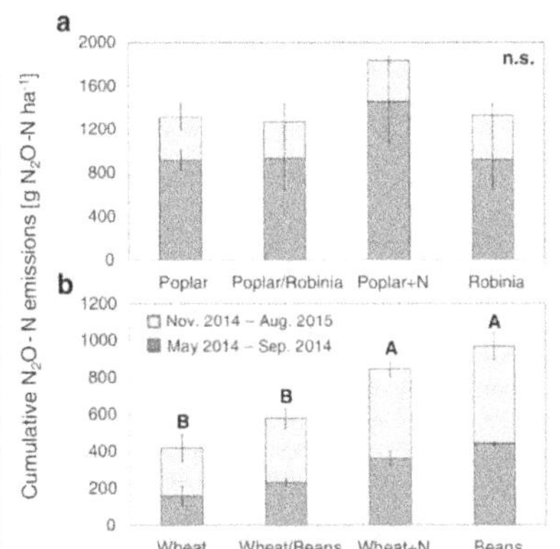

Fig. 4 Cumulative mean N_2O emission rates of the non-fertilized mono-crops (wheat (WT), faba bean (FB), poplar (PL), and Robinia (RB)), N-fertilized mono-crops (80 kg N ha^{-1} in the form of calcium-ammonium-nitrate; wheat (NWT) and poplar (NPL)) and non-fertilized mixed crops (wheat with faba bean (WFB) and poplar with Robinia (PRB)). Upper graphs (**a**) show the agro-forestry treatments, and lower graphs (**b**) show the crop domains during the period from May–September 2014 (vegetation period of spring crops) and November 2014 to August 2015. *Error bars* show the standard error of the mean of each treatment ($n = 3$). Means within each crop followed by the same letter (*A*, *B*) are not significantly different ($p > 0.05$). *n.s.* not significant

nodules most likely to prevent excess NO_3^- that inhibits the activity of N_2-fixing enzymes [35].

In agro-forestry treatments, both daily and cumulative N_2O emissions did not differ among each other in the 2014 growing period. Here, mean seasonal N_2O emissions (during the growing season of arable crops) were 1121 ± 161, 1102 ± 159, and 1052 ± 266 g N_2O–N ha^{-1} in PL, PRB, and RB treatments (no significant difference), respectively. The cumulative mean N_2O fluxes during the 2014 growth period were considerably higher in agro-forestry than in arable crop treatments. During the first year, young plantations in agro-forestry domains generally have limited N and water uptake, while wheat and faba bean as arable crops are at their most productive growth stage specifically during May and June (growth rates are almost at their maximum during this period). Here, soil conditions seem to be more favorable specifically for denitrification in agro-forestry treatments than in soils planted with arable crops that may explain large N_2O emissions [27]. The cumulative mean N_2O emissions in agro-forestry treatments were about fivefold higher than in both WT and WFB treatments, and the latter was still more than twofold higher when compared to FB plots. In line with the present data set, authors reported large N_2O fluxes after conversion of pastures lands [7] or grasslands [36]. Here, authors attributed large N_2O fluxes to the soil disturbance associated with tillage and cultivation that accelerate soil organic matter decomposition and microbial activity (nitrification and denitrification) leading to N_2O emissions. In the present experiment, soil tillage has been done almost at the same time for all treatments. Thus, we may speculate that the difference in N_2O emissions when comparing arable land to agro-forestry was mainly due to differences in water and nutrient uptake of plant species. Water and nutrient demand of young plantations during the first year are generally low which in return may cause more favorable conditions specifically for denitrification and N_2O losses from denitrification. Overall, we can therefore summarize that direct plant effect seems to be one of the key variables that regulates N_2O losses from soils. Zona et al. (2013) concluded that vegetation uptake of NO_3^- together with water ultimately may reduce the anaerobic volume of soils and may lower both denitrification rate and product stoichiometry of denitrification (lower $N_2O/N_2O + N_2$ ratio; meaning reduced N_2O and enhanced N_2 production) in agricultural soils [8, 17].

Effect of mineral N supply on N_2O emissions

Although it was not the main goal of the present experiment to study the effect of mineral-N addition on N_2O fluxes, we added fertilizer N to mono-crop wheat (80 kg N ha^{-1}; calcium-ammonium-nitrate) in parallel plots to be able to compare N_2O fluxes from soils planted with N_2-fixing plants (faba bean mono-culture or faba bean intercropped with wheat) with fertilized and non-fertilized wheat soils. N fertilization during the first year of new agro-forestry plantations is also not a common practice. However, N doses similar to the arable treatments were applied at the same date in order to be able to gain better scientific knowledge about the dominant factors regulating N_2O fluxes in agro-forest ecosystems. Expectedly, in all N-fertilizer treatments, N_2O fluxes increased immediately after fertilizer application, however, only in 2014 (wet early summer) but no response observed in 2015 (dry early summer). The latter clearly suggests that environmental factors specifically soil moisture was the most dominant factor in 2014 that leads to relatively high N_2O fluxes and without sufficient moisture or rainfall, fertilizer application alone does not affect N_2O fluxes significantly, e.g., in 2015. In line with the present study, authors also reported that the application of organic or inorganic fertilizers affects N_2O fluxes only in wet seasons but not in dry years [10, 17, 19].

Overall, the application of nitrogenous fertilizer affected N_2O fluxes predominantly in cropland soils and had limited impact in agro-forestry soils. Here, cumulative N_2O fluxes (from May to September 2014) were 46 and 121 % higher in fertilized than in non-fertilized agro-forestry (non-fertilized poplar vs. fertilized poplar) or in cropland treatments (non-fertilized wheat vs. fertilized wheat), respectively. The latter clearly suggests that N_2O fluxes were more dependent on soil mineral-N content in arable crops than in agro-forestry most likely due to greater competition between plants and N_2O-producing soil microorganisms in cropland than in agro-forestry soils. Fertilizer-derived cumulative N_2O emissions (emission factor) during the period from April to December 2014 were 0.21 and 0.45 % of applied N in wheat and poplar soils, respectively. Measured emission factors were significantly lower than what IPCC predicts (1 %; [4]). However, the latter was similar to what we reported in our previous study for central and northern Germany ([12, 19]). For operational reasons in the present IPCC protocol, the N_2O emission factor was set to 1 % for all fertilizer N regardless of crop or soil type. Large variations in N_2O emissions from different agricultural systems due to differences in management, climate, and soil type are very well known. Low N_2O emission factors in the present experiment and in our previous reports suggest that in the future, different emission factors should be considered at least for different crops or regions that account for their different risks of N_2O emissions.

Cumulative mean N_2O emission during the growing season was still 31 % higher ($p < 0.05$) in FB than in NWT treatment. In a review study, Rochette and Jansen [26] summarized that legumes can increase N_2O emissions during growth compared to evenly fertilized arable crops most likely due to the N release from the root exudates and decomposition of crop residues. Our study clearly agrees with Rochette and Jansen [26] and many others (e.g., [7, 27]) that growing legumes as mono-crop can increase N_2O fluxes compared to N-fertilized arable crops. On the other hand, seasonal N_2O fluxes were 35 % lower in WFB (wheat mixed intercropped with faba bean) than in NWT (wheat as mono-crop) treatment. The latter suggests that using legume crops as intercrop or mixed crop in wheat may significantly mitigate fertilizer-derived N_2O fluxes. However, surely more research is needed to upscale current findings due to the complexity and variability of N_2O fluxes in complex agricultural systems, e.g., mixed cropping systems.

Conclusions

The N_2O emission from soils is variable in space and time, thus measuring and quantifying variance in N_2O emissions is rather difficult, and there are only few field experiments available allowing long-term comparison of various plant species (crops, legumes, and agro-forests) simultaneously with fertilizer effects. We see three take home messages:

- Currently, biogas production from energy crops is mainly based on anaerobic fermentation of mono-crops; however, high-yielding mono-crops require high N-fertilizer input that increases the risk of N_2O losses. Present study clearly showed that mixed intercropping agricultural systems (legume and non-legume plant species) may significantly lower (about 35 %) N_2O losses compared to the N-fertilized mono-crops.
- Cumulative N_2O emissions in agro-forestry soils were about 2–5-fold higher than in cropland soils. Soil conditions in agro-forestry treatments seem to be more favorable specifically for denitrification (due to limited water and nutrient uptake of young plantations during their initial growth stage) than soils planted with arable crops that may be responsible for large N_2O emissions.
- Cumulative mean N_2O emissions during the growth period of annual crops were 31 % higher ($p < 0.05$) in soils planted with faba bean than in N-fertilized wheat. We can conclude that legumes (when grown alone) can produce substantial N_2O emissions most likely due to enhanced denitrification activity in their rhizosphere due to N-rich root exudates/dead organic matter.

Competing interests

The authors declare that they have no competing interests.

Authors' contributions

MS contributed to the acquisition of funding, supervised the field work, perform the statistics, and drafted the paper (30 %). CW as a MSc student run the experiment (70 %; gas sampling, soil sampling, and analysis), collected and analyzed the data, and drafted the paper (10 %). AI as a PhD student contributed to the field experiment (crop management, soil sampling, and data analysis), and drafted the paper (5 %). JI contributed to the acquisition of funding and design of the experiment and drafted the paper (5 %). HS contributed to the acquisition of funding, participated to the design of the experiment, and drafted the paper (5 %). CK drafted the paper and contributed to the planning of the experiment. SK as a PhD student contributed to the field experiment (crop management, soil sampling, and analysis), supervised the MSc student, and drafted the paper (25 %). All authors read and approved the final manuscript.

Acknowledgements

This study was funded by the Federal Ministry of Education and Research (FKZ 031A351A). IMPAC[3] is a project of the Centre of Biodiversity and sustainable Land Use at the University of Gottingen.

Author details

[1]Institute of Applied Plant Nutrition, University of Gottingen, Carl Sprengel-Weg 1, Gottingen 37075, Germany. [2]Present Address: Department of Plant Nutrition and Soil Science, University of Harran, SanliUrfa TR 63000, Turkey. [3]Department Crop Science: Grassland Science, University of Gottingen, Gottingen 37075, Germany. [4]Department Biodiversity and Sustainable Land Use, University of Gottingen, Gottingen 37077, Germany. [5]Department Crop Science, Plant Nutrition and Soil Science, University of Gottingen, Gottingen 37075, Germany.

Received: 11 August 2015 Accepted: 8 December 2015
Published online: 16 January 2016

References

1. Ravishankara AR, Daniel JS, Portmann RW (2009) Nitrous oxide (N2O): the dominant ozone-depleting substance emitted in the 21st Century. Science 326:123–125. doi:10.1126/science.1176985
2. Smith KA, Mosier AR, Crutzen PJ, Winiwarter W (2012) The role of N2O derived from crop-based biofuels, and from agriculture in general, in Earth's climate. Philos Trans R Soc Lond B Biol Sci 367:1169–1174. doi:10.1098/rstb.2011.0313
3. Fluckiger J, Monnin E, Stauffer B, Schwander J, Stocker TF. (2002) High-resolution Holocene N2O ice core record and its relationship with CH4 and CO2. Glob Biogeochem Cycles 16:1010. doi:10.1029/2001GB001417
4. IPCC Fourth Assessment Report: Climate Change 2007 (AR4). https://www.ipcc.ch/publications_and_data/ar4/syr/en/contents.html. Accessed 5 Aug 2015
5. Kroeze C, Mosier A, Bouwman L (1999) Closing the global N2O budget: a retrospective analysis 1500–1994. Glob Biogeochem Cycles 13:1–8. doi:10.1029/1998GB900020
6. Davidson E (1992) Sources of nitric oxide and nitrous-oxide following wetting of dry soil. Soil Sci Soc Am J 56:95–102
7. Zona D, Janssens IA, Gioli B, Jungkunst HF, Serrano MC, Ceulemans R. (2013) N2O fluxes of a bio-energy poplar plantation during a two years rotation period. GCB Bioenergy 5:536–547. doi:10.1111/gcbb.12019
8. Senbayram M, Chen R, Mühling KH, Dittert K (2009) Contribution of nitrification and denitrification to nitrous oxide emissions from soils after application of biogas waste and other fertilizers. Rapid Commun Mass Spectrom RCM 23:2489–2498. doi:10.1002/rcm.4067
9. Bergstermann A, Cardenas L, Bol R, Gilliam L, Goulding K, Meijide A, et al. (2011) Effect of antecedent soil moisture conditions on emissions and isotopologue distribution of N2O during denitrification. Soil Biol Biochem 43:240–250. doi:10.1016/j.soilbio.2010.10.003
10. Dobbie KE, Smith KA (2001) The effects of temperature, water-filled pore space and land use on N2O emissions from an imperfectly drained gleysol. Eur J Soil Sci 52:667–673. doi:10.1046/j.1365-2389.2001.00395.x
11. Bouwman AF, Boumans LJM, Batjes NH (2002) Emissions of N2O and NO from fertilized fields: summary of available measurement data. Glob Biogeochem Cycles 16:1058. doi:10.1029/2001GB001811
12. Lebender U, Senbayram M, Lammel J, Kuhlmann H (2014) Impact of mineral N fertilizer application rates on N2O emissions from arable soils under winter wheat. Nutr Cycl Agroecosystems 100:111–120. doi:10.1007/s10705-014-9630-0
13. Weier K, Doran J, Power J, Walters D (1993) Denitrification and the dinitrogen nitrous-oxide ratio as affected by soil-water, available carbon, and nitrate. Soil Sci Soc Am J 57:66–72
14. Morkved PT, Dorsch P, Henriksen TM, Bakken LR (2006) N2O emissions and product ratios of nitrification and denitrification as affected by freezing and thawing. Soil Biol Biochem 38:3411–3420. doi:10.1016/j.soilbio.2006.05.015
15. Metay A, Chapuis-Lardy L, Findeling A, Oliver R, Alves Moreira JA, Feller C. (2011) Simulating N2O fluxes from a Brazilian cropped soil with contrasted tillage practices. Agric Ecosyst Environ 140:255–263. doi:10.1016/j.agee.2010.12.012
16. Saggar S (2010) Estimation of nitrous oxide emission from ecosystems and its mitigation technologies preface. Agric Ecosyst Environ 136:189–191. doi:10.1016/j.agee.2010.01.007
17. Koester JR, Well R, Dittert K, Giesemann A, Lewicka-Szczebak D, Mühling KH, et al. (2013) Soil denitrification potential and its influence on N2O reduction and N2O isotopomer ratios. Rapid Commun Mass Spectrom 27:2363–2373. doi:10.1002/rcm.6699
18. Jungkunst HF, Fiedler S, Stahr K (2004) N2O emissions of a mature Norway spruce (Picea abies) stand in the Black Forest (southwest Germany) as differentiated by the soil pattern. J Geophys Res Atmos 109:D07302. doi:10.1029/2003JD004344
19. Senbayram M, Chen R, Wienforth B, Herrmann A, Kage H, Mühling KH, et al. (2014) Emission of N2O from biogas crop production systems in Northern Germany. Bio Energy Res 7:1223–1236. doi:10.1007/s12155-014-9456-2
20. Drury CF, Reynolds WD, Tan CS, McLaughlin NB, Yang XM, Calder W, et al. (2014) Impacts of 49–51 years of fertilization and crop rotation on growing season nitrous oxide emissions, nitrogen uptake and corn yields. Can J Soil Sci 94:421–433. doi:10.4141/CJSS2013-101
21. Herrmann A (2013) Biogas production from maize: current state, challenges and prospects. 2. Agronomic and environmental aspects. Bio Energy Res 6:372–387. doi:10.1007/s12155-012-9227-x
22. Cherubini F, Jungmeier G (2010) LCA of a biorefinery concept producing bioethanol, bioenergy, and chemicals from switchgrass. Int J Life Cycle Assess 15:53–66. doi:10.1007/s11367-009-0124-2
23. Smeets EMW, Bouwmanw LF, Stehfest E, Van Vuuren DP, Posthuma A. (2009) Contribution of N2O to the greenhouse gas balance of first-generation biofuels. Glob Chang Biol 15:1–23. doi:10.1111/j.1365-2486.2008.01704.x
24. Whitmore AP, Schroder JJ (2007) Intercropping reduces nitrate leaching from under field crops without loss of yield: a modelling study. Eur J Agron 27:81–88. doi:10.1016/j.eja.2007.02.004
25. Beaudette C, Bradley RL, Whalen JK, McVetty PBE, Vessey K, Smith DL. (2010) Tree based intercropping does not compromise canola (Brassica napus L.) seed oil yield and reduces soil nitrous oxide emissions. Agric Ecosyst Environ 139:33–39. doi:10.1016/j.agee.2010.06.014
26. Rochette P, Janzen HH (2005) Towards a revised coefficient for estimating N2O emissions from legumes. Nutr Cycl Agroecosystems 73:171–179. doi:10.1007/s10705-005-0357-9
27. Liu XP, Zhang WJ, Hu CS, Tang XG (2014) Soil greenhouse gas fluxes from different tree species on Taihang Mountain, North China. Biogeosciences 11:1649–1666. doi:10.5194/bg-11-1649-2014
28. Mayer J, Buegger F, Jensen ES, Schloter M, Heß J. (2003) Estimating N rhizodeposition of grain legumes using a N-15 in situ stem labelling method. Soil Biol Biochem 35:21–28. doi:10.1016/S0038-0717(02)00212-2
29. Breitenbeck G, Bremner J (1989) Ability of free-living cells of bradyrhizobium-japonicum to denitrify in soils. Biol Fertil Soils 7:219–224. doi:10.1007/BF00709652
30. Müller E, Rottmann N, Bergstermann A, Wildhagen, H, Joergensen RG. (2011) Soil CO2 evolution rates in the field—a comparison of three methods. Arch Agron Soil Sci 57:597–608. doi:10.1080/03650340.2010.485984
31. Hutchinson GL, Mosier AR (1981) Improved soil cover method for field measurement of nitrous oxide fluxes1. Soil Sci Soc Am J 45:311. doi:10.2136/sssaj1981.03615995004500020017x
32. Xiang S-R, Doyle A, Holden PA, Schimel JP (2008) Drying and rewetting effects on C and N mineralization and microbial activity in surface and subsurface California grassland soils. Soil Biol Biochem 40:2281–2289. doi:10.1016/j.soilbio.2008.05.004
33. Unkovich M (2012) Nitrogen fixation in Australian dairy systems: review and prospect. Crop Pasture Sci 63:787–804. doi:10.1071/CP12180
34. Jeuffroy MH, Baranger E, Carrouee B, de Chezelles E, Gosme M, Hénault C, et al. (2013) Nitrous oxide emissions from crop rotations including wheat, oilseed rape and dry peas. Biogeosciences 10:1787–1797. doi:10.5194/bg-10-1787-2013
35. Rosen A, Lindgren PE, Ljunggren H (1996) Denitrification by Rhizobium meliloti.1. Studies of free-living cells and nodulated plants. Swed J Agric Res 26:105–113
36. Gelfand I, Zenone T, Jasrotia P, Chen J, Hamilton SK, Robertson GP (2011) Carbon debt of conservation reserve program (CRP) grasslands converted to bioenergy production. Proc Natl Acad Sci U S A 108:13864–13869. doi:10.1073/pnas.1017277108

Chapter 5:
Remotely assessed vegetation development and water use of arable and grassland intercropping with legumes

Annika Meißner, Juliane Streit, Dirk Koops, Johannes Isselstein, Klaus Dittert

Manuscript draft

Remotely assessed vegetation development and water use of arable and grassland intercropping with legumes

Meißner, Annika[1,2*], Streit, Juliane[2,3], Koops, Dirk[4], Isselstein, Johannes[4], Dittert, Klaus[5]

[1]Institute of Applied Plant Nutrition, University of Goettingen, Carl-Sprengel-Weg 1, D-37075 Goettingen, Germany

[2]Center of Biodiversity and Sustainable Land Use, University of Goettingen, Grisebachstr. 6, D-37077 Goettingen, Germany

[3]Division of Agronomy, Department of Crop Sciences, University of Goettingen, Von-Siebold-Str. 8, D-37075 Goettingen, Germany

[4]Division of Grassland Science, Department of Crop Sciences, University of Goettingen, Von-Siebold-Str. 8, D-37075 Goettingen, Germany

[5]Division of Plant Nutrition and Crop Physiology, Department of Crop Sciences, University of Goettingen, Carl-Sprengel-Weg 1, D-37075 Goettingen, Germany

*** Correspondence:**
Annika Meißner, M.Sc.
annika.meissner@mein.gmx

Abstract

Including legumes in cropping systems improves the sustainability, nutrient availability and productivity of agricultural systems. Intercropping of legumes and non-legumes also comprises these advantages and further has the potential of complementary resource use in terms of light and water. However, the availability of water resources is unevenly distributed throughout the vegetation period leading to low yield stability. For early evaluations of the crop stands, remote sensing approaches were implied to monitor critical phases during the vegetation.

At this background, we set up a factorial split-plot experiment with pure and intercropped stands of legumes and non-legumes with different genotypes in arable land and grassland. Arable crops were winter faba bean (*Vicia faba* L.) and winter wheat (*Triticum aestivum* L.); grassland crops were white clover (*Trifolium repens* L.), perennial ryegrass (*Lolium perenne* L.) and chicory (*Cichorium intybus* L.). In additional non-legume grassland plots, 240 kg nitrogen fertilizer per ha were applied. The experiment was set up at two contrasting study sites in Germany to compare different growth conditions. Drone-based remote sensing of canopy surface temperature via thermography and the normalized difference vegetation index (NDVI) via spectral images was applied. Continuous measurements on a weekly basis were conducted over a period of three years (2015 - 2017).

Results showed that intercropping in both land-use systems as well as application of nitrogen fertilizer in grassland increased NDVI and decreased canopy surface temperature. This indicates an improvement in productivity and water use by species in mixtures and by enhanced nitrogen availability. Only small genotype differences within winter faba bean and white clover were detected in terms of NDVI and canopy temperature. The significance of these differences was dependent on the crop stand. The performance of intercropping was rather affected by environmental conditions such as air temperature and rainfall as well as the accompanying species. In grassland, chicory significantly reduced the surface temperature and thus increased the water consumption compared to ryegrass, both in pure and intercropped stands. This difference was observed on the basis of a single day measurement, as the canopy surface temperature was strongly influenced by daily fluctuations in growth conditions.

1 Introduction

In the last decades, crop rotations have become less diverse so that the sustainability of agricultural systems is often questioned (Stein and Steinmann, 2018). Legumes can increase the sustainability of cropping systems due to several positive benefits, such as an increased nitrogen availability, improved soil structure and lower greenhouse gas emissions (Jensen et al., 2012; Peoples et al., 2009). One of the main grassland legume species in Europe is white clover with its high protein content and good digestibility (Lüscher et al., 2014). Among arable crops, faba bean gains more interest as it has the potential to replace soybean as protein source in livestock feed production (Jensen et al., 2010). Here, breeding of winter-hardy faba bean cultivars is a promising option to increase the productivity.

The yield of grain legumes is mostly affected by yearly fluctuating growth conditions, e.g. water availability (Watson et al., 2017). Especially under ongoing shifts in global climate with less seasonal water availability, water limitations in spring and summer lead to low yield stability (Brouder and Volenec, 2008). Consequently, drought is a major factor negatively influencing growth and productivity of crops (Zampieri et al., 2017).

In this context, intercropping with legumes has been widely encouraged in terms of sustainability, as it bears the potential of increased yields compared to pure stands due to more efficient use of growth parameters (Hauggaard-Nielsen et al., 2001). Deep rooting species such as chicory can contribute to these synergistic effects by their capability to exploit water resources from deep soil layers which are not available for accompanying crops such as white clover and perennial ryegrass (Skinner, 2008). Including such deep rooting species therefore enhances the water availability in the system and promotes more stable yields (Weißhuhn et al., 2017).

Yield stability can be achieved, besides by the cropped species itself, by the precise management of agricultural land under consideration of spatial heterogeneity within the fields. Here, novel non-invasive remote sensing techniques gain increasing relevance as they are able to rapidly measure canopies on a large scale (Candiago et al., 2015). Using unmanned aerial vehicles (UAV) with lightweight cameras, estimations of biomass production, species composition as well as drought stress are possible (Bendig et al., 2014; Möckel et al., 2016; Smigaj et al., 2017). In contrast, conventional phenotyping of crop stands or genotypes is costly and time consuming (Mahlein, 2016). UAV-based techniques provide additional information and increase the efficiency in phenotyping. Spectral and thermal imaging within the breeding process therefore have the potential to improve the selection for stress adapted genotypes (Leucker et al., 2017; Thomas et al., 2017).

Applications of such sensor techniques depend on specific spectral features of the plants under certain conditions (Behmann et al., 2015). One possibility of differentiation between genotypes and crop stands is the Normalized Difference Vegetation Index (NDVI), reflecting photosynthetic activity and biomass parameters. In addition, thermography can reveal differences in stomatal closure and water use of crops such as faba bean (Ivushkin et al., 2017; Khan et al., 2010; Ortiz-Bustos et al., 2017) as transpiration and leaf temperature are negatively correlated (Lindenthal et al., 2005).

Differences in physiology, water use and yield production of crops occur due to spatial and temporal variations in growth conditions as a consequence of shifts in global climate (Bakhsh et al., 2000). Furthermore, genotypes of e.g. faba bean differ in several drought tolerance related parameters (Khan et al., 2007). This genotypic variation might be altered in intercropping systems due to competitive or synergistic situations. Especially responses to the availability of nitrogen, either by fertilizer application or by biological fixation, lead to changes in the reflectance patterns (Li et al., 2014).

In order to evaluate the crop development and the water use over the vegetation period, we set up a field experiment at two contrasting experimental sites for the domains arable land and grassland in order to compare pure and intercropped stands. In these stands, winter faba bean and winter wheat as well as white clover, perennial ryegrass and chicory were investigated, whereby each legume was represented by different genotypes. We hypothesize that (1) different species and thus intercropping of species have different spectral and thermal properties; (2) the genotypes of the particular species respond differently; (3) nitrogen application alters these properties in non-legumes of grassland; (4) general fertility and water availability of the field site increases NDVI and decreases canopy temperature.

2 Material and methods

Study Sites

Field experiments were conducted in the period of 2014 to 2017 at two experimental stations of the Georg-August University of Goettingen in central Germany with identical setups. The experimental stations were Reinshof on the study site 'Pfingstanger' (N 51°29′5′′, E 9°55′23′′, 157 m asl.) and Deppoldshausen (N 51° 34′54′′, E 9° 58′3′′, 342 m asl.). The study sites differed in their chemical and physical soil properties (Table 1). According to the FAO scheme the soil type of the study site Reinshof is classified as Gleyic Fluvisol and consists of 21 % clay, 68 % silt, 11 % sand and 2.24 % total organic carbon (C) and 2.81 % humus with the average pH of 7.0 (range 6.4 - 7.5) while the land value ranged between 81 and 89 (Table 1). The study site Deppoldshausen is classified as Calcaric Leptosol and contains 43 % clay, 55 % silt, 2 % sand and 3.18 % total organic C and 3.29 % humus with the average pH of 7.3, while the land value ranged from 38 to 48 (Table 1).

At both study sites, weather conditions were recorded by custom-made weather stations (Adolf Thies GmbH & Co. KG, Göttingen, Germany) as well as by the Deutscher Wetterdienst (DWD) at the close-by weather station 'Göttingen'. Overall, Reinshof had higher monthly air temperatures than Deppoldshausen (Fig.1). Reduced rainfall in May and June 2015 led to natural drought in comparison to the long-time average (Fig.2). After high temperatures and rainfall in June 2016, drought events occurred in the following months at both sites. High temperatures were also observed in spring 2017 with average rainfall, followed by drastically increased rainfall in summer.

Table 1: Soil properties of the two experimental sites Reinshof and Deppoldshausen

Site	Soil type (FAO)	Major composition	Total organic C	Humus	Average pH
Reinshof	Gleyic Fluvisol	21 % clay 68 % silt 11 % sand	2.24 %	2.81 %	7.0
Deppoldshausen	Calcaric Leptosol	43 % clay 55 % silt 2 % sand	3.18 %	3.29 %	7.3

Experimental design and treatments

The field trial was consisted of the domains arable land and grassland. Each domain had four replicates in a factorial split-plot design with the same set-up at both study sites. While grassland was permanently established in summer 2014, arable blocks were autumn-sown and rotated with winter rye in a four-year crop rotation. Cropping was performed in a row design.

Intercropping of legumes and non-legumes was compared to their respective pure stands. In arable land, winter faba bean (*Vicia faba* L.) and winter wheat (*Triticum aestivum* L.) were grown. In grassland, white clover (*Trifolium repens* L.) and the non-legumes perennial ryegrass (*Lolium perenne* L.) and chicory (*Cichorium intybus* L.) were grown as pure stands as well as in binary and three species intercrops. Eight different genotypes of the legume and three different non-legume partners were cropped alone or in substitutive mixtures (Table 2). Additional plots of the non-legumes in grassland received 240 kg nitrogen (N) fertilizer per ha divided into four applications per year.

In arable land, plots were sown after plowing and harrowing with a seeding density of 40 seeds/m² for winter faba bean and 320 seeds/m² for winter wheat. In intercropped plots, seeding density was halved for each species. Arable plot size was in total 3 m * 9 m with 12 rows, where in case of intercropping species were arranged in alternating rows (row distance 22.5 cm). Herbicides were used when necessary to eradicate emerging weeds while additional hand weeding was performed. Insecticides were applied against aphid pests. At both sites, harvests were performed in August with a combine harvester in a core area of 1.5 m * 7 m of the plots.

Table 2: Genotypes of the different species in arable land and grassland with their respective origin.

Arable land		**Grassland**		
Winter faba bean	Winter wheat	White clover	Perennial ryegrass	Chicory
S_004	Genius	EGB PX 90305	Elp 060687	Puna II
S_062	Boxer	EGB PX 90312		
S_069	Hybery	EGB PX 90702		
S_265		EGB PX 90710		
Hiverna/2		EGB PX 90913		
Côte d'Or/1		EGB PX 90914		
WAB-Fam157		EGB PX 90915		
WAB-EP98-267		EGB PX 90909		
Link, Univ. Goettingen;	SaatenUnion/ Nordsaat (G./H.)	DSV	DSV	British Seed
NPZ Lembke	BayWa (B.)			Houses

In the grassland, plots of 3 m * 5 m area (row distance 17.5 cm) were established, in which intercropped plots were arranged in alternating rows. Seeding density was 1000 seeds/m² in pure stands, while intercropped stands were equaled to the same amount of seeds. In intercropped plots, seeding densities were reduced to 400 seeds/m² for white clover and

600 seeds/m² for perennial ryegrass and chicory. When both non-legume partners were included, seeding densities were 300 seeds/m² each. In binary mixtures of non-legumes, seeding densities of both species were halved to 500 seeds/m² for perennial ryegrass and chicory. Pest control was regularly undertaken against mice and slugs. All plots were cut four times a year with a rhythm of approximately six weeks to simulate a regularly used grassland system. The first cut was performed in May and the last at end of September/beginning of October. Nitrogen fertilizer application was performed in March with 80 kg N/ha and after the first and the second cut with 60 kg N/ha as well as after the third cut with 40 kg N/ha.

NDVI, thermal imaging and image analysis

Remote sensing measurements were carried out in three consecutive growing seasons (2015 until 2017). The unmanned aerial vehicle (UAV) system Raptor (EagleLive Systems GmbH, Römerstein, Germany) was mounted with cameras for aerial images of the field trial. The UAV system was a quadcopter (70 cm * 70 cm * 35 cm) equipped with the gimbal system Direct Drive 2D (EagleLive Systems GmbH, Römerstein, Germany) for camera attachment and stabilization. Attached to the gimbal was a Hero4 camera (GoPro Inc., CA, USA) for normal RGB images as well as a spectral camera ADC Micro (Tetracam Inc., CA, USA) with three multispectral bands of green (520-600 nm), red (630-690 nm) and near infrared light (760-900 nm). In addition, a thermal camera (PI 400, Optris GmbH, Berlin, Germany) was mounted to the UAV. Aerial images were taken on sunny days with clear-sky conditions at +/- 3 h from solar noon at a height of 75 m (grassland) and 100 m (arable land).

Spectral images were analyzed for the Normalized Difference Vegetation Index (NDVI) with the software PixelWrench2 version 1.2.3.9 (Tetracam Inc., CA, USA) according to the following equation

$$\mathrm{NDVI} = \frac{\mathrm{NIR} - \mathrm{RED}}{\mathrm{NIR} + \mathrm{RED}}$$

where RED is the reflectance at 680 nm and NIR is the reflectance at 800 nm. In the red region of the visible light maximal absorption and thus minimal reflection occurs, indicating a high chlorophyll activity and a good nutritional supply (Reckleben, 2014). The high reflection in the near infrared (NIR) region of the spectrum is derived from scattering at cell walls and other cell structures (Pinter et al., 2003). Therefore, absorption and scattering of light in the abovementioned wavelengths is promoted by dense and healthy vegetation.

At the beginning of the image analysis, colors were processed to create monochromatic images. Index values were then obtained by marking the inner parts of the plots without border effects. For these areas, average NDVI values were calculated and integrated to an average for the four replicates.

Thermal images were analyzed using the software PI Connect (Optris GmbH, Berlin, Germany). Same as for the NDVI, non-affected parts of the plots were marked for calculating a mean canopy surface temperature per treatment. Here, energy emission in the near infrared region of 800 to 1400 nm is proportional to the surface temperature of canopies (Pinter et al., 2003). As the canopy surface temperature is highly affected by daily fluctuations in weather conditions, it was corrected by the air temperature at the time of measurement. Therefore, the difference between the canopy temperature and the air temperature (dT) was calculated as follows

$$\mathrm{dT} = \mathrm{T(canopy)} - \mathrm{T(air)}$$

Soil temperature

Soil temperature was measured between the sowing rows at 10 cm soil depth using mini-dataloggers (DS1921G, Thermochron iButton, maxim integrated, WI, USA) in 4 plots per block for each species in pure stand and intercropping, respectively. Measuring periods for arable land were from April 2015 to August 2015 and from November 2015 to August 2016. Measurements in grassland took place from April 2015 to December 2015 and from April 2016 to October 2016. Data were recorded every 4 hours with a resolution of 0.5 °C. Data for correlations were calculated as daily average.

Statistics

Statistical analyzes were performed with R version 3.4.4 (R Core Team, 2018). The R package *nlme* (Pinheiro et al., 2018) was used to calculate linear mixed effects models with block and year as random effects and treatment and site as fixed effects. Differences between treatments were evaluated by the Tukey test at $p < 0.05$ with the package *emmeans* (Lenth, 2018). Graphs were produced with R as well as with Microsoft Excel. Linear regressions were used to evaluate correlations between the variables canopy surface temperature and soil temperature.

3 Results

The development of the NDVI values throughout the growing periods of 2015 until 2017 generally showed clear differences between the crop stands in both domains. Arable crop stands including winter faba bean, both pure stand and intercropping, had generally higher NDVI than pure stands of winter wheat (Fig.3). Consequently, the intercropping of both species had increased NDVI compared to the calculated theoretical mean. In March and April of 2017 however, wheat had higher NDVI than the other crop stands at Reinshof. At Deppoldshausen, all NDVI values increased during May.

The temperature difference between canopy and air (dT) of arable crops highly varied in the course of the years (Fig.4). There were few differences between the crop stands, showing higher temperatures of pure wheat stands in comparison to crop stands including faba bean.

The NDVI of crop stands in grassland representing the crop development were generally decreased by the biomass harvests (Fig.5). Pure stands of white clover have mostly highest NDVI values, especially in 2016 and 2017. The NDVI values of intercropped stands are close to those of pure white clover or on the same level but higher than the NDVI of unfertilized non-legumes. The N fertilizer applications markedly increased NDVI of the non-legumes compared to the unfertilized treatment. Subsequently, in 2015 and in parts of the seasons of 2016 and 2017 the N-treatments reached highest NDVI.

The temperature development of grassland crops followed similar courses as the arable crops, and was thus very little affected by the biomass harvests (Fig.6). Pure stands of white clover and intercropping had similar canopy temperatures while the average of pure stands including ryegrass and chicory showed higher dT. Here, the application of nitrogen led to lower dT of the non-legumes in 2016, whereas fertilization had no effect in 2017.

The different genotypes of the species were compared based on the measurement of June 18th 2016. For arable crops, there were no significant differences in NDVI between the faba bean genotypes, irrespective whether cropped alone or in mixture with wheat (Fig.7). The NDVI values of the two study sites ranged on the same level (between 0.69 and 0.74). Pure stands of winter wheat had significantly lower NDVI, ranging between 0.50 and 0.58. For the canopy surface temperature on the contrary, all treatments differed significantly comparing Reinshof (23.4 to 25.8 °C) and Deppoldshausen (16.0 to 17.7 °C). At Reinshof, the faba bean genotypes in pure stands significantly differed in their surface temperature. Genotypes Vf5 and Vf7 had higher temperatures (24.5 °C and 24.4 °C) and genotypes Vf4 and Vf6 had lower temperatures (23.5 °C and 23.6 °C). There was also significant variability between the genotypes of winter wheat. In contrast, at Deppoldshausen there were no significant differences between the genotypes of the species.

For grassland crops measured on June 18th 2016, there was in general no significant difference between the NDVI values for any of the treatments at the two study sites (Fig.8). The white clover genotypes Tr3, Tr5 and Tr6 showed in pure stands at Reinshof significantly lower NDVI than intercropped with ryegrass. For genotype Tr6 this was the case in Deppoldshausen as well. Nitrogen fertilizer application in all non-legume treatments led to clear and significant increases in NDVI from 0.56 to 0.72 on average of both sites. In accordance with the arable crops, the canopy surface temperature of the treatments showed significant differences between Reinshof (25.4 to 29.8 °C) and Deppoldshausen (16.3 to 18.9 °C). At both study sites, there were no significant differences between the white clover genotypes in pure stands. Surface temperatures of non-legumes in pure stands were significantly decreased by nitrogen fertilizer by on average 2.5 °C at Reinshof and 1.4 °C at Deppoldshausen. The share of chicory decreased the temperatures proportionally. Similarly, in most of the intercropped stands at Reinshof, chicory significantly decreased the surface temperature compared to ryegrass by 0.8 °C on average.

The canopy surface temperature of arable crops was positively correlated with the soil temperature (R^2 = 0.663) (Fig.9). Grassland crops showed a similar positive correlation between canopy surface temperature and soil temperature with the same coefficient (R^2 = 0.663) (Fig.10). Both correlations over all treatments of the respective domain were significant.

4 Discussion

The water use efficiency of crops is affected by stomatal adjustment and transpiration but is more strongly determined by physiological and biomass-forming processes (Brueck, 2008). Those processes of yield formation and transpiration can be evaluated by drone-based methods. It is crucial to know when to apply these remote methods for screening the crop stands to get an adequate impression of the performance of the plants (Negin and Moshelion, 2017). In intercropping of plant species, the heterogeneity and high variation of the system demand for long-term evaluations to be able to consider the complexity of species interaction. In this context, this study evaluated the vegetation development of pure and intercropped stands during three consecutive vegetation periods. The results corresponding to the abovementioned hypotheses are discussed in the following chapter.

Soil properties do not have a great impact on different productivity and water use

The two study sites Reinshof and Deppoldshausen differ remarkably in their soil properties and climatic conditions, resulting in differing prerequisites for plant production and yield potential. Grain legumes are known to need high fertile fields due to their sensitivity to varying growth conditions between the years (Reckling et al., 2015). Consequently, faba bean has a high yield potential on clay soils with adequate precipitation (Reckling et al., 2016). This demand is met at Reinshof, where water availability is high. On the contrary, none of the crop stands containing faba bean had significantly higher NDVI at Reinshof. This could be explained by an observation of Abdelhamid et al. (2011) who stated that yield production of faba bean is impaired if upper soil layers are dried, even though they are able to reach deep soil layers. Among arable crops, only the wheat genotype Ta3 showed a significant difference in NDVI between the study sites, suggesting that this genotype of winter wheat is more suitable for fertile sites than for marginal sites.

Similarly, all crop stands in grassland did not show significant differences between the study sites. Though chicory with its deep taproot was hypothesized to benefit from the deep soil profile at Reinshof. At Deppoldshausen there is a shallow soil profile of maximum 40 cm depth. Here, the soil volume for root distribution is considerably limited and thus reduces available water and nutrient resources. There are contrasting findings in literature regarding the impact of soil properties on nutrient uptake. Carlsson et al. (2017) found higher resource use efficiency and nitrogen availability for intercropping compared to fertilized grass in pure stands at marginal sites. Contrastingly, Suter et al. (2015) showed that nitrogen gain in intercropping is not affected by the fertility of the site but rather correlated with minimum site temperature. The latter findings can be supported by the present results, showing greater impact of weather conditions in contrast to soil properties.

Different water availability at the study sites probably led to overall higher transpiration at Reinshof, indicated by lower dT values. However, these results were masked by high daily fluctuations. This is why on the single day measurement on June 18th 2016, the opposite trend was shown. Though on that day, dT was significantly higher indicating lower transpiration at Reinshof, the whole course of vegetation period suggested higher transpiration at Reinshof. For the method of canopy surface temperature, weather conditions such as the air temperature have a greater impact than soil properties. Single day measurements are thus less reliable than for NDVI.

Weather conditions determine the performance in different years

Warmer air temperatures at Reinshof generally enable a longer vegetation period and a higher yield potential. This was clearly visible in the NDVI values of arable crops in spring

2017, which increased earlier at Reinshof than at Deppoldshausen, indicative of an earlier vegetation start. Nevertheless, comparing the NDVI development over the investigated years, there were clear differences between the years resulting from varying environmental conditions. Also, the canopy surface temperature is only indirectly affected by the soil type. It is rather influenced by temperature and humidity as lower potential biomass production increases the canopy temperature (Lenthe et al., 2007). The present results even indicate that the canopy temperature is more influenced by environmental conditions than by crop stand or genotype. This is in line with studies of Pinter et al. (2003) who discussed that several abiotic factors additionally affect the transpiration and thus plant water status and yield potential, as spectral properties and plant status are generally related.

The weather conditions in 2017 were more favorable for wheat than for faba bean, revealed by higher NDVI of wheat in spring 2017. Crop development of wheat was probably faster than of faba bean as ground cover and plant height determine reflectance pattern in imaging techniques (Bendig et al., 2014; Vaesen et al., 2001). In that year, wheat had a bigger influence on the performance of the intercropping system than in the other years. Here, intercropping of arable crops had NDVI values close to the theoretical mean, which was comparably lower than in 2015 and 2016. These differences were also reflected by the canopy temperatures, showing lower transpiration of wheat in 2016 whereas dT was similar for all crop stands in 2017. This influence from wheat also led to an earlier senescence rate of intercropping than in the other years. The environmental conditions in 2017 were in contrast to the conditions in 2016, which were beneficial for faba bean instead. In 2016, water availability was much higher compared to water shortage in spring 2017. This led to drastic height growth of faba bean in 2016, while in 2017 the height of both species was rather equal (Siebrecht-Schöll, 2019). Decreased height and leaf area of faba bean in intercropping result in higher light interception (Bilalis et al., 2010), further promoting the growth of wheat.

In-seasonal changes in environmental conditions can be measured via NDVI as anomalies in the vegetation development, indicative of yield variations (Rembold et al., 2013). It was shown by Hou et al. (2015) that there is a considerable relation between NDVI and air temperature rather than precipitation. Decreases in the course of the presented measurements were observable in May 2016 and May 2017. However, both were probably induced by lower-than-average rainfall and not the air temperature. This is why we conclude that variability in air temperature did not determine the productivity in our experiments. This contradiction can be explained by a study of Karnieli et al. (2010) who found out that the relationship between NDVI and environmental temperature highly depends on the observation period: In early and late vegetation stages, NDVI is limited by solar radiation and thus air temperature due to impaired photosynthesis (Hou et al., 2015). In the mid-season on the other hand, radiation is higher and the water availability limits the plant growth as we could show in our experiment.

Nitrogen application in grassland increases productivity and water use of non-legumes

With remote sensing, it was clearly visible when N fertilizer was applied in grassland as NDVI of the non-legumes increased drastically after each application. Consequently, shortly before the biomass harvests, the N fertilizer application showed significant enhancing impacts on NDVI as well as declining effects on canopy surface temperature. The observed lower values of unfertilized non-legumes, i.e. winter wheat and perennial ryegrass, can be explained by lower values as consequence to the abiotic stress of N deficiency and early senescence of leaves (Carter and Knapp, 2001). This effect was slightly more pronounced at Deppoldshausen due to lower overall fertility of the site. This monitoring is possible because of the good correlation between leaf N concentration and the leaf reflectance (Pavuluri et al., 2015). That illustrates how remote sensing techniques and the exact management of farm inputs lead to better exploitation of crop yield potentials and higher profitability (Mulla, 2013).

The effects of nitrogen fertilizer in ryegrass and chicory on NDVI were dependent on the year and its environmental conditions. In 2015, NDVI values of the fertilized treatments were among the highest, indicating nitrogen as limiting resource. In contrast in 2017, those treatments rarely reached the NDVI level of intercropping and pure stands of white clover. These different effects can be explained by the weather which was warm with sufficient rainfall in summer and autumn 2015, resulting in favorable growth conditions so that the applied nitrogen could easily be taken up by the plants. Nitrogen uptake in return promotes productivity which is also translated into root growth and then again better water acquisition (Mullan and Reynolds, 2010). In 2017 in contrast, the weather was either cold or dry, except for a warm and moist June, the only month were NDVI values of the fertilized non-legumes were high. During the rest of the year, biomass production was impaired as a consequence of unfavorable growth conditions. This was also visible in the canopy temperature, where fertilized non-legumes showed transpiration rates similar to pure white clover and ryegrass in 2016, whereas in 2017 the transpiration of fertilized and unfertilized non-legumes was similarly low. Hence it can be concluded that water use by transpiration is reduced in seasons with unfavorable growth conditions due to lower water availability by rainfall and lower air temperature. Also, reduced transpiration induced by lack of nitrogen points to impaired photosynthesis and thus yield reductions (Negin and Moshelion, 2017).

As discussed above, many NDVI values of fertilized non-legumes were higher than those of intercropping. This is in line with studies on perennial grassland and annual cover crops, where it was shown that biomass of high yielding grass monocultures is higher compared to non-fertilized mixtures (Carlsson et al., 2017; Finney et al., 2016). This is in contrast to a long-term diversity experiment with grassland crops (*Jena experiment*), where it was concluded that diversity characteristics lead to equal biomass production as intensive management strategies (Weigelt et al., 2009). One of those characteristics is the increased N_2 fixation in intercropping systems. In grassland intercrops, white clover is relying to more than 80 % on N_2 fixation, while in pure stands the proportion is 60-80 % (Høgh-Jensen and Schjoerring, 1997). Yet, biological N_2 fixation is energy consuming so that easy attainable N resources are usually preferred. Accordingly, Nyfeler et al. (2009) showed that little nitrogen fertilizer application in grassland mixtures had a comparable effect as high amounts of N fertilizer in pure grass stands. Therefore, it can be concluded that in the present experiment, the performance of the intercropped treatments could be further improved by little amounts of N fertilizer.

Genotype-specific responses

For the comparison of the winter faba bean and white clover genotypes, the method of the peak NDVI was chosen (Rembold et al., 2013; Tucker et al., 1981). The peak was chosen in a year of favorable growth conditions at a time point of high production rates. In arable land this was at full flowering of winter faba bean and in grassland this was shortly before the second biomass harvest. The method uses the NDVI peak during the growing season which is most related to biomass of wheat as well as grasses. Because of this relationship, most reliable conclusions can be drawn regarding differences between the genotypes.

Only few significant differences between the genotypes were found. An explanation could be that the reflectance patterns of the plants depend on various factors such as chemical composition (e.g. leaf pigments) and leaf structure (Mahlein, 2016) which might hide genotypic effects. Another reason could be too low sensitivity of the imaging technique (Negin and Moshelion, 2017) because of errors induced by varying leaf angles and mutual shading (Mullan and Reynolds, 2010). Though NDVI is related to fractional vegetation cover and leaf area index (LAI), at high LAI the relation between both parameters gets worse (Carlson and Ripley, 1997). Then the reflection of NIR is scattered within the crop stand, NIR signal is lost and the dense vegetation is underestimated by NDVI (Ollinger, 2011). This saturation effect might have restricted the differentiation of genotypes via NDVI. Additionally, canopy surface temperature was strongly affected by environmental conditions as shown by the relation between the canopy temperature and the soil temperature. In Deppoldshausen, the marginal study site, canopy temperatures also revealed heterogeneity in the field which was higher compared to the fertile study site Reinshof. Such variability determines the accuracy of thermal images (Chaerle et al., 2007).

Nevertheless, some variation could be explained by genotypic differences which were visible comparing the treatments on June 18th 2016. The winter faba bean genotypes differed in their transpiration, evaluated by the canopy temperature. Genotypes Vf4 and Vf6 can be characterized as high transpiring, whereas genotypes Vf5 and Vf7 had rather low transpiration. Though these genotypic characteristics were only visible at Reinshof and could not be supported by NDVI, they were stable for pure as well as intercropped stands. Therefore, it can be concluded that within the breeding process, selection for the best performing genotype can be performed in both crop stands. According to a definition by Romano et al. (2011), Vf4 and Vf6 can be considered as drought tolerant due to their lower canopy temperatures. Here, the preservation of physiological processes through stomatal conductivity is regarded as favorable. The plants then sustain productivity and yield formation even though water resources are short.

In contrast to winter faba bean, genotypes of white clover showed significant differences at both study sites. The NDVI revealed the highest productive genotype Tr4, both in pure as well as intercropped stands. Therefore, genotype Tr4 is recommended for further breeding, independently of the crop stand. Considering only pure stands, genotypes Tr3, Tr5 and Tr6 in contrast can be characterized by low productivity. This behavior was not visible in intercropped stands and indicates improved performance in intercropped stands in comparison to pure stands. In terms of the canopy temperature at Reinshof, there were additional mixing effects: Tr1 had in intercropping with ryegrass a comparably low transpiration, while Tr5 and Tr7 had increased transpiration in comparison to the other genotypes. This altered water use behavior could affect the root/shoot ratio, so that the high transpiring crops produce a deeper root system to the disadvantage of reduced production of aboveground biomass (Marshall et al., 2016). This would explain why there was no relationship between the ranking after NDVI and canopy temperature.

Higher-than-average performance of intercropped stands

In both land-use systems, the theoretical NDVI as average of the pure stands was calculated. That theoretical average assumes zero interaction effects between the species in intercropping. We could observe positive interaction effects as intercropping at both study sites was generally higher than the calculated average. That development indicates the advantage of ameliorated yield stability induced by enhanced biodiversity on the field. This effect is independent of variability in growth conditions as discussed above and also reviewed by Frison et al. (2011). The contrary plant architecture of winter faba bean and white clover compared to winter wheat and ryegrass might add to the facilitation in the intercropping system. However, horizontally orientated leaves of faba bean and white clover have higher NIR reflectance than upright orientated leaves (Asner, 1998). Due to this technical aspect, conclusions in terms of yield prediction might be limited.

Intercropping of winter faba bean and winter wheat benefit from nitrogen fixation

The comparably higher NDVI values of intercropped arable crops were most often similar to pure faba bean stands. This observed increase in productivity of intercropping systems is generally considered as result of resource complementarity (Bulson et al., 1997). It indicates the often stated transfer of nitrogen of the legume to the non-legume, supporting the companion crop in growth and development (e.g. Paynel and Cliquet, 2003; Xiao et al., 2004; Meng et al., 2015). Faba bean provide additional nitrogen, resulting in significantly increased grain protein of both faba bean and wheat (Bulson et al., 1997). Moreover, light interception of faba bean-wheat intercrops is higher than of species in pure stands leading to vigorous canopy development (Haymes and Lee, 1999). As a consequence, photosynthesis and biomass production is increased, resulting in rapid ground cover and reduced soil evaporation (Mullan and Reynolds, 2010).

During 2015 and 2016, it was visible that a later decrease in NDVI of intercropped faba bean and wheat was indicative of slower ripening and delayed leaf senescence compared to pure stands of wheat. Here, the difference in the development of faba bean and wheat was decremented due to a stay-green effect in wheat induced by the biological N_2 fixation of faba bean. However, this positive effect was not observable in 2017, displaying a clear dependency on year-to-year fluctuations in growth conditions, as discussed above. Correspondingly, towards the end of the vegetation period canopy temperatures showed a stronger increase in 2016 than in 2017. Such increases in canopy temperature are due to lack of natural cooling of senescent leaf tissue (Lindenthal et al., 2005).

The canopy surface temperature did not show as pronounced effects as the NDVI. In contrast to NDVI, the canopy temperature is strongly affected by several factors such as (micro-)

climate and plant growth status, leading to short-term changes in the canopy temperature (Lenthe et al., 2007). Present results also indicate a dependence on the soil temperature. Therefore, thermography is suitable to relatively compare the evapotranspiration of crop stands, whereas absolute values are not reliable enough (Craparo et al., 2017). Another aspect is that upper leaves have greater contribution to the recorded values, whereas lower leaves within the canopy have other drying and transpiration characteristics (Lenthe et al., 2007). This especially accounts for the horizontally orientated leaves of faba bean, whereas stands of wheat have a higher radiation transmission. That might be the reason why only on few measurement dates wheat had higher canopy temperatures than faba bean.

White clover and chicory define the performance in grassland

In intercropping treatments of grassland crops, NDVI as well as transpiration was relatively increased compared to the respective pure stands of white clover and unfertilized non-legumes. It can thus be concluded that the integration of legumes into grassland systems improves productivity under low-input conditions (Rochon et al., 2004). This can be explained by N transfer from white clover to perennial ryegrass and chicory which leads to increased resource use efficiency. In previous studies, it was described that up to one third of N in ryegrass is derived from the N_2 fixation of white clover (Høgh-Jensen and Schjoerring, 1997). In some parts of the evaluated seasons, this effect was even more effective than mineral N fertilizer.

Similar to arable crops, the legume white clover has mostly highest NDVI values, particularly in 2016 and 2017. This might be due to lateral growth of the shoots, leading to fully covered soil surface and foliage aggregation. Consequently, in combination with the horizontally orientated leaf angle, NIR reflectance of pure white clover stands is highest (Biewer et al., 2009). Adverse leaf orientation was probably also the reason for similar canopy temperature of intercropping and pure white clover stands. Thermal imaging is not affected by the leaf architecture (Barkla and Rhodes, 2017) but complex interactions of biochemical aspects and leaf structure as well as mutual shading might have masked the effect of the crop stand in the light reflection of the canopy (Mahlein, 2016).

The lateral growth of white clover also leads to more remaining biomass and leaf area on the field after harvests. This is why after the biomass harvests in the present experiment NDVI values are less decreased in crop stands containing white clover compared to crop stands of vertically grown non-legumes, as observable especially at Deppoldshausen in 2017. The years before, this effect was not as clear, which might be due to the dynamics in species composition and shifted dominance of the species in permanently established grassland.

In addition to comparisons between genotypes, on June 18th 2016, also crop stands with non-legumes were more precisely evaluated at this time point of the growing season. Here, clear differences between perennial ryegrass and chicory were visible. Though NDVI values did not differ, the canopy temperature was significantly reduced in crop stands of chicory compared to perennial ryegrass. The increased transpiration results from a deeper rooting system of chicory which improves water acquisition (Skinner, 2008). This beneficial effect was also visible in some intercropping treatments with white clover and perennial ryegrass.

For such crop stands higher water availability is of particular advantage under dry conditions (Cranston, 2017).

5 Conclusion

Drone-based sensor systems open up possibilities for low labor but high frequent continuous measurements of plant growth in the field. In close surveillance, these techniques can help to evaluate plant responses to weather and growth conditions and to detect critical periods during the vegetation with resource deficiencies. Gerhards et al. (2017) rise the question of at which developmental stage water deficit is detectable. At this question our results might contribute as we surveyed wide parts of the vegetation period. We could show that the crop stands were clearly affected by lower-than-average rainfall. Furthermore, the performance of the investigated intercropping systems varied with species composition, genotypes and site-specific weather conditions. As genotypic variations were low, which is why further research or increased number of candidates is necessary to identify suitable genotypes as basis for breeding. Overall, with the applied remote sensing approaches, we could demonstrate that productivity and water use throughout the season can be improved by intercropping of legumes and non-legumes. As a consequence of the continuous development of such techniques, the possibilities of application in terms of crop stand monitoring increase. Therefore, remote sensing techniques can be implemented on the farm level as well.

6 Acknowledgments

We thank Regina Martsch for the attentive maintenance of the arable crops in field experiments and Horst-Henning Steinmann for coordination of the project. Furthermore, we thank for the work done by Christian Wenthe and Ronny Schmidt who contributed in data acquisition and image analysis. Funded by the Federal Ministry of Education and Research (FKZ 031A351A). IMPAC³ is a project of the Center of Biodiversity and sustainable Land Use at the University of Goettingen.

Figures

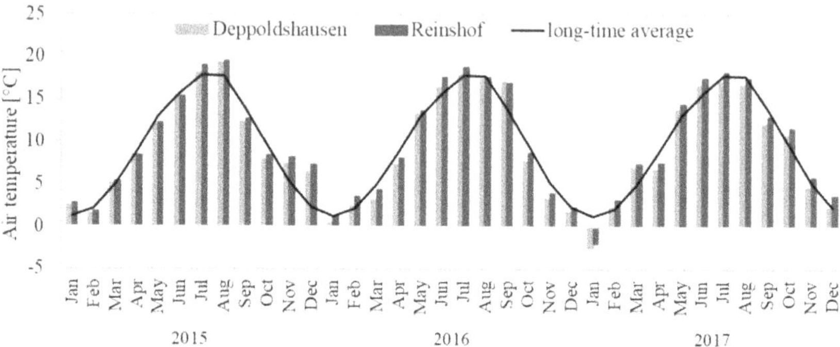

Fig.1: Monthly mean air temperature of the years 2015 to 2017. Air temperature of the field sites Reinshof and Deppoldshausen are compared to the long-time average, calculated over 30 years (1987-2017).

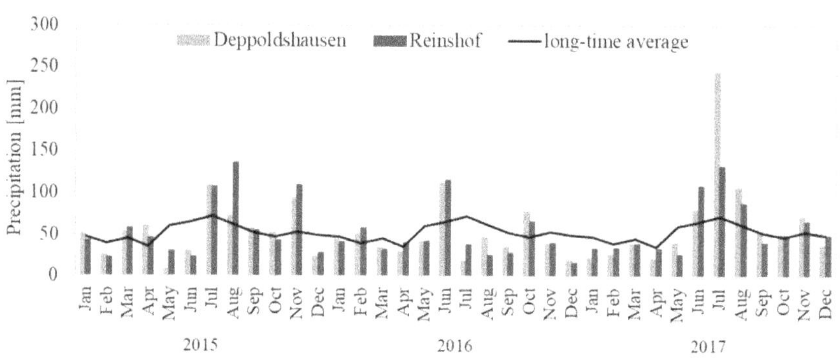

Fig.2: Monthly sum of precipitation of the years 2015 to 2017. Precipitation of the field sites Reinshof and Deppoldshausen are compared to the long-time average, calculated over 30 years (1987-2017).

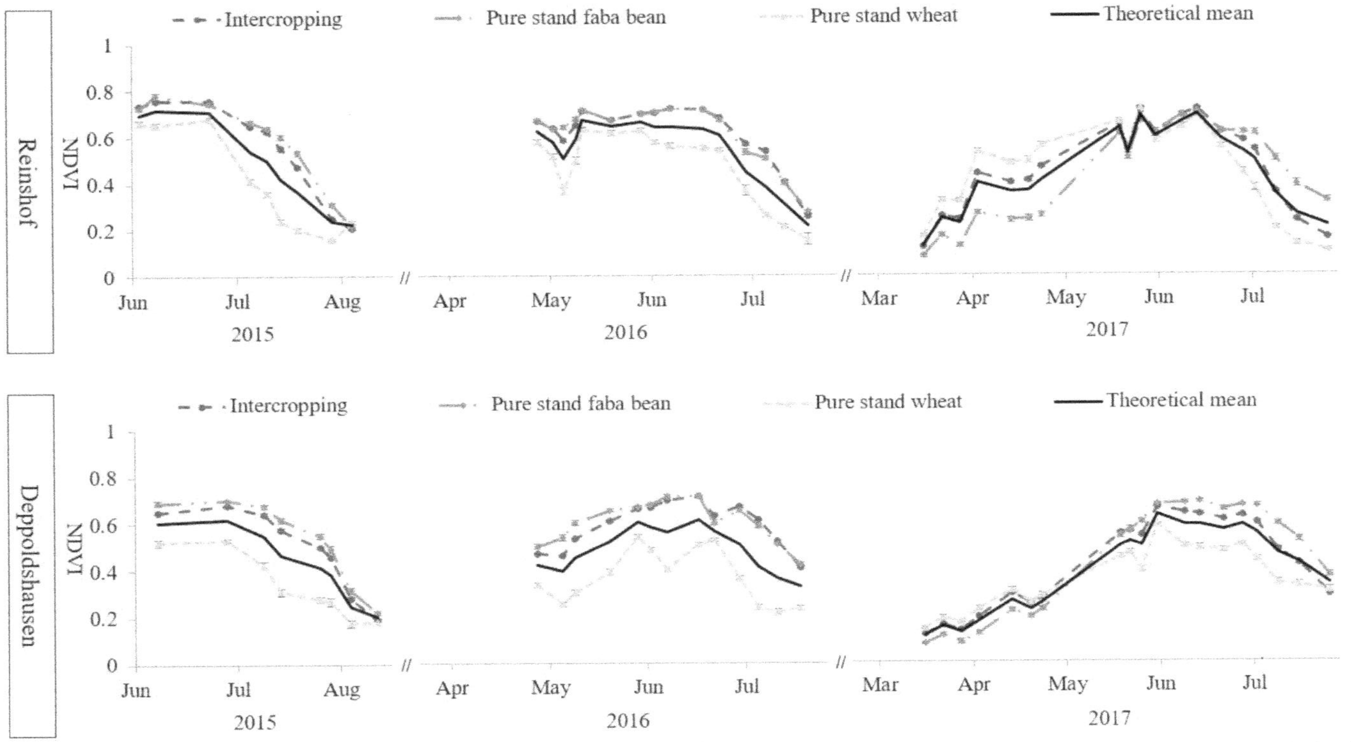

Fig.3: NDVI of crop stands in arable land in the years 2015 to 2017 at the field sites Reinshof and Deppoldshausen. The theoretical mean was calculated as average of the pure stands of winter faba bean and winter wheat. Means were calculated over all investigated genotypes per species. Error bars indicate standard error.

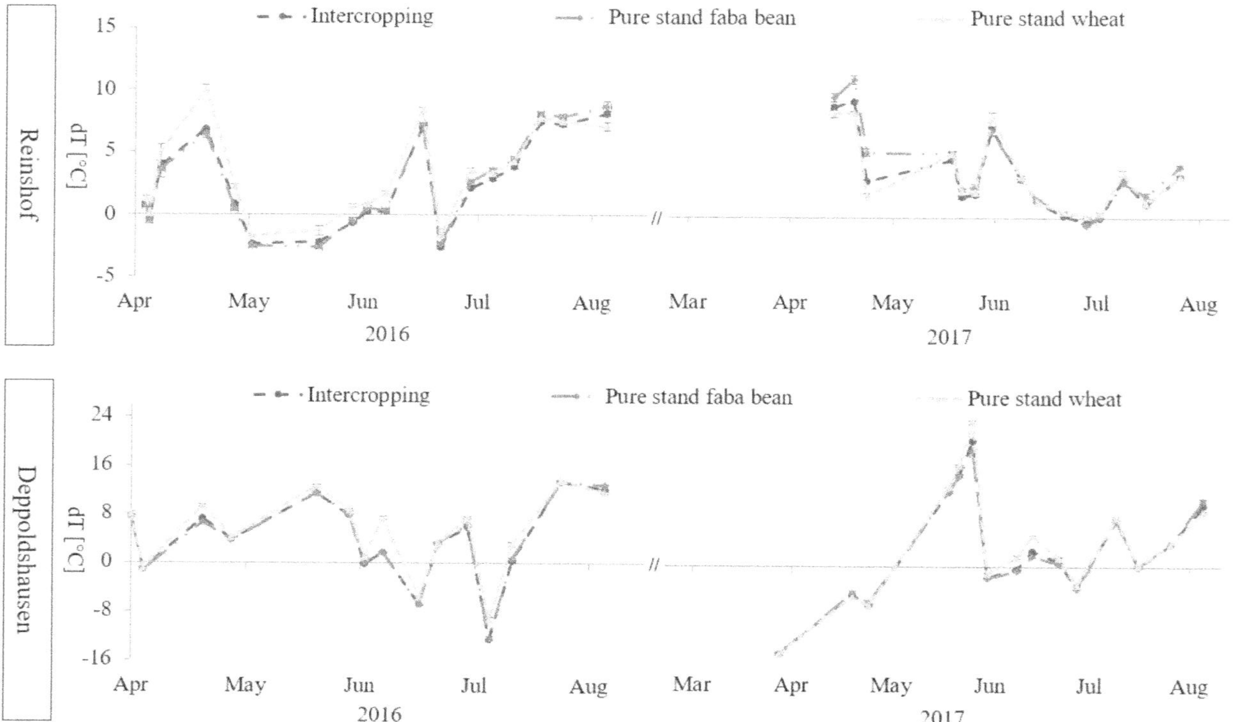

Fig.4: Temperature difference between canopy and air (dT) of crop stands in arable land in the years 2016 and 2017 at the field sites Reinshof and Deppoldshausen. Means were calculated over all investigated genotypes per species. Error bars indicate standard error.

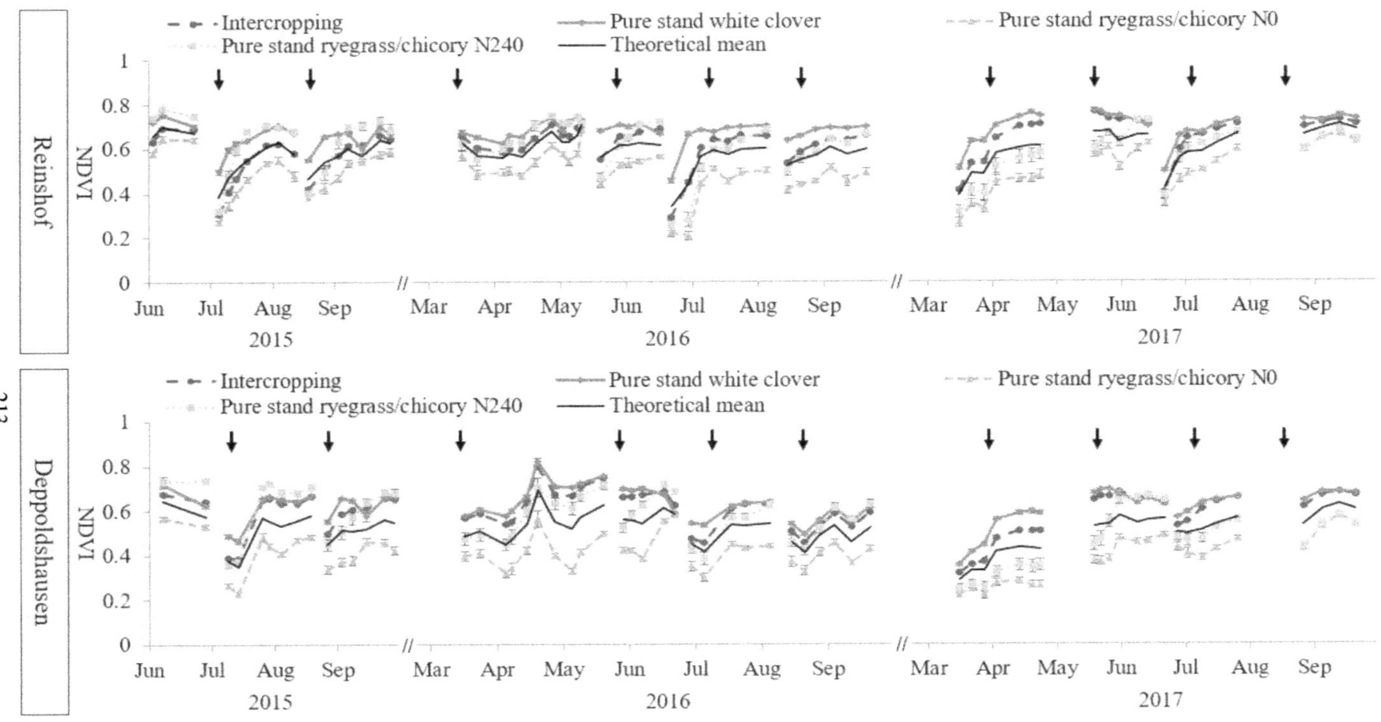

Fig.5: NDVI of crop stands in grassland in the years 2015 to 2017 at the field sites Reinshof and Deppoldshausen. Interruptions in the lines indicate biomass harvests. Means were calculated over all investigated genotypes of white clover as well as all stands of perennial ryegrass, the binary mixture of perennial ryegrass & chicory and chicory. N0: no nitrogen fertilizer; N240: 240 kg N/ha, arrows indicate partial applications. Error bars indicate standard error.

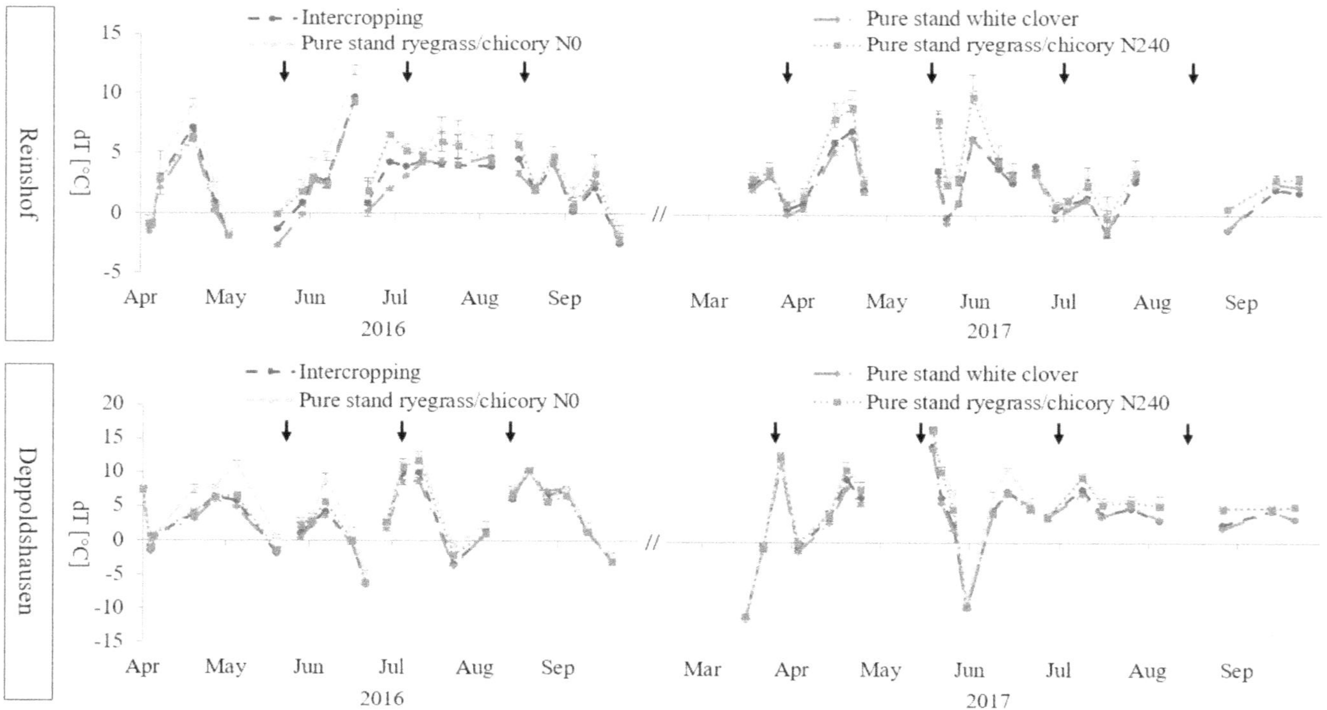

Fig.6: Temperature difference between canopy and air (dT) of crop stands in grassland in the years 2016 and 2017 at the field sites Reinshof and Deppoldshausen. Interruptions in the lines indicate biomass harvests. Means were calculated over all investigated genotypes of white clover as well as all stands of perennial ryegrass, the binary mixture of perennial ryegrass & chicory and chicory. N0: no nitrogen fertilizer; N240: 240 kg N/ha, arrows indicate partial applications. Error bars indicate standard error.

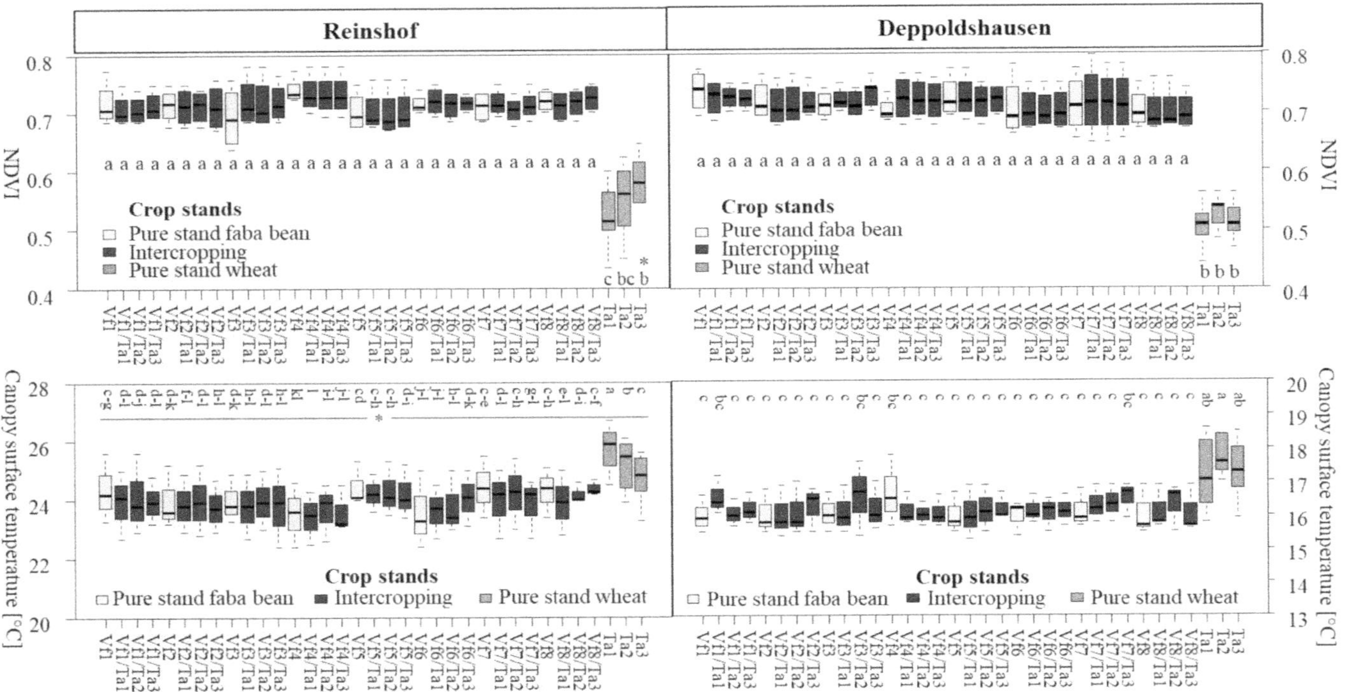

Fig.7: NDVI and canopy surface temperature of arable land on June 18th 2016 at the field sites Reinshof and Deppoldshausen. Crop stands include different genotypes of winter faba bean (Vf) and winter wheat (Ta). Different letters indicate significant differences at $p < 0.05$, LSD-test. Stars indicate significant differences of the treatments between Reinshof and Deppoldshausen.

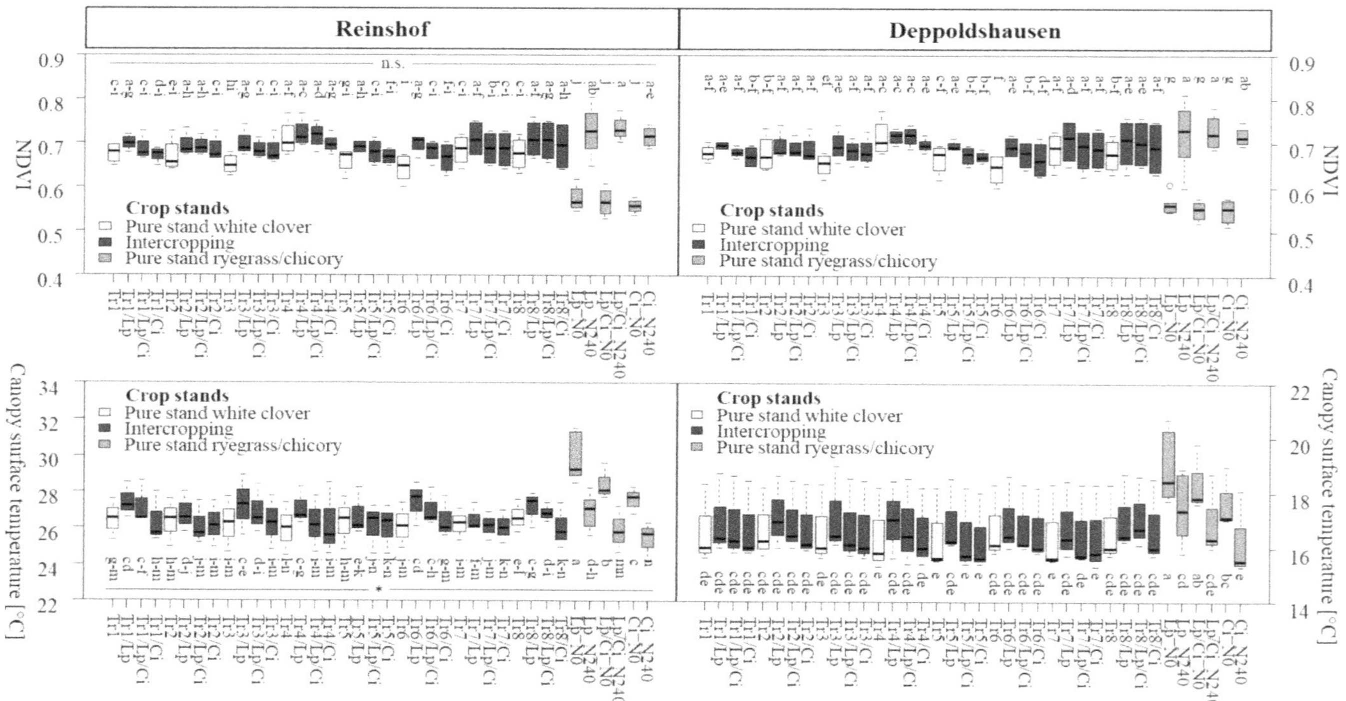

Fig.8: NDVI and canopy surface temperature of grassland on June 18th 2016 at the field sites Reinshof and Deppoldshausen. Crop stands include different genotypes of white clover (Tr) as well as the non-legumes perennial ryegrass (Lp), a binary mixture of perennial ryegrass/chicory (Lp/Ci) and chicory (Ci). N0: no nitrogen fertilizer, N240: 240 kg N/ha. Different letters indicate significant differences at $p < 0.05$, LSD-test. Stars indicate significant differences of the treatments between Reinshof and Deppoldshausen.

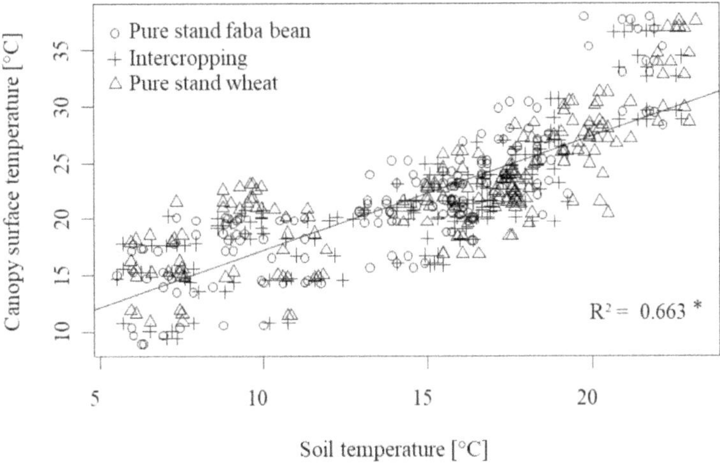

Fig.9: Linear relation between soil temperature and canopy surface temperature of arable crops. The star indicates significance of the correlation at $p < 0.05$ after Pearson product-moment correlation.

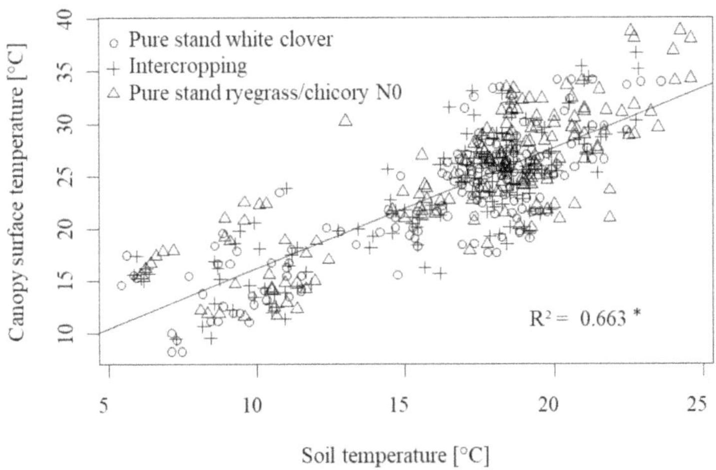

Fig.10: Linear relation between soil temperature and canopy surface temperature of grassland crops. Pure stands ryegrass/chicory contain either ryegrass, chicory or a mixture of both. N0: no nitrogen fertilizer. The star indicates significance of the correlation at $p < 0.05$ after Pearson product-moment correlation.

7 References

Abdelhamid, M.T., Palta, J.A., Veneklaas, E.J., Atkins, C., Turner, N.C., Siddique, K.H.M., 2011. Drying the surface soil reduces the nitrogen content of faba bean (Vicia faba L.) through a reduction in nitrogen fixation. Plant Soil 339, 351–362. https://doi.org/10.1007/s11104-010-0586-9

Asner, G.P., 1998. Biophysical and Biochemical Sources of Variability in Canopy Reflectance. Remote Sens. Environ. 64, 234–253. https://doi.org/10.1016/S0034-4257(98)00014-5

Bakhsh, A., Jaynes, D.B., Colvin, T.S., Kanwar, R.S., 2000. Spatio-temporal analysis of yield variability for a corn-soybean field in Iowa. Trans. ASAE 43, 31–38. https://doi.org/10.13031/2013.2684

Barkla, B.J., Rhodes, T., 2017. Use of infrared thermography for monitoring crassulacean acid metabolism. Funct. Plant Biol. 44, 46–51. https://doi.org/10.1071/FP16210

Behmann, J., Mahlein, A.-K., Rumpf, T., Römer, C., Plümer, L., 2015. A review of advanced machine learning methods for the detection of biotic stress in precision crop protection. Precis. Agric. 16, 239–260. https://doi.org/10.1007/s11119-014-9372-7

Bendig, J., Bolten, A., Bennertz, S., Broscheit, J., Eichfuss, S., Bareth, G., 2014. Estimating Biomass of Barley Using Crop Surface Models (CSMs) Derived from UAV-Based RGB Imaging. Remote Sens. 6, 10395–10412. https://doi.org/10.3390/rs61110395

Biewer, S., Fricke, T., Wachendorf, M., 2009. Determination of Dry Matter Yield from Legume–Grass Swards by Field Spectroscopy. Crop Sci. 49, 1927. https://doi.org/10.2135/cropsci2008.10.0608

Bilalis, D., Papastylianou, P., Konstantas, A., Patsiali, S., Karkanis, A., Efthimiadou, A., 2010. Weed-suppressive effects of maize–legume intercropping in organic farming. Int. J. Pest Manag. 56, 173–181. https://doi.org/10.1080/09670870903304471

Brouder, S.M., Volenec, J.J., 2008. Impact of climate change on crop nutrient and water use efficiencies. Physiol. Plant. 133, 705–724. https://doi.org/10.1111/j.1399-3054.2008.01136.x

Brueck, H., 2008. Effects of nitrogen supply on water-use efficiency of higher plants. J. Plant Nutr. Soil Sci. 171, 210–219. https://doi.org/10.1002/jpln.200700080

Bulson, H.A.J., Snaydon, R.W., Stopes, C.E., 1997. Effects of plant density on intercropped wheat and field beans in an organic farming system. J. Agric. Sci. 128, 59–71.

Candiago, S., Remondino, F., De Giglio, M., Dubbini, M., Gattelli, M., 2015. Evaluating Multispectral Images and Vegetation Indices for Precision Farming Applications from UAV Images. Remote Sens. 7, 4026–4047. https://doi.org/10.3390/rs70404026

Carlson, T.N., Ripley, D.A., 1997. On the relation between NDVI, fractional vegetation cover, and leaf area index. Remote Sens. Environ. 62, 241–252. https://doi.org/10.1016/S0034-4257(97)00104-1

Carlsson, G., Mårtensson, L.-M., Prade, T., Svensson, S.-E., Jensen, E.S., 2017. Perennial species mixtures for multifunctional production of biomass on marginal land. GCB Bioenergy 9, 191–201. https://doi.org/10.1111/gcbb.12373

Carter, G.A., Knapp, A.K., 2001. Leaf Optical Properties in Higher Plants: Linking Spectral Characteristics to Stress and Chlorophyll Concentration. Am. J. Bot. 88, 677. https://doi.org/10.2307/2657068

Chaerle, L., Hagenbeek, D., Vanrobaeys, X., Van Der Straeten, D., 2007. Early detection of nutrient and biotic stress in Phaseolus vulgaris. Int. J. Remote Sens. 28, 3479–3492. https://doi.org/10.1080/01431160601024259

Cranston, L., 2017. Herb and clover mixes for New Zealand sheep and cattle grazing systems. Livestock 22, 86–89. https://doi.org/10.12968/live.2017.22.2.86

Craparo, A.C.W., Steppe, K., Van Asten, P.J.A., Läderach, P., Jassogne, L.T.P., Grab, S.W., 2017. Application of thermography for monitoring stomatal conductance of Coffea arabica under different shading systems. Sci. Total Environ. 609, 755–763. https://doi.org/10.1016/j.scitotenv.2017.07.158

Finney, D.M., White, C.M., Kaye, J.P., 2016. Biomass Production and Carbon/Nitrogen Ratio Influence Ecosystem Services from Cover Crop Mixtures. Agron. J. 108, 39–52. https://doi.org/10.2134/agronj15.0182

Frison, E.A., Cherfas, J., Hodgkin, T., 2011. Agricultural Biodiversity Is Essential for a Sustainable Improvement in Food and Nutrition Security. Sustainability 3, 238–253. https://doi.org/10.3390/su3010238

Gerhards, M., Rock, G., Schlerf, M., Udelhoven, T., 2017. Water Stress Detection using Temperature, Emissivity, and Reflectance. Presented at the EGU General Assembly Conference Abstracts, p. 14906.

Hauggaard-Nielsen, H., Ambus, P., Jensen, E.S., 2001. Interspecific competition, N use and interference with weeds in pea–barley intercropping. Field Crops Res. 70, 101–109. https://doi.org/10.1016/S0378-4290(01)00126-5

Haymes, R., Lee, H.., 1999. Competition between autumn and spring planted grain intercrops of wheat (Triticum aestivum) and field bean (Vicia faba). Field Crops Res. 62, 167–176. https://doi.org/10.1016/S0378-4290(99)00016-7

Høgh-Jensen, H., Schjoerring, J.K., 1997. Interactions between white clover and ryegrass under contrasting nitrogen availability: N2 fixation, N fertilizer recovery, N transfer and water use efficiency. Plant and Soil 187–199.

Hou, W., Gao, J., Wu, S., Dai, E., 2015. Interannual Variations in Growing-Season NDVI and Its Correlation with Climate Variables in the Southwestern Karst Region of China. Remote Sens. 7, 11105–11124. https://doi.org/10.3390/rs70911105

Ivushkin, K., Bartholomeus, H., Bregt, A.K., Pulatov, A., Bui, E.N., Wilford, J., 2017. Infrared thermal remote sensing for soil salinity assessment on landscape scale. Presented at the EGU General Assembly Conference Abstracts, p. 11888.

Jensen, E.S., Peoples, M.B., Boddey, R.M., Gresshoff, P.M., Hauggaard-Nielsen, H., J.R. Alves, B., Morrison, M.J., 2012. Legumes for mitigation of climate change and the provision of feedstock for biofuels and biorefineries. A review. Agron. Sustain. Dev. 32, 329–364. https://doi.org/10.1007/s13593-011-0056-7

Jensen, E.S., Peoples, M.B., Hauggaard-Nielsen, H., 2010. Faba bean in cropping systems. Field Crops Res., Faba Beans in Sustainable Agriculture 115, 203–216. https://doi.org/10.1016/j.fcr.2009.10.008

Karnieli, A., Agam, N., Pinker, R.T., Anderson, M., Imhoff, M.L., Gutman, G.G., Panov, N., Goldberg, A., 2010. Use of NDVI and Land Surface Temperature for Drought Assessment: Merits and Limitations. J. Clim. 23, 618–633. https://doi.org/10.1175/2009JCLI2900.1

Khan, H.R., Link, W., Hocking, T.J., Stoddard, F.L., 2007. Evaluation of physiological traits for improving drought tolerance in faba bean (Vicia faba L.). Plant Soil 292, 205–217. https://doi.org/10.1007/s11104-007-9217-5

Khan, H.R., Paull, J.G., Siddique, K.H.M., Stoddard, F.L., 2010. Faba bean breeding for drought-affected environments: A physiological and agronomic perspective. Field Crops Res. 115, 279–286. https://doi.org/10.1016/j.fcr.2009.09.003

Lenth, R.V., 2018. emmeans: Estimated Marginal Means, aka Least-Squares Means.

Lenthe, J.-H., Oerke, E.-C., Dehne, H.-W., 2007. Digital infrared thermography for monitoring canopy health of wheat. Precis. Agric. 8, 15–26. https://doi.org/10.1007/s11119-006-9025-6

Leucker, M., Wahabzada, M., Kersting, K., Peter, M., Beyer, W., Steiner, U., Mahlein, A.-K., Oerke, E.-C., 2017. Hyperspectral imaging reveals the effect of sugar beet quantitative trait loci on Cercospora leaf spot resistance. Funct. Plant Biol. 44, 1–9. https://doi.org/10.1071/FP16121

Li, F., Mistele, B., Hu, Y., Chen, X., Schmidhalter, U., 2014. Reflectance estimation of canopy nitrogen content in winter wheat using optimised hyperspectral spectral indices and partial least squares regression. Eur. J. Agron. 52, 198–209. https://doi.org/10.1016/j.eja.2013.09.006

Lindenthal, M., Steiner, U., Dehne, H.-W., Oerke, E.-C., 2005. Effect of Downy Mildew Development on Transpiration of Cucumber Leaves Visualized by Digital Infrared Thermography. Phytopathology 95, 233–240. https://doi.org/10.1094/PHYTO-95-0233

Lüscher, A., Mueller-Harvey, I., Soussana, J.F., Rees, R.M., Peyraud, J.L., 2014. Potential of legume-based grassland-livestock systems in Europe: a review. Grass Forage Sci. 69, 206–228. https://doi.org/10.1111/gfs.12124

Mahlein, A.-K., 2016. Plant Disease Detection by Imaging Sensors – Parallels and Specific Demands for Precision Agriculture and Plant Phenotyping. Plant Dis. 100, 241–251. https://doi.org/10.1094/PDIS-03-15-0340-FE

Marshall, A.H., Collins, R.P., Humphreys, M.W., Scullion, J., 2016. A new emphasis on root traits for perennial grass and legume varieties with environmental and ecological benefits. Food Energy Secur. 5, 26–39. https://doi.org/10.1002/fes3.78

Meng, L., Zhang, A., Wang, F., Han, X., Wang, D., Li, S., 2015. Arbuscular mycorrhizal fungi and rhizobium facilitate nitrogen uptake and transfer in soybean/maize intercropping system. Front. Plant Sci. 6. https://doi.org/10.3389/fpls.2015.00339

Möckel, T., Dalmayne, J., Schmid, B., Prentice, H., Hall, K., 2016. Airborne Hyperspectral Data Predict Fine-Scale Plant Species Diversity in Grazed Dry Grasslands. Remote Sens. 8, 133. https://doi.org/10.3390/rs8020133

Mulla, D.J., 2013. Twenty five years of remote sensing in precision agriculture: Key advances and remaining knowledge gaps. Biosyst. Eng., Special Issue: Sensing Technologies for Sustainable Agriculture 114, 358–371. https://doi.org/10.1016/j.biosystemseng.2012.08.009

Mullan, D.J., Reynolds, M.P., 2010. Quantifying genetic effects of ground cover on soil water evaporation using digital imaging. Funct. Plant Biol. 37, 703. https://doi.org/10.1071/FP09277

Negin, B., Moshelion, M., 2017. The advantages of functional phenotyping in pre-field screening for drought-tolerant crops. Funct. Plant Biol. 44, 107. https://doi.org/10.1071/FP16156

Nyfeler, D., Huguenin-Elie, O., Suter, M., Frossard, E., Connolly, J., Lüscher, A., 2009. Strong mixture effects among four species in fertilized agricultural grassland led to persistent and consistent transgressive overyielding. J. Appl. Ecol. 46, 683–691. https://doi.org/10.1111/j.1365-2664.2009.01653.x

Ollinger, S.V., 2011. Sources of variability in canopy reflectance and the convergent properties of plants: Tansley review. New Phytol. 189, 375–394. https://doi.org/10.1111/j.1469-8137.2010.03536.x

Ortiz-Bustos, C.M., Pérez-Bueno, M.L., Barón, M., Molinero-Ruiz, L., 2017. Use of Blue-Green Fluorescence and Thermal Imaging in the Early Detection of Sunflower Infection by the Root Parasitic Weed Orobanche cumana Wallr. Front. Plant Sci. 8. https://doi.org/10.3389/fpls.2017.00833

Pavuluri, K., Chim, B.K., Griffey, C.A., Reiter, M.S., Balota, M., Thomason, W.E., 2015. Canopy spectral reflectance can predict grain nitrogen use efficiency in soft red winter wheat. Precis. Agric. 16, 405–424. https://doi.org/10.1007/s11119-014-9385-2

Paynel, F., Cliquet, J.B., 2003. N transfer from white clover to perennial ryegrass, via exudation of nitrogenous compounds. Agronomie 23, 503–510. https://doi.org/10.1051/agro:2003022

Peoples, M.B., Brockwell, J., Herridge, D.F., Rochester, I.J., Alves, B.J.R., Urquiaga, S., Boddey, R.M., Dakora, F.D., Bhattarai, S., Maskey, S.L., Sampet, C., Rerkasem, B., Khan, D.F., Hauggaard-Nielsen, H., Jensen, E.S., 2009. The contributions of nitrogen-fixing crop legumes to the productivity of agricultural systems. Symbiosis 1–17.

Pinheiro, J., Bates, D., DebRoy, S., Sarkar, D., R Core Team, 2018. nlme: Linear and Nonlinear Mixed Effects Models.

Pinter, P.J., Jr., Hatfield, J.L., Schepers, J.S., Barnes, E.M., Moran, M.S., Daughtry, C.S.T., Upchurch, D.R., 2003. Remote Sensing for Crop Management. Photogramm. Eng. Remote Sens. 69, 647–664. https://doi.org/10.14358/PERS.69.6.647

R Core Team, 2018. R: A language and environment for statistical computing.

Reckleben, Y., 2014. Sensoren für die Stickstoffdüngung–Erfahrungen in 12 Jahren praktischem Einsatz. J. Kult. 66, 42–47.

Reckling, M., Bergkvist, G., Watson, C.A., Stoddard, F.L., Zander, P.M., Walker, R.L., Pristeri, A., Toncea, I., Bachinger, J., 2016. Trade-Offs between Economic and Environmental Impacts of Introducing Legumes into Cropping Systems. Front. Plant Sci. 7. https://doi.org/10.3389/fpls.2016.00669

Reckling, M., Döring, T.F., Stein-Bachinger, K., Bloch, R., Bachinger, J., 2015. Yield stability of grain legumes in an organically managed monitoring experiment. Aspects of Applied Biology 57–61. https://doi.org/10.13140/rg.2.1.1122.0966

Rembold, F., Atzberger, C., Savin, I., Rojas, O., 2013. Using Low Resolution Satellite Imagery for Yield Prediction and Yield Anomaly Detection. Remote Sens. 5, 1704–1733. https://doi.org/10.3390/rs5041704

Rochon, J.J., Doyle, C.J., Greef, J.M., Hopkins, A., Molle, G., Sitzia, M., Scholefield, D., Smith, C.J., 2004. Grazing legumes in Europe: a review of their status, management, benefits, research needs and future prospects. Grass Forage Sci. 59, 197–214. https://doi.org/10.1111/j.1365-2494.2004.00423.x

Romano, G., Zia, S., Spreer, W., Sanchez, C., Cairns, J., Araus, J.L., Müller, J., 2011. Use of thermography for high throughput phenotyping of tropical maize adaptation in water stress. Comput. Electron. Agric. 79, 67–74. https://doi.org/10.1016/j.compag.2011.08.011

Siebrecht-Schöll, D.J., 2019. Breeding analysis of eight winter faba bean genotypes for mixed cropping with winter wheat. Georg-August-Universität Göttingen, Göttingen. http://hdl.handle.net/21.11130/00-1735-0000-0005-128E-7

Skinner, R.H., 2008. Yield, Root Growth, and Soil Water Content in Drought-Stressed Pasture Mixtures Containing Chicory. Crop Sci. 48, 380. https://doi.org/10.2135/cropsci2007.04.0201

Smigaj, M., Gaulton, R., Suarez, J.C., Barr, S.L., 2017. Use of Miniature Thermal Cameras for Detection of Physiological Stress in Conifers. Remote Sens. 9, 957. https://doi.org/10.3390/rs9090957

Stein, S., Steinmann, H.-H., 2018. Identifying crop rotation practice by the typification of crop sequence patterns for arable farming systems – A case study from Central Europe. Eur. J. Agron. 92, 30–40. https://doi.org/10.1016/j.eja.2017.09.010

Suter, M., Connolly, J., Finn, J.A., Loges, R., Kirwan, L., Sebastià, M.-T., Lüscher, A., 2015. Nitrogen yield advantage from grass–legume mixtures is robust over a wide range of legume proportions and environmental conditions. Glob. Change Biol. 21, 2424–2438. https://doi.org/10.1111/gcb.12880

Thomas, S., Mahlein, A.-K., Kuska, M.T., Rascher, U., Wahabzada, M., 2017. Observation of plant–pathogen interaction by simultaneous hyperspectral imaging reflection and transmission measurements (No. FZJ-2017-01496). Pflanzenwissenschaften. https://doi.org/10.1071/FP16127

Tucker, C.J., Holben, B.N., Elgin, J.H., McMurtrey, J.E., 1981. Remote sensing of total dry-matter accumulation in winter wheat. Remote Sens. Environ. 11, 171–189. https://doi.org/10.1016/0034-4257(81)90018-3

Vaesen, K., Gilliams, S., Nackaerts, K., Coppin, P., 2001. Ground-measured spectral signatures as indicators of ground cover and leaf area index: the case of paddy rice. Field Crops Res. 69, 13–25. https://doi.org/10.1016/S0378-4290(00)00129-5

Watson, C.A., Reckling, M., Preissel, S., Bachinger, J., Bergkvist, G., Kuhlman, T., Lindström, K., Nemecek, T., Topp, C.F.E., Vanhatalo, A., Zander, P., Murphy-Bokern, D., Stoddard, F.L., 2017. Grain Legume Production and Use in European Agricultural Systems, in: Advances in Agronomy. Elsevier, pp. 235–303. https://doi.org/10.1016/bs.agron.2017.03.003

Weigelt, A., Weisser, W.W., Buchmann, N., Scherer-Lorenzen, M., 2009. Biodiversity for multifunctional grasslands: equal productivity in high-diversity low-input and low-diversity high-input systems. Biogeosciences 6, 1695–1706. https://doi.org/10.5194/bg-6-1695-2009

Weißhuhn, P., Reckling, M., Stachow, U., Wiggering, H., 2017. Supporting Agricultural Ecosystem Services through the Integration of Perennial Polycultures into Crop Rotations. Sustainability 9, 2267. https://doi.org/10.3390/su9122267

Xiao, Y., Li, L., Zhang, F., 2004. Effect of root contact on interspecific competition and N transfer between wheat and fababean using direct and indirect 15N techniques. Plant Soil 262, 45–54. https://doi.org/10.1023/B:PLSO.0000037019.34719.0d

Zampieri, M., Ceglar, A., Dentener, F., Toreti, A., 2017. Wheat yield loss attributable to heat waves, drought and water excess at the global, national and subnational scales. Environ. Res. Lett. 12, 064008. https://doi.org/10.1088/1748-9326/aa723b

Chapter 6: Epilogue

Water availability is an ever-changing factor in crop growth conditions and water resources in soil vary on a daily basis (Boyer and Westgate, 2004). In addition to the predicted shifts in global climate, year-to-year variations in precipitation patterns will lead to extreme weather events even under comparably stable climatic conditions (Sun et al., 2018). Consequently, seasons with limited water resources will endanger productivity and thus yield stability of agricultural systems. These less favorable growth conditions may induce higher competition in legume-based intercropping. However, the productivity of intercropping can still be comparably higher than that of pure stands (Zając et al., 2013). Within these systems, the partitioning of water resources is complex, resulting in both increasing and decreasing effects on the water use efficiency (WUE; Ren et al., 2016). In this context, the concise findings of this thesis regarding WUE of intercropping are discussed in the following chapter.

6.1 Intercropping with arable crops

By using spectral and thermal screening of crop stands in the field, we showed that intercropping of winter faba bean and winter wheat has positive effects on the vegetation development (chapter 5). Furthermore, we found that intercropping of these species has the potential of improving WUE under water deficit conditions (chapter 2.1). This was, however, dependent on the combined effects of genotypic responses and different cropping systems. In agricultural practice, the appropriate choice of variety has therefore to be made considering the cropping system and drought expectations. This is of particular importance for intercropping with winter forms of faba bean and wheat. Autumn-sown faba bean is often negatively affected by frost and drought events leading to decreased yield advantages compared to spring-sown faba bean (Neugschwandtner et al., 2015). Acreage of faba bean has declined in Europe over the last decades due to low and instable yields of available varieties (Rubiales and Mikic, 2015; Reckling et al., 2016b). Here, breeding of suitable winter faba bean varieties with high WUE in intercropping is necessary and could be based on the methods applied in chapter 2.1.

One important factor in improving the water use efficiency of a crop stand is to minimize evaporation from the soil (Brueck, 2008; Sadras and Angus, 2006). This can be achieved in intercropping systems with legumes due to higher soil coverage (Bilalis et al., 2010). On the contrary, it was shown in chapter 3 that evapotranspiration and thus the amount of water used by intercropped stands was increased. Nevertheless, WUE of intercropping was similar to pure legumes as the CO_2 assimilation was also enhanced, displaying no negative effect of intercropping. This is in line with a study of Caviglia et al. (2004), who observed wheat and soybean intercropping to use greater amounts of the available water than the respective pure stands. Due to a similarly increased radiation use efficiency, intercropping was able to take up bigger portions of the available water and translate it into biomass. This strategy would be rather water spending but results in highest yields under dry conditions unless soil water resources are depleted. Accordingly, the target strategy is to increase productivity at the same level of water use instead of crops that save water but produce less.

Additionally, intercropping of winter faba bean and winter wheat proved to be more efficient in using nitrogen (N) resources. It could be shown in chapter 4 that nitrous oxide (N_2O) emissions are lower when compared to crop stands with higher N availability by either fertilizer application or biological N_2 fixation. Despite the lower N availability, results of net ecosystem exchange of CO_2 (chapter 3) and NDVI (chapter 5) indicated a comparably high productivity at simultaneously reduced N_2O emissions in the mixed cropping field plots. This can be explained by the preferential use of soil mineral N by winter faba bean (Schwenke et al., 2015) and the N transfer of biologically fixed N to winter wheat (Xiao et al., 2004). Nitrogen losses are usually most affected by soil conditions and soil management, showing similar effects in cropping systems with and without legumes (Reckling et al., 2016a). However, intercropping of legumes and non-legumes, as shown in this thesis, can reduce N losses due to the complementary use of soil mineral N and biologically fixed N.

6.2 Intercropping with grassland crops

Like in the arable system, intercropped white clover with perennial ryegrass increased the water use as estimated by thermal images, both near the ground and remotely assessed (chapter 3 and 5). The WUE was enhanced compared to unfertilized non-legumes but did not exceed the WUE of pure white cover. In contrast to a study by Angus et al. (2001), who stated that clover-based pastures use water resources to a lesser extent than other crops, we found that pure white clover used highest amounts of water. Nevertheless, white clover roots lead to more soil aggregation than perennial ryegrass which promotes the water storage ability of soil as well as hydraulic conductivity (Marshall et al. 2016). This circumstance might have led to the observed carbon isotope discrimination ($\delta^{13}C$) in chapter 3. Using this method, WUE of pure white clover was highest which might be partly due to limited gas exchange via stomata. The difference between the latter and other methods such as thermal imaging might derive from different spatial and temporal scales measured (Jákli et al., 2018).

Deep rooting species like chicory and faba bean are able to acquire more water resources from deeper soil layers than white clover and perennial ryegrass (Weißhuhn et al., 2017). Chicory is therefore able to maintain transpiration within first periods of water deficit (Vandoorne et al., 2012). Accordingly, we could show that including chicory in grassland systems has a beneficial effect on the use of water resources (chapter 3). Therefore, the species mixed with white clover in intercropped stands had greater influence on the performance of the system than the white clover genotype (chapter 5). Reasons are reviewed in an article of Díaz and Cabido (2001), who concluded that positive influences of intercropping systems are not determined by the number of species but rather by their traits and trait expressions. In grassland species, root systems are less pronounced at high harvestable biomass production aboveground (Bessler et al., 2009). In this situation, an additional species like chicory results in higher tolerance of the cropping system towards water limitations. Adding the ability of water transfer from chicory to accompanying crops (Skinner, 2008), intercropping systems including chicory have the properties to use water resources efficiently at a high productivity level of the system. Here, N fertilizer applied to non-legumes only shows advantages over these intercropped stands as well as pure stands of white clover if the weather conditions of the respective year are favorable (chapter 5). From our present results, it can therefore be concluded that intercropping, though rather water demanding, improves the overall performance of the agro-ecosystem by enhanced water acquisition and higher productivity.

6.3 Drought stress tolerance

The relation between water use efficiency and drought stress tolerance is to save water without compromising on yield (Farooq et al., 2009). With regard to drought stress tolerance, several parameters such as gas exchange via stomata, WUE, RWC and carbon isotope discrimination are studied in the literature as reviewed by Negin and Moshelion (2017). Several of them were investigated in the present research to achieve a holistic impression on the dynamics of water use in intercropping systems. However, the interpretation of the results in terms of positive or negative effects on the water use of the cropping system has to be clarified.

There are several contradictory definitions of drought stress tolerance in literature. Drought tolerance in faba bean can be either characterized by low plant height, small thick leaves and few small stomata (Amede et al., 1999; Link et al., 1999) or by tall growth, large leaves, high transpiration per leaf area, increased root length and root surface (Nerkar et al., 1981; Belachew et al., 2018). Some authors report negative correlations between yield and drought tolerance of faba bean varieties (Abdellatif et al., 2012). In other studies, maintenance of stable and high yields under water deficit through constantly high yield parameters are stated as most important aspects of drought tolerance (Mwanamwenge et al., 1999). Accordingly, there exist two contrasting definitions of drought tolerance: (i) water saving mechanisms, which are induced by fluctuations in soil water potential (Senbayram et al., 2015) and (ii) maintained water use for yield production even under water limitations. In both situations, efficient use of available water resources is of major importance.

Nevertheless, increased WUE under water limiting conditions, e.g. by stomatal adjustment or root growth, is often accompanied by reduced yield production (Tambussi et al., 2007). This is a constant challenge of plant breeding for water limited environments. Reducing the amount of unproductive water losses, such as evaporation from soil, is therefore an opportunity to increase WUE without compromising on yield (Tambussi et al., 2007). To achieve this, early vigor and fast initial growth is crucial for soil cover. This especially accounts for intercropping systems such as mixtures of maize and common bean (Walker and Ogindo, 2003). Here, the authors found lower evaporation due to soil cover but higher transpiration rates of the intercropped treatments, which is in line with the findings of chapter 3.

The best strategy to survive drought events is also determined by the duration of the water deficit period. This aspect also affected the associated bacterial community (chapter 2.2). Intensive water use even under water limited conditions is often acceptable as severe droughts happen rarely in European climates and water deficit usually occurs in a range where plant survival is not endangered (Tardieu, 1996). Therefore, Negin and Moshelion

(2017) propose the term drought tolerance as dynamic 'risk management' to optimize yield parameters.

After water limited periods, it is crucial that plants are able to fully recover their physiological activities like carbon fixation (Zarafshar et al., 2014). The ability to recover after drought stress is also called the resilience of a system (Negin and Moshelion, 2017). The extent and rate of recovery in gas exchange for photosynthesis and transpiration determine the yield production after drought. In the greenhouse experiment on water deficit (chapter 2.1), it could be shown that both tested genotypes of winter faba bean were able to fully recover at re-irrigation in the investigated parameters. Nonetheless, there were clear differences between genotypes and crop stands during the phase of water limitation, indicating that different characteristics are reasonable to account for a drought tolerant cultivar.

At early stages of water deficit, faba bean reduce stomatal conductance and net photosynthesis, although water is still available in deeper rooting zones (Abdelhamid et al., 2011). This results in reduced root, nodule and shoot biomass, indicating yield loss even under mild water limiting conditions. Reductions in tap root length and lateral root length however depend on genotype x treatment interactions (Belachew et al., 2018). Therefore, faba bean need to activate mechanisms to expand the rooting system under developing water deficit to capture available water resources. Stomatal characteristics can then be evaluated by leaf temperature and carbon isotope discrimination (Khan et al., 2010) as demonstrated in chapter 3. The WUE is increased when transpiration per carbon assimilation is reduced. Some water loss, however, needs to be accepted in order to maintain stomatal conductance for photosynthetic processes and biomass production, which indicates yield stability even under drought conditions (Serraj and Sinclair, 2002).

Stomatal adjustment is related to turgor pressure and several osmotically active solutes (Bajji et al., 2001). Increases in such solutes are usually considered as drought stress responses. In contrast to this most widespread opinion, it was concluded in chapter 2.1 that an accumulation of proline rather acts as stress signal and is not accumulated in drought tolerant plants. This corresponds to findings of Amede et al. (2003), who proved the hypothesis, that a drought-induced solute accumulation is not particularly a consequence of high drought tolerance. Similarly, Serraj and Sinclair (2002) reviewed in their article that an accumulation of osmolytes does not necessarily lead to better drought adaptation. Instead, it is a general concentration effect. Thus, selection in terms of drought stress should not only be based on avoidance of cell dehydration but rather on an active metabolism, continued growth, and yield formation (Tardieu, 1996).

6.4 Breeding for intercropping

The overall aim of crop breeding for drought tolerance is to create varieties which are able to withstand unfavorable conditions in terms of limited water resources throughout the entire vegetation period (Joshi et al., 2016), resulting in stable yields. We could observe in the field experiments, that this aspect is very important as all three investigated years (2015-2017) had different environmental conditions with high variability in arable crop yields (chapter 3; Siebrecht-Schöll, 2019). Breeding for enhanced drought tolerance and high water use efficiency however is usually accompanied by reductions in yield as current cultivars realize high productivity by the intensive use of water resources (Negin and Moshelion, 2017). Thus, as discussed above, selection of varieties demands attention on the maintenance of productivity, supported by the efficient use of available water. This was realized over the last decades in Australian wheat varieties where water use efficiency could be enhanced, driven by increased yields at similar evapotranspiration levels (Sadras and Lawson, 2013). In this context, intercropping of species can develop its advantages if breeding for highly productive and simultaneously drought tolerant cultivars is conducted in intercropped stands.

6.4.1 Genotypes of winter faba bean

Similar to our findings in the greenhouse experiment (chapter 2.1), Mwanamwenge et al. (1998) found a faba bean genotype which they characterized as drought tolerant due to its ability of highest yield production even under water limitations. The genotype was early flowering, similar to the genotypes S_062 and WAB-Fam157 used in our field experiments. Genotype S_062 accordingly proved to be drought tolerant in pure stands in the pot experiment. Nevertheless, the better suitable winter faba bean genotype for intercropping was S_004. The different performance of S_004 and S_062 in intercropping highlights, that mixtures of faba bean and wheat can also result in competitive situations with less yield production (Fan et al., 2006). It is therefore of major importance to choose varieties with facilitating abilities and overall good performance in intercropping systems.

These different responses of the genotypes could not be verified in the field under natural conditions. There, the faba bean genotypes S_004 and S_062 both showed highest water use estimated by carbon isotope discrimination independent of the crop stand (Tab.6.1), whereas the water use estimated by remote sensing was intermediate. The NDVI did not reveal differences between winter faba bean genotypes (chapter 5) in contrast to other methods such as $\delta^{13}C$ (chapter 3) and WUE in the greenhouse (chapter 2.1). In this case, other remote sensing techniques might be more sensitive to small differences between genotypes, providing additional information to detections by the breeder's eye (Mullan and Reynolds, 2010). Furthermore, the spectral reflectance of plants is affected by size and orientation of the leaves as well as their spatial distribution within the canopy (Ollinger, 2011). Diverse leaf size and plant height of the investigated winter faba bean genotypes probably led to the observed differences. Moreover, the relation between carbon isotope fractionation and grain yield of crops is complex and highly depending on the environment (Tambussi et al., 2007).

Among winter faba bean genotypes, S_265 could be characterized as high transpiring both by remote and physiological methods. In contrast, Hiverna/2 and WAB-Fam157 showed comparatively low transpiration. These differences were most pronounced at the fertile study site Reinshof. Obviously, genotypic variability did not appear at the marginal study site Deppoldshausen. This shows that, regardless of the crop stand, winter faba bean can realize its potential only at fertile sites, resulting in less importance for the choice of variety at marginal sites. A reason for that might be that none of the genotypes at Deppoldshausen was able to reach its yield potential due to an overall reduced production under these growth conditions with less water availability (Link et al., 1999). In contrast at Reinshof, genotypic performance split up according to the yield potential. Thus genotype x environment interactions have to be taken into account in breeding processes (Link et al., 1996). In order to make use of genotypic variability, breeders need to test promising lines at suitable field sites with favorable growth conditions.

Tab.6.1: Evaluation of the water use of eight winter faba bean (Vf) genotypes over both crop stands regarding results from the difference between canopy and air temperature (dT) via remote sensing and carbon isotope discrimination ($\delta^{13}C$).

Data derived from remote sensing	Data derived from carbon isotope discrimination: high water use	intermediate	low water use	
high water use	Vf4: S_265	Vf6: Côte d'Or/1		low dT at Reinshof --> high transpiration; NDVI was similar
intermediate	Vf1: S_004 Vf2: S_062	Vf3: S_069 Vf8: WAB-EP98-267		
low water use			Vf5: Hiverna/2 Vf7: WAB-Fam157	high dT at Reinshof --> low transpiration; NDVI was similar
	low $\delta^{13}C$ at both sites --> Stomata open		high $\delta^{13}C$ at both sites --> Stomata closed	

Selection should aim for an active metabolism with higher transpiration rates as displayed by genotype S_265. These characteristics are desirable for a potential higher grain yield. Consequently, this genotype of winter faba bean is recommended for breeding and should be further evaluated under greenhouse conditions. Information from greenhouse and field conditions need then to be consolidated with yield parameters to get comprehensive information about the performance of the crop stand with this genotype.

6.4.2 Genotypes of white clover

For white clover genotypes, the classification was strongly dependent on the method applied for water use estimations. While carbon isotope discrimination as well as canopy temperature did not reveal significant variation between the genotypes of white clover, only by NDVI on June 18th 2016 the genotypes could be differentiated (Tab.6.2). Based on these results, the white clover genotype EGB PX 90710 showed the best performance, while most of the other genotypes ranged on similar intermediate levels. Three genotypes can be characterized as inferior when pure stands are regarded only. Their intercropping range on higher, hence intermediate levels. This differentiation was possible at both study sites. In contrast to winter faba bean, there was no intensification in genotypic variability at the fertile study site Reinshof. Either the investigated genotypes do not show variability in their water use due to low genetic distance or the applied methods were not suitable to detect such intraspecific variation. A reason might be that genotypic variation appeared during winter (Ledgard et al., 1990) which was not covered in our period of measurements. Furthermore, white clover is in contrast to ryegrass generally able to realize its photosynthetic potential due to the positioning of young leaves near the top of the sward (Dennis and Woledge, 1982). This effect might mask genotypic variability due to an overall good performance of white clover in intercropping.

Tab.6.2: Evaluation of the productivity of eight white clover (Tr) genotypes over both crop stands regarding results from the NDVI via remote sensing. Genotypes in grey color were categorized based on the performance in pure stands.

Data derived from remote sensing		
high NDVI at both sites	intermediate	low NDVI at both sites
Tr4: EGB PX 90710	Tr1: EGB PX 90305 Tr2: EGB PX 90312 Tr7: EGB PX 90915 Tr8: EGB PX 90909	Tr3: EGB PX 90702 Tr5: EGB PX 90913 Tr6: EGB PX 90914

6.5 Technical aspects

The complex interactions of the component crops in permanently established grassland systems led to difficulties in obtaining reliable results by remote sensing approaches as well as ground-truth methods as genotypic variation was low. Furthermore, thermography and spectral reflectance are limited to general evaluations of the plant status with regard to transpiration and biomass production, and thus cannot differentiate between resource limitations (Lenthe et al., 2007). We observed high variability between measurement dates instead and some block effect at Deppoldshausen, especially in the canopy surface temperature. Zhao et al. (2017) also stated that different flights of airborne vehicles achieved different NDVI values within a day. The authors therefore propose an effect of solar motion on data stability. Consequently, we conclude that more detailed investigations need to be done regarding daily fluctuations in spectral patterns of canopies.

The combination of the applied ground-truth and remote techniques with other sensor systems might further improve the knowledge of genotype-specific performances. The obtained results already showed that information on the physiological level can be improved by airborne canopy monitoring as remote approaches can reduce errors derived from small scale differences within the canopy. Higher developed techniques, such as hyperspectral imaging, have the potential of highly specified detection of impaired biomass production under unfavorable growth conditions such as drought stress (Römer et al., 2012). With those techniques, differences between white clover genotypes might be visible. In literature, there are different opinions regarding the number of spectral bands needed for accurate estimation of biomass, ranging from four to up to 22 spectral wavebands (Psomas et al., 2011; Thenkabail et al., 2004). Compared to results based on vegetation indices, better results may be achieved with multi- or hyperspectral image analysis. The application of indices based on e.g. the red-edge reflection provides useful information about the N status of plants (Morier et al., 2015), which can be used to identify N usage and transfer from legumes to non-legumes in intercropping systems. However, yet it is still a challenge to manage the quantity of data that is easily generated by novel remote sensing techniques and to relate these data to the diverse characteristics of plant physiology (Mahlein, 2016). Beside the demand for automated and professional analysis of big data sets, these advanced technologies will support breeding processes as well as sustainable, resource- and cost-efficient agriculture in the future.

6.6 Concluding remarks

The recent climatic variability demands for resilient cropping systems and specific drought adapted varieties (Frison et al., 2011). The results of this thesis suggest that intercropping of legumes and non-legumes could enhance the water use efficiency compared to conventional low-input systems. However, due to the complexity of the system containing different species, there was variability in the responses of the crop stands which arises the need for more detailed investigations.

Apart from the discussed advantages, challenges of intercropping systems such as nutrient supply and weed control exist (Malézieux et al. 2009; Weißhuhn et al. 2017). The adoption of intercropping by farmers is to date still restricted by technical barriers (Lemken et al., 2017). However, it was shown by Bulson et al. (1997) that intercropping of faba bean and wheat can be harvested with a conventional combine harvester and biomass harvests in grasslands neither demand for special machineries. The harvested goods of faba bean-wheat intercrops can easily be separated by sieving due to different grain size (Haymes and Lee, 1999). Nevertheless, farmers usually respond in their management practices rather to subsidies than to environmental needs (Auerswald et al., 2018).

Future challenges are therefore not only to improve the efficiency of production systems and to deal with environmental alterations in terms of climate and diseases (Cazenave, 2018), but also to set up supportive politics to raise the attractiveness of sustainable agricultural systems. Then, legumes such as winter faba bean can become key components in future cropping systems when the yield variability will be stabilized (Jensen et al., 2010). Hence, intercropping systems and breeding for efficient and tolerant species for these specific environments will lead to robust cropping systems with stable productivity. But as there are more questions to address, the complexity of those systems requests combined approaches, as within the IMPAC^3 project, in order to answer research questions in a holistic kind of way (Cazenave, 2018).

In closing, interdisciplinary approaches will deepen our understanding on complex interactions of intra- and interspecific competition and facilitation as well as environmental impacts. This knowledge will foster breeding strategies for resource efficient cultivars adapted to interactions in intercropping systems. Then, and if technical barriers are overcome, intercropping is a promising management strategy for farmers who aim at high production rates and efficient resource use in low-input agriculture.

References

Abdelhamid, M.T., Palta, J.A., Veneklaas, E.J., Atkins, C., Turner, N.C., Siddique, K.H.M., 2011. Drying the surface soil reduces the nitrogen content of faba bean (Vicia faba L.) through a reduction in nitrogen fixation. Plant Soil 339, 351–362. https://doi.org/10.1007/s11104-010-0586-9

Abdellatif, K.F., A. El Absawy, E.S., M. Zakaria, A., 2012. Drought Stress Tolerance of Faba Bean as Studied by Morphological Traits and Seed Storage Protein Pattern. J. Plant Stud. 1. https://doi.org/10.5539/jps.v1n2p47

Amede, T., Kittlitz, E.V., Schubert, S., 1999. Differential Drought Responses of Faba Bean (Vicia faba L.) Inbred Lines. J. Agron. Crop Sci. 183, 35–45. https://doi.org/10.1046/j.1439-037x.1999.00310.x

Amede, T., Schubert, S., Stahr, K., 2003. Mechanisms of drought resistance in grain legumes I: Osmotic adjustment. SINET Ethiop. J. Sci. 26, 37–46.

Angus, J.F., Gault, R.R., Peoples, M.B., Stapper, M., Herwaarden, A.F. van, 2001. Soil water extraction by dryland crops, annual pastures, and lucerne in south-eastern Australia. Aust. J. Agric. Res. 52, 183–192. https://doi.org/10.1071/ar00103

Auerswald, K., Fischer, F.K., Kistler, M., Treisch, M., Maier, H., Brandhuber, R., 2018. Behavior of farmers in regard to erosion by water as reflected by their farming practices. Sci. Total Environ. 613–614, 1–9. https://doi.org/10.1016/j.scitotenv.2017.09.003

Bajji, M., Lutts, S., Kinet, J.-M., 2001. Water deficit effects on solute contribution to osmotic adjustment as a function of leaf ageing in three durum wheat (Triticum durum Desf.) cultivars performing differently in arid conditions. Plant Sci. 160, 669–681. https://doi.org/10.1016/S0168-9452(00)00443-X

Belachew, K.Y., Nagel, K.A., Fiorani, F., Stoddard, F.L., 2018. Diversity in root growth responses to moisture deficit in young faba bean (Vicia faba L.) plants. PeerJ 6, e4401. https://doi.org/10.7717/peerj.4401

Bessler, H., Temperton, V.M., Roscher, C., Buchmann, N., Schmid, B., Schulze, E.-D., Weisser, W.W., Engels, C., 2009. Aboveground overyielding in grassland mixtures is associated with reduced biomass partitioning to belowground organs. Ecology 90, 1520–1530. https://doi.org/10.1890/08-0867.1

Bilalis, D., Papastylianou, P., Konstantas, A., Patsiali, S., Karkanis, A., Efthimiadou, A., 2010. Weed-suppressive effects of maize–legume intercropping in organic farming. Int. J. Pest Manag. 56, 173–181. https://doi.org/10.1080/09670870903304471

Boyer, J.S., Westgate, M.E., 2004. Grain yields with limited water. J. Exp. Bot. 55, 2385–2394. https://doi.org/10.1093/jxb/erh219

Brueck, H., 2008. Effects of nitrogen supply on water-use efficiency of higher plants. J. Plant Nutr. Soil Sci. 171, 210–219. https://doi.org/10.1002/jpln.200700080

Bulson, H.A.J., Snaydon, R.W., Stopes, C.E., 1997. Effects of plant density on intercropped wheat and field beans in an organic farming system. J. Agric. Sci. 128, 59–71.

Caviglia, O.., Sadras, V.., Andrade, F.., 2004. Intensification of agriculture in the south-eastern Pampas. Field Crops Res. 87, 117–129. https://doi.org/10.1016/j.fcr.2003.10.002

Cazenave, A.-B., 2018. Facing changes today for an agriculture tomorrow. Adv Agr Environ Sci. 1, 00004. https://doi.org/10.13140/RG.2.2.15062.96322

Dennis, W.D., Woledge, J., 1982. Photosynthesis by White Clover Leaves in Mixed Clover/Ryegrass Swards. Ann. Bot. 49, 627–635. https://doi.org/10.1093/oxfordjournals.aob.a086290

Díaz, S., Cabido, M., 2001. Vive la différence: plant functional diversity matters to ecosystem processes. Trends Ecol. Evol. 16, 646–655. https://doi.org/10.1016/S0169-5347(01)02283-2

Fan, F., Zhang, F., Song, Y., Sun, J., Bao, X., Guo, T., Li, L., 2006. Nitrogen Fixation of Faba Bean (Vicia faba L.) Interacting with a Non-legume in Two Contrasting Intercropping Systems. Plant Soil 283, 275–286. https://doi.org/10.1007/s11104-006-0019-y

Farooq, M., Kobayashi, N., Wahid, A., Ito, O., Basra, S.M.A., 2009. RETRACTED: Strategies for Producing More Rice with Less Water, in: Advances in Agronomy. Elsevier, p. e1. https://doi.org/10.1016/S0065-2113(08)00806-7

Frison, E.A., Cherfas, J., Hodgkin, T., 2011. Agricultural Biodiversity Is Essential for a Sustainable Improvement in Food and Nutrition Security. Sustainability 3, 238–253. https://doi.org/10.3390/su3010238

Haymes, R., Lee, H.., 1999. Competition between autumn and spring planted grain intercrops of wheat (Triticum aestivum) and field bean (Vicia faba). Field Crops Res. 62, 167–176. https://doi.org/10.1016/S0378-4290(99)00016-7

Jákli, B., Hauer-Jákli, M., Böttcher, F., Meyer zur Müdehorst, J., Senbayram, M., Dittert, K., 2018. Leaf, canopy and agronomic water-use efficiency of field-grown sugar beet in response to potassium fertilization. J. Agron. Crop Sci. 204, 99–110. https://doi.org/10.1111/jac.12239

Jensen, E.S., Peoples, M.B., Hauggaard-Nielsen, H., 2010. Faba bean in cropping systems. Field Crops Res., Faba Beans in Sustainable Agriculture 115, 203–216. https://doi.org/10.1016/j.fcr.2009.10.008

Joshi, R., Wani, S.H., Singh, B., Bohra, A., Dar, Z.A., Lone, A.A., Pareek, A., Singla-Pareek, S.L., 2016. Transcription Factors and Plants Response to Drought Stress: Current Understanding and Future Directions. Front. Plant Sci. 7. https://doi.org/10.3389/fpls.2016.01029

Khan, H.R., Paull, J.G., Siddique, K.H.M., Stoddard, F.L., 2010. Faba bean breeding for drought-affected environments: A physiological and agronomic perspective. Field Crops Res. 115, 279–286. https://doi.org/10.1016/j.fcr.2009.09.003

Ledgard, S.F., Brier, G.J., Upsdel, M.P., 1990. Effect of clover cultivar on production and nitrogen fixation in clover-ryegrass swards under dairy cow grazing. N. Z. J. Agric. Res. 33, 243–249. https://doi.org/10.1080/00288233.1990.10428416

Lemken, D., Spiller, A., von Meyer-Höfer, M., 2017. The Case of Legume-Cereal Crop Mixtures in Modern Agriculture and the Transtheoretical Model of Gradual Adoption. Ecol. Econ. 137, 20–28. https://doi.org/10.1016/j.ecolecon.2017.02.021

Lenthe, J.-H., Oerke, E.-C., Dehne, H.-W., 2007. Digital infrared thermography for monitoring canopy health of wheat. Precis. Agric. 8, 15–26. https://doi.org/10.1007/s11119-006-9025-6

Link, W., Abdelmula, A.A., Kittlitz, E. von, Bruns, S., Riemer, H., Stelling, D., 1999. Genotypic variation for drought tolerance in Vicia faba. Plant Breed. 118, 477–484.

Link, W., Schill, B., von Kittlitz, E., 1996. Breeding for wide adaptation in faba bean. Euphytica 92, 185–190. https://doi.org/10.1007/BF00022844

Mahlein, A.-K., 2016. Plant Disease Detection by Imaging Sensors – Parallels and Specific Demands for Precision Agriculture and Plant Phenotyping. Plant Dis. 100, 241–251. https://doi.org/10.1094/PDIS-03-15-0340-FE

Malézieux, E., Crozat, Y., Dupraz, C., Laurans, M., Makowski, D., Ozier-Lafontaine, H., Rapidel, B., Tourdonnet, S., Valantin-Morison, M., 2009. Mixing plant species in cropping systems: concepts, tools and models. A review. Agron. Sustain. Dev. 29, 43–62. https://doi.org/10.1051/agro:2007057

Marshall, A.H., Collins, R.P., Humphreys, M.W., Scullion, J., 2016. A new emphasis on root traits for perennial grass and legume varieties with environmental and ecological benefits. Food Energy Secur. 5, 26–39. https://doi.org/10.1002/fes3.78

Morier, T., Cambouris, A.N., Chokmani, K., 2015. In-Season Nitrogen Status Assessment and Yield Estimation Using Hyperspectral Vegetation Indices in a Potato Crop. Agron. J. 107, 1295. https://doi.org/10.2134/agronj14.0402

Mullan, D.J., Reynolds, M.P., 2010. Quantifying genetic effects of ground cover on soil water evaporation using digital imaging. Funct. Plant Biol. 37, 703. https://doi.org/10.1071/FP09277

Mwanamwenge, J., Loss, S.., Siddique, K.H.., Cocks, P.., 1999. Effect of water stress during floral initiation, flowering and podding on the growth and yield of faba bean (Vicia faba L.). Eur. J. Agron. 11, 1–11. https://doi.org/10.1016/S1161-0301(99)00003-9

Mwanamwenge, J., Loss, S.P., Siddique, K.H.M., Cocks, P.S., 1998. Growth, seed yield and water use of faba bean (Vicia faba L.) in a short-season Mediterranean-type environment. Aust. J. Exp. Agric. 38, 171. https://doi.org/10.1071/EA97098

Negin, B., Moshelion, M., 2017. The advantages of functional phenotyping in pre-field screening for drought-tolerant crops. Funct. Plant Biol. 44, 107. https://doi.org/10.1071/FP16156

Nerkar, Y.S., Wilson, D., Lawes, D.A., 1981. Genetic variation in stomatal characteristics and behaviour, water use and growth of five Vicia faba L. genotypes under contrasting soil moisture regimes. Euphytica 30, 335–345.

Neugschwandtner, R., Ziegler, K., Kriegner, S., Wagentristl, H., Kaul, H.-P., 2015. Nitrogen yield and nitrogen fixation of winter faba beans. Acta Agric. Scand. Sect. B — Soil Plant Sci. 65, 658–666. https://doi.org/10.1080/09064710.2015.1042028

Ollinger, S.V., 2011. Sources of variability in canopy reflectance and the convergent properties of plants: Tansley review. New Phytol. 189, 375–394. https://doi.org/10.1111/j.1469-8137.2010.03536.x

Psomas, A., Kneubühler, M., Huber, S., Itten, K., Zimmermann, N.E., 2011. Hyperspectral remote sensing for estimating aboveground biomass and for exploring species richness patterns of grassland habitats. Int. J. Remote Sens. 32, 9007–9031. https://doi.org/10.1080/01431161.2010.532172

Reckling, M., Bergkvist, G., Watson, C.A., Stoddard, F.L., Zander, P.M., Walker, R.L., Pristeri, A., Toncea, I., Bachinger, J., 2016a. Trade-Offs between Economic and Environmental Impacts of Introducing Legumes into Cropping Systems. Front. Plant Sci. 7. https://doi.org/10.3389/fpls.2016.00669

Reckling, M., Hecker, J.-M., Bergkvist, G., Watson, C.A., Zander, P., Schläfke, N., Stoddard, F.L., Eory, V., Topp, C.F.E., Maire, J., Bachinger, J., 2016b. A cropping system assessment framework—Evaluating effects of introducing legumes into crop rotations. Eur. J. Agron. 76, 186–197. https://doi.org/10.1016/j.eja.2015.11.005

Ren, Y., Liu, J., Wang, Z., Zhang, S., 2016. Planting density and sowing proportions of maize–soybean intercrops affected competitive interactions and water-use efficiencies on the Loess Plateau, China. Eur. J. Agron. 72, 70–79. https://doi.org/10.1016/j.eja.2015.10.001

Römer, C., Wahabzada, M., Ballvora, A., Pinto, F., Rossini, M., Panigada, C., Behmann, J., Léon, J., Thurau, C., Bauckhage, C., Kersting, K., Rascher, U., Plümer, L., 2012. Early drought stress detection in cereals: simplex volume maximisation for hyperspectral image analysis. Funct. Plant Biol. 39, 878. https://doi.org/10.1071/FP12060

Rubiales, D., Mikic, A., 2015. Introduction: Legumes in Sustainable Agriculture. Crit. Rev. Plant Sci. 34, 2–3. https://doi.org/10.1080/07352689.2014.897896

Sadras, V.O., Angus, J.F., 2006. Benchmarking water-use efficiency of rainfed wheat in dry environments. Aust. J. Agric. Res. 57, 847. https://doi.org/10.1071/AR05359

Sadras, V.O., Lawson, C., 2013. Nitrogen and water-use efficiency of Australian wheat varieties released between 1958 and 2007. Eur. J. Agron. 46, 34–41. https://doi.org/10.1016/j.eja.2012.11.008

Schwenke, G.D., Herridge, D.F., Scheer, C., Rowlings, D.W., Haigh, B.M., McMullen, K.G., 2015. Soil N2O emissions under N2-fixing legumes and N-fertilised canola: A reappraisal of emissions factor calculations. Agric. Ecosyst. Environ. 202, 232–242. https://doi.org/10.1016/j.agee.2015.01.017

Senbayram, M., Tränkner, M., Dittert, K., Brück, H., 2015. Daytime leaf water use efficiency does not explain the relationship between plant N status and biomass water-use efficiency of tobacco under non-limiting water supply. J. Plant Nutr. Soil Sci. 178, 682–692. https://doi.org/10.1002/jpln.201400608

Serraj, R., Sinclair, T.R., 2002. Osmolyte accumulation: can it really help increase crop yield under drought conditions? Plant Cell Environ. 25, 333–341.

Siebrecht-Schöll, D.J., 2019. Breeding analysis of eight winter faba bean genotypes for mixed cropping with winter wheat. Georg-August-Universität Göttingen, Göttingen. http://hdl.handle.net/21.11130/00-1735-0000-0005-128E-7

Skinner, R.H., 2008. Yield, Root Growth, and Soil Water Content in Drought-Stressed Pasture Mixtures Containing Chicory. Crop Sci. 48, 380. https://doi.org/10.2135/cropsci2007.04.0201

Sun, F., Roderick, M.L., Farquhar, G.D., 2018. Rainfall statistics, stationarity, and climate change. Proc. Natl. Acad. Sci. 201705349. https://doi.org/10.1073/pnas.1705349115

Tambussi, E. a., Bort, J., Araus, J. l., 2007. Water use efficiency in C3 cereals under Mediterranean conditions: a review of physiological aspects. Ann. Appl. Biol. 150, 307–321. https://doi.org/10.1111/j.1744-7348.2007.00143.x

Tardieu, F., 1996. Drought perception by plants Do cells of droughted plants experience water stress? Plant Growth Regul. 20, 93–104. https://doi.org/10.1007/BF00024005

Thenkabail, P.S., Enclona, E.A., Ashton, M.S., Van Der Meer, B., 2004. Accuracy assessments of hyperspectral waveband performance for vegetation analysis applications. Remote Sens. Environ. 91, 354–376. https://doi.org/10.1016/j.rse.2004.03.013

Vandoorne, B., Beff, L., Lutts, S., Javaux, M., 2012. Root Water Uptake Dynamics of var. Under Water-Limited Conditions. Vadose Zone J. 11, 0. https://doi.org/10.2136/vzj2012.0005

Walker, S., Ogindo, H.O., 2003. The water budget of rainfed maize and bean intercrop. Phys. Chem. Earth Parts ABC 28, 919–926. https://doi.org/10.1016/j.pce.2003.08.018

Weißhuhn, P., Reckling, M., Stachow, U., Wiggering, H., 2017. Supporting Agricultural Ecosystem Services through the Integration of Perennial Polycultures into Crop Rotations. Sustainability 9, 2267. https://doi.org/10.3390/su9122267

Xiao, Y., Li, L., Zhang, F., 2004. Effect of root contact on interspecific competition and N transfer between wheat and fababean using direct and indirect 15N techniques. Plant Soil 262, 45–54. https://doi.org/10.1023/B:PLSO.0000037019.34719.0d

Zając, T., Oleksy, A., Stokłosa, A., Klimek-Kopyra, A., Kulig, B., 2013. The development competition and productivity of linseed and pea-cultivars grown in a pure sowing or in a mixture. Eur. J. Agron. 44, 22–31. https://doi.org/10.1016/j.eja.2012.08.001

Zarafshar, M., Akbarinia, M., Askari, H., Hosseini, S.M., Rahaie, M., Struve, D., Striker, G.G., 2014. Morphological, physiological and biochemical responses to soil water deficit in seedlings of three populations of wild pear (Pyrus boisseriana).

Zhao, T., Stark, B., Chen, Y., Ray, A.L., Doll, D., 2017. Challenges in Water Stress Quantification Using Small Unmanned Aerial System (sUAS): Lessons from a Growing Season of Almond. J. Intell. Robot. Syst. 88, 721–735. https://doi.org/10.1007/s10846-017-0513-x

Summary

Under low input conditions, legume-based intercropping systems have the potential to achieve increased yield production in comparison to pure stands. This is due to complementary resource use. A more efficient use of water resources as a result of species intercropping is of particular importance, as changes in precipitation patterns towards extreme weather events are likely to occur. The performance of the intercropping system depends on various factors such as environmental conditions (e.g. air temperature and precipitation) and genotypic characteristics (e.g. transpiration). However, breeding of novel cultivars, like winter types of faba bean, is performed in pure stands and the performance in intercropping systems is not considered. The suitability of a specific genotype for intercropping is governed by complex interactions, as genotypes with best ranking in pure stands may not necessarily perform as strongly as in intercropping systems.

In this thesis, legumes and non-legumes of arable, grassland and woody crops were tested in intercropped and pure stands, with special focus on their water use efficiency and the suitability of various genotypes for intercropping. Due to the complexity of such intercropping systems, it is necessary to assess different parameters for water use efficiency, involving both greenhouse and field conditions. Physiological and microbial aspects (i.e. bacterial and fungal communities) of winter faba bean-winter wheat intercropping in the greenhouse provide an in-depth insight into how the crops respond to a period of water deficit. On a larger scale, field experiments allow us to analyze and understand more complex growth conditions such as weather events and variations in soil structure.

In greenhouse experiments, two winter faba bean (*Vicia faba* L.) genotypes differing in growth habitus and maturing time were investigated in pure stands and intercropped with winter wheat (*Triticum aestivum* L.). The crop stands were cultivated in soil substrate and a period of water deficit was mimicked by reducing irrigation, followed by re-irrigation to adequate water contents in soil. Parameters such as the gas exchange, relative water content of leaves (RWC; %), plant biomass production and water use efficiency (WUE; calculated as total aboveground biomass per water consumption (kg/L) during a growth period) were studied. A strong interrelation of genotype and crop stand in all investigated parameters was observed. The water deficit resulted in reduced biomass and WUE of winter faba bean for one genotype in pure stands and for the other genotype in intercropping. This observation was confirmed by other results such as RWC, net CO_2 assimilation and evapotranspiration. These findings indicate the preservation of photosynthetic processes under water deficit of the adapted genotype.

Within this pot experiment, it was shown that the water deficit but also the crop stand affected the active bacterial community. In the leaf endosphere, bacterial diversity and richness was reduced under the condition of the water deficit. However, the results were dependent on crop species, genotype and plant compartment and differed in the duration of the water deficit period.

Another experiment with the same set-up showed that intercropping in combination with the inoculation of an entomopathogenic fungus acting as biological pesticide significantly affects the microbial diversity and community structure. These effects were dependent on the crop species as well as the plant compartment such as the endosphere of the plants and the rhizosphere of the soil. The results could partly be explained by changes in the C/N ratio of the soil. That shows that there are complex interactions between plants, associated microorganisms and their environment, influencing the cropping system's productivity.

Following that, field experiments were conducted. Within arable, grassland and woodland crops, the examination of the crop stands focused on key indicators of the water use as well as the sustainability in terms of greenhouse gas emissions. Additional drone-based remote sensing approaches were implemented in pure and intercropped stands for the observation of the vegetation development. Therefore, crop stands of winter faba bean with winter wheat in arable land as well as white clover (*Trifolium repens* L.) with perennial ryegrass (*Lolium perenne* L.) and chicory (*Cichorium intybus* L.) in grassland were investigated. In grassland, additional plots of the non-legumes were fertilized with 240 kg N/ha. In order to compare the performance under varying growth conditions, two contrasting sites were chosen, i.e. Reinshof as a fertile site and Deppoldshausen as a marginal site, both located near Göttingen in central Germany. Plants were exposed to regular weather and soil moisture conditions without artificial interference in three consecutive years. Measurements were performed throughout the growing cycles in the months of March to September of the years 2015 until 2017.

Among the evaluated parameters, there were the discrimination of carbon isotopes ($\delta^{13}C$) in plant tissues as estimator for intrinsic WUE as well as the instantaneous WUE as ratio of net CO_2 exchange and evapotranspiration. Results show that crop stands including N_2-fixing legumes, i.e. pure stands and intercropping, had improved WUE and productivity compared to pure non-legumes. Here, nitrogen fertilizer increased the use of water resources, but did not affect the WUE. Due to different plant characteristics in water acquisition, grassland crop stands including chicory led to higher evapotranspiration and lower $\delta^{13}C$ compared to stands including ryegrass. The impact of the different growth conditions at the study sites was low. Intercropping with legumes thus had significant advantages in WUE compared to pure winter wheat or perennial ryegrass.

Intercropping of legumes and non-legumes also showed advantages in regard to the sustainability of agricultural ecosystems. In both arable as well as woodland crop stands, which included black locust (*Robinia pseudoacacia*) and poplar (*Populus* sp.), the emissions of nitrous oxide (N_2O) were investigated in order to develop concepts for reduced greenhouse gas emissions. In agro-forestry, the N_2O emissions were generally high due to the initially slow growth phase which resulted in a low nitrogen uptake by the plants. In arable crops, intercropping of faba bean and wheat resulted in significantly reduced N_2O emissions in comparison to pure stands of faba bean as well as wheat stands with additional nitrogen fertilizer. As a result, intercropping with legumes contributed to diminished greenhouse gas emissions of agro-ecosystems.

Drone-based remote monitoring of the crop stands in arable land and grassland highlighted the vegetation development in terms of photosynthetic activity and biomass production by the Normalized Difference Vegetation Index (NDVI). Additionally, water use and transpiration of the crops was estimated based on the canopy surface temperature measured via thermography. The results indicated that intercropping as well as the application of nitrogen fertilizer in grassland improve productivity and water use. However, the canopy surface temperature was strongly influenced by daily fluctuations in environmental conditions. Only few differences could be observed between leguminous genotypes. In this instance, the accompanying species had greater impact on the performance of the intercropped stand, which was particularly visible in grassland intercropping. Differing from ryegrass, chicory significantly reduced the surface temperature while it increased the water use.

From the greenhouse experiment it can be concluded that the performance and drought tolerance of intercropping systems can be improved depending on the choice of the winter faba bean genotype. Though observed differences between genotypes in the field were less significant, some genotypes could be identified for further breeding for intercropping systems. All in all, intercropping with legumes improves the water use efficiency and the general performance and sustainability of the agro-ecosystem. Consequently, intercropping is a promising farming practice to meet future challenges in agriculture.

Further publications

Peer-reviewed articles

Cabeza R, Koester B, Liese R, **Lingner A**, Baumgarten V, Dirks J, Salinas-Riester G, Pommerenke C, Dittert K, Schulze J (2014) An RNA sequencing transcriptome analysis reveals novel insights into molecular aspects of the nitrate impact on the nodule activity of *Medicago truncatula*. Plant physiology, 164(1), 400-411.

Cabeza R, **Lingner A**, Liese R, Sulieman S, Senbayram M, Tränkner M, Dittert K, Schulze J (2014) The Activity of Nodules of the Supernodulating Mutant Mt_{sunn} Is not Limited by Photosynthesis under Optimal Growth Conditions. Int. J. Mol. Sci. 15, 6031–6045. https://doi.org/10.3390/ijms15046031

Cabeza RA, Liese R, **Lingner A**, von Stieglitz I, Neumann J, Salinas-Riester G, Pommerenke C, Dittert K, Schulze J (2014) RNA-seq transcriptome profiling reveals that *Medicago truncatula* nodules acclimate N_2 fixation before emerging P deficiency reaches the nodules. J. Exp. Bot. 65, 6035–6048. https://doi.org/10.1093/jxb/eru341

Cabeza RA, Liese R, Fischinger SA, Sulieman S, Avenhaus U, **Lingner A**, Hein H, Koester B, Baumgarten V, Dittert K, Schulze J (2015) Long-term non-invasive and continuous measurements of legume nodule activity. Plant J. 81, 637–648. https://doi.org/10.1111/tpj.12751

Avenhaus U, Cabeza RA, Liese R, **Lingner A**, Dittert K, Salinas-Riester G, Pommerenke C, Schulze J (2016) Short-Term Molecular Acclimation Processes of Legume Nodules to Increased External Oxygen Concentration. Front. Plant Sci. 6. https://doi.org/10.3389/fpls.2015.01133

Talks

Lingner A, Dittert K, Senbayram M (2016) Water use efficiency and drone-based spectral screening of productivity in mixed cropping systems. Second International Legume Society Conference (ILS), 11.-14.10.2016, Tróia, Portugal.

Awarded an honorable mention.

Lingner A, Dittert K (2017) Remote sensing of productivity and water use in legume-based mixed cropping systems. 60. Jahrestagung der Gesellschaft für Pflanzenbauwissenschaften e.V. (GPW), 26.-28.09.2017, Witzenhausen, Germany.

Lingner A, Dittert K, Pfeiffer B (2018) The cropping system matters – Contrasting responses of winter faba bean genotypes to drought stress. EUCARPIA Symposium on Breeding for Diversification, 19.-21.02.2018, Witzenhausen, Germany.

Poster

Lingner A, Dittert K, Steinmann H, Isselstein J, Link W, Senbayram M (2015) Traits for water and nutrient use in mixed cropping systems. Tagung der Deutschen Gesellschaft für Pflanzenernährung (DGP), 17.-18.09.2015, Göttingen, Germany.

Lingner A, Dittert K, Senbayram M (2016) Traits for water use in mixed cropping systems. PLANT 2030 Status Seminar, 14.-16.03.2016, Potsdam, Germany.

Lingner A, Siebrecht D, Senbayram M, Link W, Dittert K (2016) Productivity of legume-based mixed cropping systems as estimated by NDVI imaging. Tagung der Deutschen Gesellschaft für Pflanzenernährung (DGP), 28.-30.09.2016, Hohenheim, Germany.

Lingner A, Dittert K (2016) Spectral screening of legume-based mixed cropping systems using NDVI imaging. Internationale Pflanzenbautagung, 17.-18.11.2016, Bernburg, Germany.

Lingner A, Pfeiffer B, Dittert K (2017) Winter faba bean growth in pure and mixed stands under water deficit conditions in the greenhouse. PLANT 2030 Status Seminar, 20.-22.02.2017, Potsdam, Germany.

Including a short poster presentation.

Lingner A, Pfeiffer B, Dittert K (2017) Performance and drought stress response of winter faba bean genotypes in mixed cropping with winter wheat. 18th International Plant Nutrition Colloquium (IPNC), 19.-24.08.2017, Copenhagen, Denmark.

Including a short poster presentation.

Lingner A, Dittert K, Pfeiffer B (2018) The cropping system matters – Contrasting responses of winter faba bean genotypes to drought stress. PLANT 2030 Status Seminar, 05.-07.02.2018, Potsdam, Germany.

Acknowledgments

Danksagung

Die Promotionszeit war für mich ein Abschnitt meines Lebens, in dem ich vieles lernen und erfahren durfte, so dass ich mich professionell und persönlich weiterentwickeln konnte. Darum möchte ich an dieser Stelle allen danken, die mir dabei geholfen, mich unterstützt, gefördert und begleitet haben.

Während dieser Zeit hatte ich mehrere Betreuer, denen ich verschiedenste Dinge zu verdanken habe. Alles begann mit Joachim Schulze, der während meines Masterstudiums ein Potential in mir erkannt hat, das ich damals noch nicht gesehen habe. Diese Unterstützung hat es mir erst ermöglicht, mit einer Doktorarbeit zu beginnen und den Weg einer Wissenschaftlerin einzuschlagen. Mehmet Senbayram danke ich, dass er ebenso Vertrauen in mich gesetzt hat und mich als Doktorandin übernommen und gestärkt hat. Durch ihn hatte ich die Chance, sowohl im IMPAC³-Projekt als auch am IAPN mit vielen netten Kollegen zusammenzuarbeiten. Nachdem Mehmet die Universität Göttingen verlassen hat, war ich sehr dankbar, dass Klaus Dittert die Lücke des Erstbetreuers ausgefüllt hat und ich somit meine Doktorarbeit weiterführen konnte. Für das Vertrauen in mich und die Möglichkeit, eigenständige Entscheidungen treffen zu können und Verantwortung zu übernehmen bin ich sehr dankbar. So konnte ich auch in diesem Abschnitt Neues lernen und mich persönlich stärken. Außerdem danke ich Johannes Isselstein und Andrea Carminati dafür, dass sie sich ganz unkompliziert dazu bereit erklärt haben, Zweit- und Drittprüfer zu werden und mich dahingehend trotz knapper Zeitressourcen betreuen und fördern.

Ein besonderer Dank gilt Birgit Pfeiffer, die maßgeblich zum Gelingen dieser Doktorarbeit beigetragen hat. Sie hat mich motiviert, mit mir diskutiert, meine Texte vielfach korrigiert, Ratschläge erteilt, Druck gemacht wo es nötig war und freundschaftliche Unterstützung geboten. Mit diesen vielfältigen Hilfestellungen hat sie meine Arbeit erleichtert und mich erheblich vorangebracht, wofür ich herzlich danken möchte.

Das übergeordnete IMPAC³-Projekt, in dem ich meine Doktorarbeit geschrieben habe, hat mich mit vielen lieben Kollegen zusammengebracht, und mir ermöglicht, Einblicke in verschiedenste Themenbereiche zu bekommen. An dieser Stelle bedanke ich mich sehr bei Horst Steinmann für die Koordination des Projektes und dass er bei Bedarf immer bereit war, Ratschläge zu erteilen und auch anderweitig zu unterstützen. Die Techniker, insbesondere Regina Martsch und Dirk Koops, haben sich auch sehr für das Projekt eingesetzt und es mir ermöglicht, Feldarbeiten und Drohnenflüge reibungslos durchzuführen und haben mir dabei nach besten Möglichkeiten hilfreich zur Seite gestanden. Vielen Dank dafür. Auch meinen Mit-Doktoranden und Post-Docs möchte ich danken, dass die Zusammenarbeit so angenehm war und wir so manche schönen Tage im Feld, auf Konferenzen und bei diversen Getränken in Pausen und nach der Arbeit verbracht haben. Es war eine tolle Zeit mit euch!

Christiane Lüers, die mit meiner Projektverwaltung einiges an Nerven brauchte, danke ich sehr für ihre Unterstützung während der gesamten Promotionszeit. Im Labor und auch im Gewächshaus hatte ich sehr viel Hilfe von Susanne Koch, Simone Urstadt, Marlies Niebuhr,

Kirsten Fladung und Ulrike Kierbaum, die immer alles möglich gemacht haben und mir damit oftmals eine große Unterstützung waren. Insbesondere bei der Herkulesaufgabe der Tausenden von Bodenproben haben sie mir tatkräftig unter die Arme gegriffen und mir mit Tipps und Ratschlägen zur Seite gestanden. Hierbei hat vor allem Ulrike mit unseren schwer verständlichen Probenbezeichnungen und Sortierungen immer die Ruhe bewahrt und die vielen Proben unermüdlich gemessen. Für die Drohnenflüge im Feld war es eine Erleichterung, als das IAPN-Auto angeschafft wurde, wobei ich Martina Renneberg sehr dankbar bin, dass sie so manche Matschspuren toleriert hat und mich auch bei anderen Dingen im Alltag immer wieder unterstützt hat. Bernd Steingrobe verdanke ich die Möglichkeit, an dem Bachelorpraktikum teilzunehmen und Erfahrungen in der Lehre zu sammeln, was mir viel Spaß gemacht hat. Bei technischen und praktischen Fragen waren mir Reinhard Hilmer und Jürgen Kobbe oft eine große Hilfe.

Ohne meine vielen fleißigen Studenten und Hiwis, die mir in vielen Bereichen und mitunter stressigen Zeiten eine wichtige Stütze waren, wäre diese Doktorarbeit in dem Rahmen auch nicht möglich gewesen. Vielen Dank an Jonas Lotze, Sascha Baun, Christian Wenthe, Nina Bacchi, Ronny Schmidt, Michaela Böhme, Phillip Walter, Isa Bulut, Salomé Gonçalves, Leonie Gollisch, Kirsten Oldenburg, Marie Bayer und all die anderen Hiwis, Studenten und Praktikanten, die mir unter die Arme gegriffen und geholfen haben. Besonders hervorheben möchte ich hierbei Ruja Mansorian, die in der ganzen Zeit seit Beginn meiner Arbeit bis zum Ende eine konstante Hilfe war, auf die ich mich immer verlassen konnte.

Den aktuellen und ehemaligen Doktoranden aus Pflanzenernährung, IAPN und Qualität pflanzlicher Produkte danke ich dafür, dass sie nicht nur nette und hilfsbereite Kollegen waren, sondern während dieser Zeit auch zu guten Freunden geworden sind. Allen Uni-internen und -externen Freunden danke ich hiermit sehr für Kaffeepausen, Mensabesuche, gemeinsame Arbeit in der SUB, offene Ohren für Sorgen und Nöte, das Teilen von Erfahrungen, Bastelaktionen, tolle Urlaube, Umzugshelfer, das Erklären von Landwirtschaft, Telefonate, lustige Fotos, zusammen Bier oder Wein trinken, gemeinsames Kochen und Backen, die Bereitstellung von Therapieschweinen, motorisierte und nicht-motorisierte Ausflüge, Spieleabende, Freundschaftsbekenntnisse, stundenlange Gespräche, Albernheiten, Unterstützung in schwierigen Phasen, Wiedersehensfreude, das Diskutieren von Fachfragen, Statistikberatung, unermüdliches und konstruktives Korrekturlesen und vieles weitere mehr.

Herzlich danken möchte ich an dieser Stelle auch Dennis Meißner, der mir auf den letzten Metern in mehrfacher Hinsicht eine große und liebevolle Motivationshilfe war. Neben vielen freudigen Momenten und dem Entdecken von Gemeinsamkeiten hat er mir ganz unkompliziert den Raum gegeben, den ich brauchte, um die Doktorarbeit zu beenden.

Zu guter Letzt möchte ich mich ganz herzlich bei meiner Familie bedanken für all den Rückhalt, den ich stets bekomme, das Verständnis, die Liebe und Geduld. Auch wenn die Doktorarbeit sprachlich und fachlich eine Herausforderung war, habt ihr mir immer gezeigt, dass ihr für mich da seid und ich auf eure Unterstützung zählen kann.

Appendix

Supplementary material for chapter 2.1

Table S.1. ANOVA table of dry matter (DM) of winter faba bean and winter wheat.

Species	Model	Factor	Df	Sum Sq	Mean Sq	F value	Pr(>F)	
Winter faba bean	DM~	Crop stand	1	0.2736	0.2736	5.06	0.0344	*
	Crop stand	Water availability	1	0.123	0.123	2.275	0.1451	
	* Water availability	Genotype	1	0.2455	0.2455	4.54	0.044	*
	* Genotype	Crop stand:Water availability	1	0.0001	0.0001	0.002	0.9614	
		Crop stand:Genotype	1	0.2994	0.2994	5.536	0.0276	*
		Water availability:Genotype	1	0.0017	0.0017	0.031	0.8613	
		Crop stand:Water availability:Genotype	1	0.3619	0.3619	6.693	0.0165	*
		Residuals	23	1.2437	0.0541			
Winter wheat	DM~	Crop stand	1	0.03007	0.03007	12.08	0.00253	**
	Crop stand	Water availability	1	0.01294	0.012936	5.197	0.03435	*
	+ Water availability	Genotype	1	0.00043	0.00043	0.173	0.68234	
	+ Genotype	Residuals	19	0.04729	0.002489			

Signif. codes: 0 '***' 0.001 '**' 0.01 '*' 0.05 '.' 0.1 ' ' 1

Table S.2. ANOVA table of leaf area (LA) of winter faba bean and winter wheat.

Species	Model	Factor	Df	Sum Sq	Mean Sq	F value	Pr(>F)	
Winter faba bean	LA~	Crop stand	1	181	181	0.114	0.73914	
	Crop stand	Water availability	1	18791	18791	11.785	0.00227	**
	* Water availability	Genotype	1	7589	7589	4.76	0.03961	*
	* Genotype	Crop stand:Water availability	1	1829	1829	1.147	0.2953	
		Crop stand:Genotype	1	4684	4684	2.938	0.09999	.
		Water availability:Genotype	1	379	379	0.238	0.63052	
		Crop stand:Water availability:Genotype	1	6351	6351	3.983	0.05794	.
		Residuals	23	36674	1595			
Winter wheat	LA~	Crop stand	1	483.8	483.8	14.403	0.00133	**
	Crop stand	Water availability	1	166.2	166.2	4.948	0.03916	*
	+ Water availability	Genotype	1	1.3	1.3	0.039	0.84504	
	+ Genotype	Residuals	18	604.6	33.6			

Signif. codes: 0 ‘***’ 0.001 ‘**’ 0.01 ‘*’ 0.05 ‘.’ 0.1 ‘ ’ 1

Table S.3. ANOVA table of water use efficiency (WUE) of winter faba bean and winter wheat.

Species	Model	Factor	Df	Sum Sq	Mean Sq	F value	Pr(>F)	
Winter faba bean	WUE~	Crop stand	1	0.711	0.711	3.202	0.08619	.
	Crop stand	Water availability	1	2.404	2.4035	10.823	0.00309	**
	* Water availability	Genotype	1	1.829	1.8288	8.235	0.00844	**
	* Genotype	Crop stand:Water availability	1	0.041	0.0413	0.186	0.67004	
		Crop stand:Genotype	1	0.071	0.0713	0.321	0.57636	
		Water availability:Genotype	1	0.005	0.0053	0.024	0.87906	
		Crop stand:Water availability:Genotype	1	1.688	1.6882	7.602	0.01097	*
		Residuals	24	5.33	0.2221			
Winter wheat	WUE~	Water availability	1	0.0136	0.01361	0.115	0.746	
	Water availability	Residuals	6	0.7083	0.11805			

Signif. codes: 0 '***' 0.001 '**' 0.01 '*' 0.05 '.' 0.1 ' ' 1

Table S.4. ANOVA table of net ecosystem exchange of CO_2 (NEE) of winter faba bean and winter wheat.

Species	Day	Model	Factor	Df	Sum Sq	Mean Sq	F value	Pr(>F)	
Winter faba bean	24	NEE~	Crop stand	1	67.36	67.36	10.341	0.0037	**
		Crop stand	Water availability	1	1.12	1.12	0.173	0.6815	
		* Water availability	Genotype	1	5.5	5.5	0.845	0.3672	
		* Genotype	Crop stand:Water availability	1	3.57	3.57	0.548	0.4663	
			Crop stand:Genotype	1	7.15	7.15	1.097	0.3053	
			Water availability:Genotype	1	0.05	0.05	0.008	0.9301	
			Crop stand:Water availability:Genotype	1	18.87	18.87	2.897	0.1017	
			Residuals	24	156.35	6.51			
	29	NEE~	Crop stand	1	7.83	7.83	20.274	0.00014	***
		Crop stand	Water availability	1	39.57	39.57	102.455	2.51E-10	***
		+ Water availability	Genotype	1	0.01	0.01	0.029	0.86631	
		+ Genotype	Crop stand:Water availability	1	0.84	0.84	2.174	0.15282	
		+ Crop stand:Water availability	Crop stand:Genotype	1	6.56	6.56	16.982	0.00036	***
		+ Crop stand:Genotype	Water availability:Genotype	1	1.25	1.25	3.244	0.08377	.
		+ Water availability:Genotype	Residuals	25	9.66	0.39			
	34	NEE~	Crop stand	1	38.74	38.74	24.496	8.90E-05	***
		Crop stand	Water availability	1	31.55	31.55	19.951	2.64E-04	***
		* Water availability	Genotype	1	0.01	0.01	0.006	0.93861	
		* Genotype	Crop stand:Water availability	1	0.76	0.76	0.481	0.49637	
			Crop stand:Genotype	1	47.69	47.69	30.159	2.68E-05	***
			Water availability:Genotype	1	1.52	1.52	0.961	0.33933	
			Crop stand:Water availability:Genotype	1	3.7	3.7	2.338	0.14272	
			Residuals	19	30.05	1.58			
	38	NEE~	Crop stand	1	3.528	3.528	5.193	0.0305	*
		Crop stand	Water availability	1	0.075	0.075	0.111	0.7417	
		+ Water availability	Genotype	1	0.464	0.464	0.682	0.4157	
		+ Genotype	Residuals	28	19.024	0.679			
Winter wheat	24	NEE~	Water availability	1	2.29	2.293	0.434	0.535	
		Water availability	Residuals	6	31.72	5.286			

29	NEE~ Water availability	Water availability	1	2.6	2.6	2.101	0.197
		Residuals	6	7.427	1.238		
34	NEE~ Water availability	Water availability	1	5.127	5.127	12.34	0.0171 *
		Residuals	5	2.078	0.416		
38	NEE~ Water availability	Water availability	1	3.921	3.921	2.977	0.135
		Residuals	6	7.904	1.317		

Signif. codes: 0 '***' 0.001 '**' 0.01 '*' 0.05 '.' 0.1 ' ' 1

Table S.5. Net ecosystem exchange of CO_2 (μmol CO_2 m^{-2} s^{-1}) of intercropping and pure stands of winter faba bean and winter wheat. Intercropping considers the canopy of both species together. WD: water deficit. Different letters in columns indicate significant differences as calculated by the Duncan-test ($p < 0.05$, n = 4).

Species	Stand	Genotype of faba bean	Water treatment	day 24 – fully irrigated			day 29 – beginning of WD			day 34 – end of WD			day 38 – re-irrigation		
				mean	SE		mean	SE		mean	SE		mean	SE	
Winter faba bean and winter wheat	Intercropping	S_004	Control	**8.2**	1.28	b	**2.9**	0.18	bc	**5.4**	0.40	ab	**1.7**	0.27	a
			Water deficit	**9.5**	0.83	b	**1.3**	0.28	e	**2.3**	0.37	c	**2.1**	0.50	a
		S_062	Control	**9.7**	2.09	b	**2.3**	0.40	cd	**1.7**	0.65	c	**2.3**	0.41	a
			Water deficit	**7.8**	0.40	b	**0.0**	0.56	f	**-0.6**	0.22	d	**1.8**	0.11	a
Winter faba bean	Pure stand	S_004	Control	**11.0**	0.63	ab	**3.2**	0.15	b	**3.3**	0.57	bc	**2.3**	0.52	a
			Water deficit	**10.6**	0.57	ab	**1.1**	0.14	e	**2.5**	1.12	c	**2.6**	0.39	a
		S_062	Control	**11.3**	0.97	ab	**4.5**	0.35	a	**6.6**	0.85	a	**2.7**	0.36	a
			Water deficit	**13.8**	2.12	a	**1.6**	0.25	de	**3.5**	0.55	bc	**3.0**	0.67	a
Winter wheat	Pure stand	-	Control	**7.1**	1.26	A	**1.8**	0.35	A	**2.4**	0.33	A	**1.6**	0.24	A
		-	Water deficit	**6.1**	1.03	A	**0.6**	0.70	A	**0.7**	0.37	B	**0.2**	0.78	A

Table S.6. ANOVA table of evapotranspiration (ET) of winter faba bean and winter wheat.

Species	Day	Model	Factor	Df	Sum Sq	Mean Sq	F value	Pr(>F)	
Winter faba bean	24	ET~	Crop stand	1	0.16	0.1605	0.761	0.392	
		Crop stand	Water availability	1	0.457	0.457	2.166	0.154	
		* Water availability	Genotype	1	0.097	0.0975	0.462	0.503	
		* Genotype	Crop stand:Water availability	1	0.069	0.069	0.327	0.573	
			Crop stand:Genotype	1	0.032	0.0315	0.149	0.703	
			Water availability:Genotype	1	0.015	0.0154	0.073	0.789	
			Crop stand:Water availability:Genotype	1	0.412	0.4122	1.954	0.175	
			Residuals	24	5.063	0.2109			
	29	ET~	Crop stand	1	0.004	0.004	0.119	0.733	
		Crop stand	Water availability	1	4.167	4.167	126.76	6.60E-12	***
		+ Water availability	Genotype	1	0.002	0.002	0.054	0.818	
		+ Genotype	Residuals	28	0.921	0.033			
	34	ET~	Crop stand	1	0.84	0.84	8.386	9.26E-03	**
		Crop stand	Water availability	1	5.637	5.637	56.258	4.31E-07	***
		* Water availability	Genotype	1	0.846	0.846	8.448	0.00905	**
		* Genotype	Crop stand:Water availability	1	0.005	0.005	0.054	0.8195	
			Crop stand:Genotype	1	2.239	2.239	22.35	1.47E-04	***
			Water availability:Genotype	1	0.471	0.471	4.699	0.0431	*
			Crop stand:Water availability:Genotype	1	0.41	0.41	4.093	0.05735	.
			Residuals	19	1.904	0.1			
	38	ET~	Crop stand	1	0.5746	0.5746	14.164	0.00079	***
		Crop stand	Water availability	1	0.0841	0.0841	2.072	0.16112	
		+ Water availability	Genotype	1	0.0901	0.0901	2.221	0.14732	
		+ Genotype	Residuals	28	1.1359	0.0406			
Winter wheat	24	ET~	Water availability	1	0.0884	0.08841	0.639	0.455	
		Water availability	Residuals	6	0.8304	0.13841			
	29	ET~	Water availability	1	0.06643	0.06643	4.861	0.0696	.
		Water availability	Residuals	6	0.08199	0.01367			
	34	ET~	Water availability	1	0.3545	0.3545	7.749	0.0318	*
		Water availability	Residuals	6	0.2745	0.0457			

38	ET~ Water availability	Water availability	1	0.0504	0.0504	2.312	0.179
		Residuals	6	0.1308	0.0218		

Signif. codes: 0 '***' 0.001 '**' 0.01 '*' 0.05 '.' 0.1 ' ' 1

Table S.7. Evapotranspiration (mmol H_2O m^{-2} s^{-1}) of intercropping and pure stands of winter faba bean and winter wheat. Intercropping considers the canopy of both species together. WD: water deficit. Different letters in columns indicate significant differences as calculated by the Duncan-test ($p < 0.05$, n = 4).

Species	Stand	Genotype of faba bean	Water treatment	day 24 – fully irrigated		day 29 – beginning of WD		day 34 – end of WD		day 38 – re-irrigation	
				mean	SE	mean	SE	mean	SE	mean	SE
Winter faba bean and winter wheat	Intercropping	S_004	Control	**1.5**	0.07 a	**1.0**	0.08 a	**1.8**	0.09 b	**1.0**	0.07 ab
			Water deficit	**2.1**	0.22 a	**0.5**	0.05 b	**0.8**	0.07 de	**1.0**	0.07 ab
		S_062	Control	**1.6**	0.23 a	**1.1**	0.15 a	**1.4**	0.04 bc	**1.0**	0.06 ab
			Water deficit	**1.7**	0.05 a	**0.4**	0.03 b	**0.5**	0.03 e	**0.7**	0.05 b
Winter faba bean	Pure stand	S_004	Control	**1.7**	0.21 a	**1.1**	0.14 a	**1.2**	0.18 cd	**1.3**	0.08 a
			Water deficit	**1.6**	0.12 a	**0.4**	0.07 b	**0.8**	0.33 de	**1.2**	0.19 a
		S_062	Control	**1.4**	0.21 a	**1.2**	0.10 a	**2.5**	0.19 a	**1.1**	0.03 a
			Water deficit	**1.8**	0.46 a	**0.4**	0.02 b	**1.1**	0.17 cd	**1.1**	0.16 a
Winter wheat	Pure stand	-	Control	**1.5**	0.22 A	**0.6**	0.06 A	**1.0**	0.15 A	**0.7**	0.09 A
		-	Water deficit	**1.3**	0.14 A	**0.4**	0.06 A	**0.6**	0.03 B	**0.5**	0.06 A

Table S.8. ANOVA table of canopy surface temperature of winter faba bean and winter wheat.

Species	Day	Model	Factor	Df	Sum Sq	Mean Sq	F value	Pr(>F)	
Winter faba bean	24	Temperature~	Crop stand	1	2.101	2.101	1.791	0.1934	
		Crop stand	Water availability	1	0.72	0.72	0.614	0.4411	
		* Water availability	Genotype	1	1.711	1.711	1.458	0.2389	
		* Genotype	Crop stand:Water availability	1	0.605	0.605	0.516	0.4796	
			Crop stand:Genotype	1	0.011	0.011	0.01	0.9228	
			Water availability:Genotype	1	1.445	1.445	1.232	0.2781	
			Crop stand:Water availability:Genotype	1	6.845	6.845	5.834	0.0237	*
			Residuals	24	28.16	1.173			
	29	Temperature~	Crop stand	1	7.291	7.291	12.879	0.00155	**
		Crop stand	Water availability	1	25.575	25.575	45.178	7.42E-07	***
		* Water availability	Genotype	1	11.812	11.812	20.866	1.37E-04	***
		* Genotype	Crop stand:Water availability	1	0.001	0.001	0.003	9.60E-01	
			Crop stand:Genotype	1	1.491	1.491	2.633	1.18E-01	
			Water availability:Genotype	1	1.005	1.005	1.776	1.96E-01	
			Crop stand:Water availability:Genotype	1	3.245	3.245	5.732	0.0252	*
			Residuals	23	13.02	0.566			
	34	Temperature~	Crop stand	1	1.565	1.565	6.879	1.42E-02	*
		Crop stand	Water availability	1	16.815	16.815	73.923	3.27E-09	***
		+ Water availability	Genotype	1	3.232	3.232	14.209	0.00081	***
		+ Genotype	Residuals	27	6.142	0.227			
	38	Temperature~	Crop stand	1	0.013	0.013	0.041	0.84159	
		Crop stand	Water availability	1	0.164	0.164	0.518	0.47796	
		+ Water availability	Genotype	1	4.872	4.872	15.37	0.00055	***
		+ Genotype	Residuals	27	8.559	0.317			
Winter wheat	24	Temperature~	Water availability	1	0.2112	0.21125	4.738	0.0724	.
		Water availability	Residuals	6	0.2675	0.04458			
	29	Temperature~	Water availability	1	0.943	0.943	1.64	0.256	
		Water availability	Residuals	5	2.874	0.5748			

34	Temperature~ Water availability	Water availability	1	0.54	0.54	2.418	0.195
		Residuals	4	0.8933	0.2233		
38	Temperature~ Water availability	Water availability	1	1.143	1.1433	1.255	0.314
		Residuals	5	4.557	0.9113		

Signif. codes: 0 '***' 0.001 '**' 0.01 '*' 0.05 '.' 0.1 ' ' 1

Table S.9. Canopy surface temperature (°C) of intercropping and pure stands of winter faba bean and winter wheat. Intercropping considers the canopy of both species together. WD: water deficit. Different letters in columns indicate significant differences as calculated by the Duncan-test ($p < 0.05$, n = 4).

Species	**Stand**	**Genotype of faba bean**	**Water treatment**	**day 24 – fully irrigated**			**day 29 – beginning of WD**			**day 34 – end of WD**			**day 38 – re-irrigation**		
				mean	SE		mean	SE		mean	SE		mean	SE	
Winter faba bean and winter wheat	Intercropping	S_004	Control	**22.6**	0.85	ab	**23.2**	0.39	d	**23.7**	0.24	e	**24.6**	0.12	ab
			Water deficit	**23.9**	0.43	a	**24.7**	0.26	bc	**25.3**	0.34	bc	**24.3**	0.17	ab
		S_062	Control	**23.4**	0.61	ab	**23.8**	0.18	cd	**24.2**	0.19	de	**24.9**	0.20	a
			Water deficit	**22.1**	0.17	b	**25.8**	0.38	ab	**25.4**	0.16	b	**25.2**	0.10	a
Winter faba bean	Pure stand	S_004	Control	**23.7**	0.47	ab	**23.1**	0.28	d	**23.7**	0.12	e	**24.5**	0.29	ab
			Water deficit	**23.8**	0.74	ab	**26.0**	0.46	a	**25.3**	0.28	bc	**24.0**	0.34	b
		S_062	Control	**22.8**	0.38	ab	**25.8**	0.35	ab	**24.7**	0.19	cd	**25.2**	0.56	a
			Water deficit	**23.8**	0.37	ab	**26.6**	0.62	a	**26.4**	0.23	a	**25.2**	0.15	a
Winter wheat	Pure stand	-	Control	**23.8**	0.13	A	**22.8**	0.30	A	**23.8**	0.20	A	**23.9**	0.57	A
		-	Water deficit	**23.5**	0.07	A	**23.6**	0.44	A	**24.4**	0.33	A	**24.7**	0.32	A

Table S.10. ANOVA table of relative water content (RWC) of winter faba bean and winter wheat.

Species	Day	Model	Factor	Df	Sum Sq	Mean Sq	F value	Pr(>F)	
Winter faba bean	29	RWC~	Crop stand	1	21.1	21.12	0.762	0.3912	
		Crop stand	Water availability	1	101.5	101.53	3.664	0.0676	.
		* Water availability	Genotype	1	75.6	75.65	2.73	0.1115	
		* Genotype	Crop stand:Water availability	1	27.8	27.75	1.001	0.3269	
			Crop stand:Genotype	1	0.4	0.4	0.015	0.9048	
			Water availability:Genotype	1	75	75.03	2.708	0.1129	
			Crop stand:Water availability:Genotype	1	44.7	44.65	1.611	0.2165	
			Residuals	24	665	27.71			
	34	RWC~	Crop stand	1	83.1	83.1	3.422	0.07721	.
		Crop stand	Water availability	1	657.1	657.1	27.046	2.84E-05	***
		* Water availability	Genotype	1	145.7	145.7	5.995	0.02239	*
		* Genotype	Crop stand:Water availability	1	18.5	18.5	0.763	0.39132	
			Crop stand:Genotype	1	33.2	33.2	1.368	0.25405	
			Water availability:Genotype	1	533.8	533.8	21.974	0.0001	***
			Crop stand:Water availability:Genotype	1	219.5	219.5	9.034	0.00631	**
			Residuals	23	558.8	24.3			
	38	RWC~	Crop stand	1	0.75	0.75	0.08	0.7794	
		Crop stand	Water availability	1	1.97	1.97	0.21	0.6504	
		+ Water availability	Genotype	1	71.14	71.14	7.576	0.0106	*
		+ Genotype	Crop stand:Genotype	1	32.68	32.68	3.481	0.0734	.
		+ Crop stand:Genotype	Residuals	26	244.15	9.39			
Winter wheat	29	RWC~	Crop stand	1	36.4	36.38	1.852	0.191	
		Crop stand	Water availability	1	30.1	30.14	1.534	0.232	
		+ Water availability	Genotype	1	39.4	39.36	2.004	0.175	
		+ Genotype	Residuals	17	334	19.64			
	34	RWC~	Crop stand	1	114.2	114.25	5.982	0.0256	*
		Crop stand	Water availability	1	0.2	0.23	0.012	0.9136	
		+ Water availability	Genotype	1	52.2	52.2	2.733	0.1166	
		+ Genotype	Water availability:Genotype	2	113.5	56.74	2.971	0.0782	.
		+ Water availability:Genotype	Residuals	17	324.7	19.1			

38	RWC~ Crop stand + Water availability + Genotype	Crop stand	1	6.82	6.82	1.369	0.2581
		Water availability	1	2	2	0.403	0.5341
		Genotype	1	41.03	41.03	8.243	0.0106 *
		Residuals	17	84.63	4.98		

Signif. codes: 0 '***' 0.001 '**' 0.01 '*' 0.05 '.' 0.1 ' ' 1

Table S.11. Leaf relative water content (%) of intercropping and pure stands of winter faba bean and winter wheat. WD: water deficit. Different letters in columns indicate significant differences as calculated by the Duncan-test ($p < 0.05$, n = 4).

Species	Stand	Genotype of faba bean	Water treatment	day 29 – beginning of WD		day 34 – end of WD		day 38 – re-irrigation	
				mean	SE	mean	SE	mean	SE
Winter faba bean	Intercropping	S_004	Control	**93.8**	2.17 a	**90.0**	1.78 ab	**94.6**	2.22 a
			Water deficit	**93.8**	1.54 a	**92.6**	3.34 a	**93.0**	0.86 ab
		S_062	Control	**95.9**	2.25 a	**97.2**	0.95 a	**88.9**	1.37 b
			Water deficit	**85.1**	3.67 b	**72.9**	4.27 c	**88.4**	1.53 b
	Pure stand	S_004	Control	**92.5**	2.20 ab	**94.7**	2.36 a	**91.7**	0.87 ab
			Water deficit	**91.5**	2.78 ab	**89.7**	1.48 ab	**91.5**	0.49 ab
		S_062	Control	**90.3**	1.46 ab	**95.1**	0.98 a	**90.8**	0.94 ab
			Water deficit	**87.9**	3.88 ab	**84.5**	2.95 b	**90.3**	3.09 ab
Winter wheat	Intercropping	S_004	Control	**90.8**	3.78 A	**99.9**	1.00 B	**100.3**	1.13 AB
			Water deficit	**94.3**	1.63 A	**104.4**	2.94 AB	**103.4**	1.71 A
		S_062	Control	**95.5**	2.04 A	**108.6**	2.90 A	**98.6**	0.43 B
			Water deficit	**95.9**	2.55 A	**102.9**	1.75 AB	**98.4**	1.10 B
	Pure stand	-	Control	**95.4**	1.04 A	**97.8**	1.89 B	**99.9**	1.06 AB
		-	Water deficit	**98.7**	0.20 A	**100.1**	2.03 B	**98.6**	0.39 B

Table S.12. ANOVA table of proline content of winter faba bean and winter wheat.

Species	Day	Model	Factor	Df	Sum Sq	Mean Sq	F value	Pr(>F)	
Winter faba bean	29	Proline~	Crop stand	1	14333	14333	4.038	0.0628	.
		Crop stand	Water availability	1	28379	28379	7.995	0.0127	*
		+ Water availability	Genotype	1	25189	25189	7.096	0.0177	*
		+ Genotype	Crop stand:Genotype	1	110495	110495	31.13	5.26E-05	***
		+ Crop stand:Genotype	Water availability:Genotype	1	5343	5343	1.505	0.2388	
		+ Water availability:Genotype	Residuals	15	53243	3550			
	34	Proline~	Crop stand	1	1514	1514	0.038	0.8484	
		Crop stand	Water availability	1	2127724	2127724	52.699	5.05E-07	***
		* Water availability	Genotype	1	201061	201061	4.98	0.03725	*
		* Genotype	Crop stand:Water availability	1	420050	420050	10.404	0.00424	**
			Crop stand:Genotype	1	398486	398486	9.87	0.00514	**
			Water availability:Genotype	1	162233	162233	4.018	0.05874	.
			Crop stand:Water availability:Genotype	1	1113351	1113351	27.575	3.87E-05	***
			Residuals	20	807497	40375			
	38	Proline~	Crop stand	1	191824	191824	21.168	0.00012	***
		Crop stand	Water availability	1	585	585	0.065	0.80154	
		+ Water availability	Genotype	1	1151	1151	0.127	0.72472	
		+ Genotype	Crop stand:Water availability	1	21613	21613	2.385	0.13559	
		+ Crop stand:Water availability	Crop stand:Genotype	1	156213	156213	17.238	0.00036	***
		+ Crop stand:Genotype	Residuals	24	217493	9062			
Winter wheat	29	Proline~	Crop stand	1	8540	8540	1.127	0.3064	
		Crop stand	Water availability	1	13813	13813	1.823	0.1984	
		+ Water availability	Genotype	1	39572	39572	5.222	0.0384	*
		+ Genotype	Water availability:Genotype	2	19145	9572	1.263	0.313	
		+ Water availability:Genotype	Residuals	14	106081	7577			
	34	Proline~	Crop stand	1	93771	93771	4.282	0.0551	.
		Crop stand	Water availability	1	6478	6478	0.296	0.594	
		+ Water availability	Genotype	1	139578	139578	6.374	0.0225	*

	+ Genotype	Water availability:Genotype	2	242287	121144	5.532	0.0149	*
	+ Water availability:Genotype	Residuals	16	350383	21899			
38	Proline~	Crop stand	1	1792	1792	0.052	0.8219	
	Crop stand	Water availability	1	19744	19744	0.575	0.4582	
	+ Water availability	Genotype	1	390413	390413	11.362	0.0034	**
	+ Genotype	Residuals	18	618493	34361			

Signif. codes: 0 '***' 0.001 '**' 0.01 '*' 0.05 '.' 0.1 ' ' 1

Table S.13. Leaf proline content (µg proline g^{-1} FW) of intercropping and pure stands of winter faba bean and winter wheat. WD: water deficit. Different letters in columns indicate significant differences as calculated by the Duncan-test ($p < 0.05$, n = 4).

Species	Stand	Genotype of faba bean	Water treatment	day 29 – beginning of WD			day 34 – end of WD			day 38 – re-irrigation		
				mean	SE		mean	SE		mean	SE	
Winter faba bean	Intercropping	S_004	Control	**292.4**	0.20	c	**398.0**	9.14	cd	**569.2**	59.22	b
			Water deficit	**414.1**	89.30	bc	**645.7**	79.13	cd	**602.6**	14.48	b
		S_062	Control	**423.9**	6.17	bc	**328.8**	18.85	d	**665.9**	47.89	ab
			Water deficit	**434.3**	43.0	b	**1690.1**	233.4	a	**785.5**	74.4	a
	Pure stand	S_004	Control	**564.0**	NA	a	**488.6**	107.17	cd	**587.9**	39.77	b
			Water deficit	**610.3**	28.15	a	**1047.3**	122.08	b	**532.2**	66.18	bc
		S_062	Control	**341.7**	7.21	bc	**633.0**	120.39	cd	**409.8**	42.48	c
			Water deficit	**374.3**	34.04	bc	**693.4**	78.10	c	**399.6**	41.86	c
Winter wheat	Intercropping	S_004	Control	**455.2**	7.41	B	**560.9**	52.97	BC	**422.1**	18.62	B
			Water deficit	**466.4**	73.42	AB	**724.1**	98.78	AB	**429.8**	340.93	B
		S_062	Control	**584.0**	39.72	AB	**355.2**	61.75	C	**846.4**	55.11	A
			Water deficit	**567.2**	54.03	AB	**481.9**	48.64	BC	**643.3**	113.52	AB
	Pure stand	-	Control	**624.0**	25.57	A	**810.9**	193.81	A	**609.2**	152.51	AB
		-	Water deficit	**496.8**	56.70	AB	**518.9**	48.44	BC	**613.1**	111.87	AB

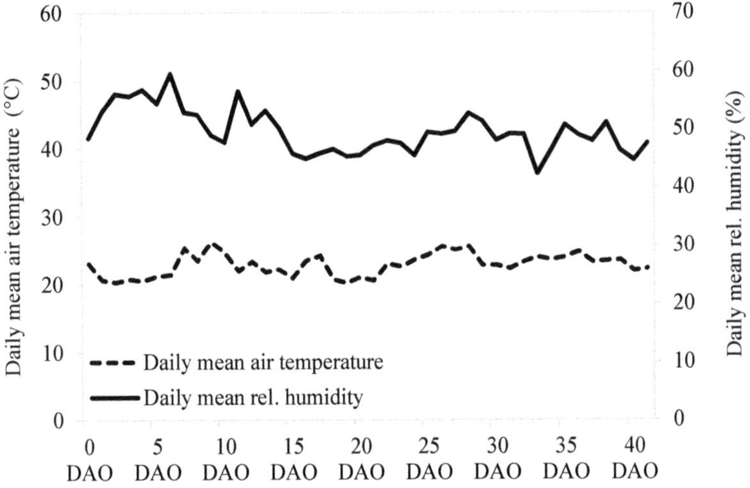

Figure S.1. Daily means of air temperature and relative humidity during the course of the experiment. DAO: days after onset of the experiment.

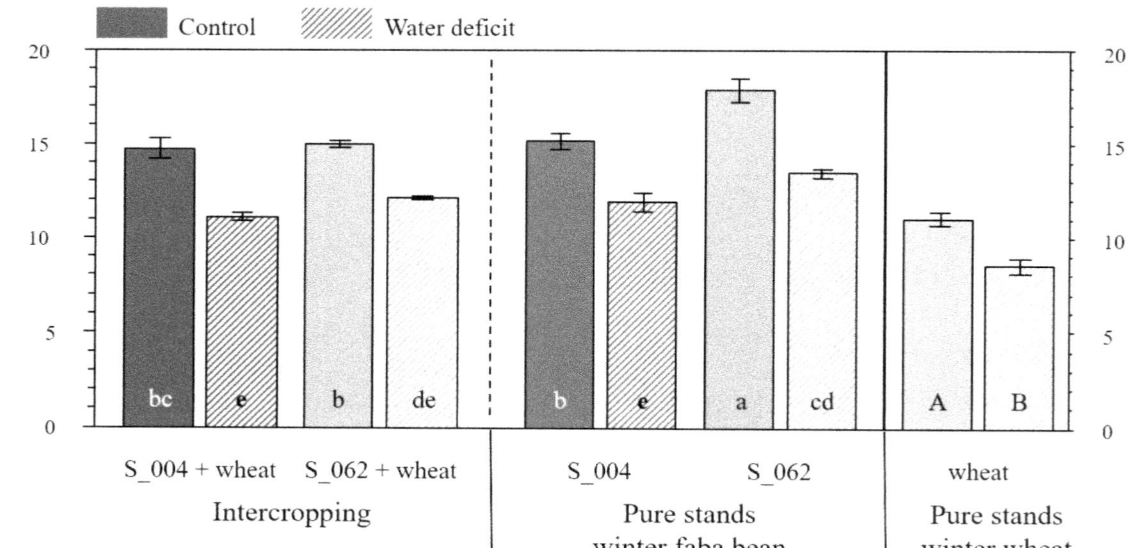

Figure S.2. Total amount of irrigated water during the whole experimental phase. Intercropping considers both species together. Error bars display the standard error. Different letters indicate significant differences as calculated by the Duncan-test ($p < 0.05$, $n = 4$).

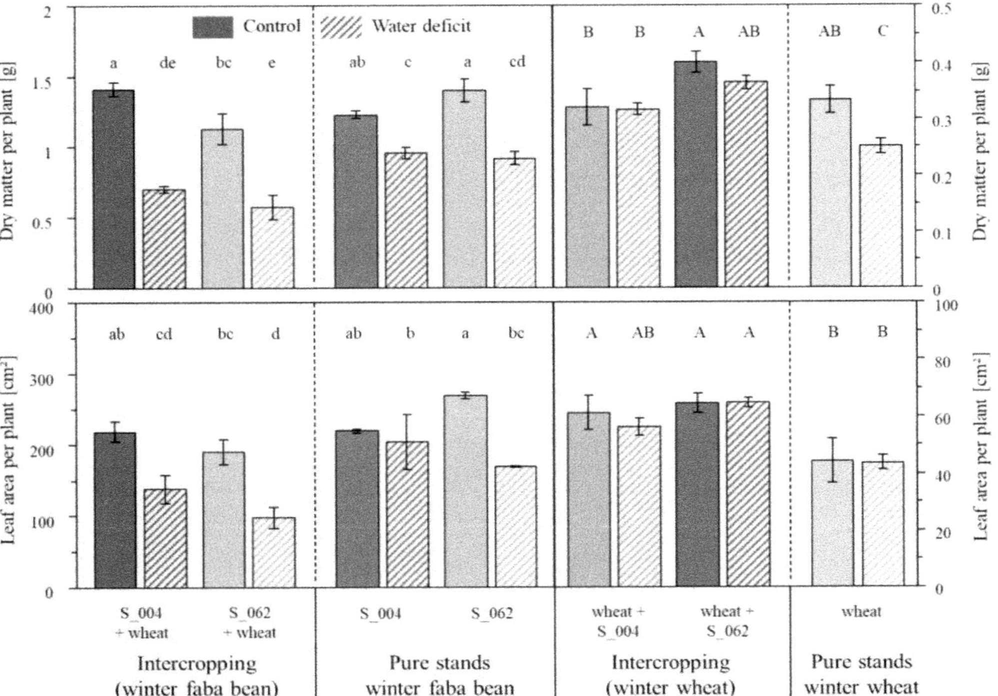

Figure S.3. Pre-experiment: Dry matter and leaf area of winter faba bean genotypes S_004 and S_062 as well as winter wheat in intercropped and pure stands at the end of the experiment. Error bars display the standard error. Different letters indicate significant differences as calculated by the Duncan-test ($p < 0.05$).

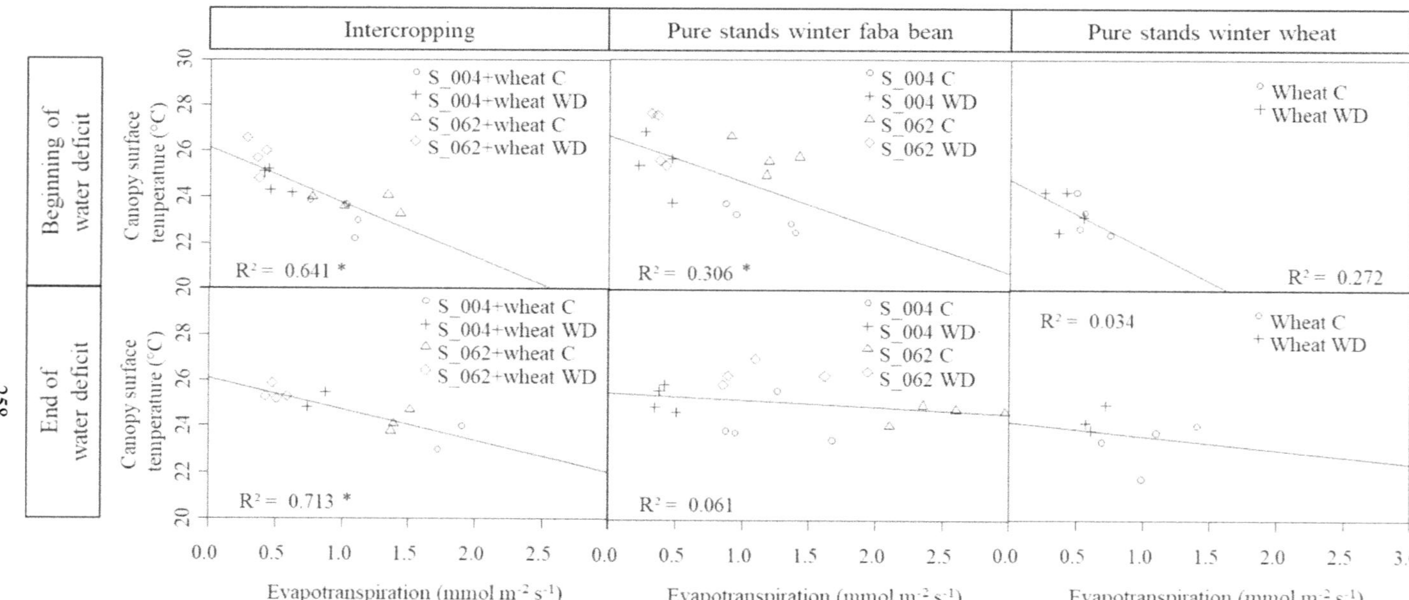

Figure S.4. Correlation between evapotranspiration and canopy surface temperature at beginning and end of water deficit. Intercropping considers both species together. C: Control. WD: Water deficit. Asterisks indicate the significance of the correlation at $p < 0.05$.

Supplementary material for chapter 2.2

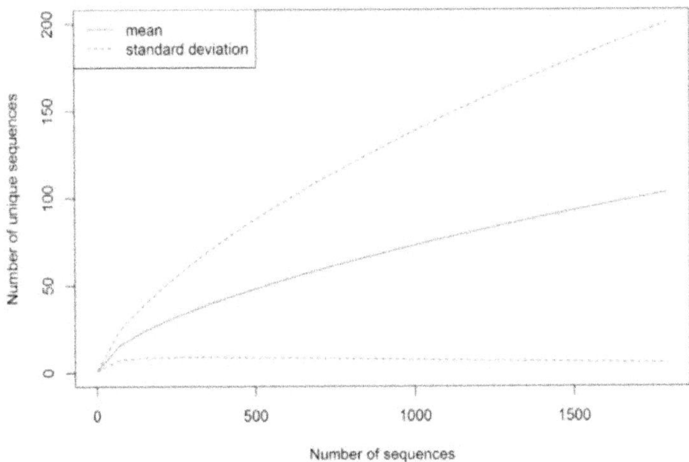

Figure S1. Rarefaction curve for bacterial leaf endophytes. Only the mean of all curves and the standard deviation are shown.

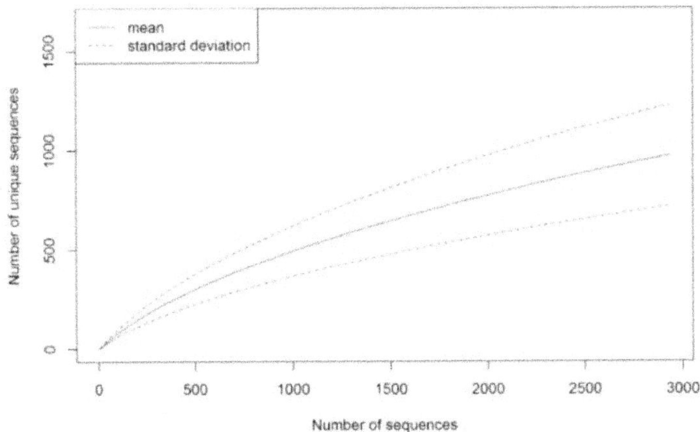

Figure S2. Rarefaction curve for bacteria in the rhizosphere. Only the mean of all curves and the standard deviation are shown.

Supplementary material for chapter 2.3

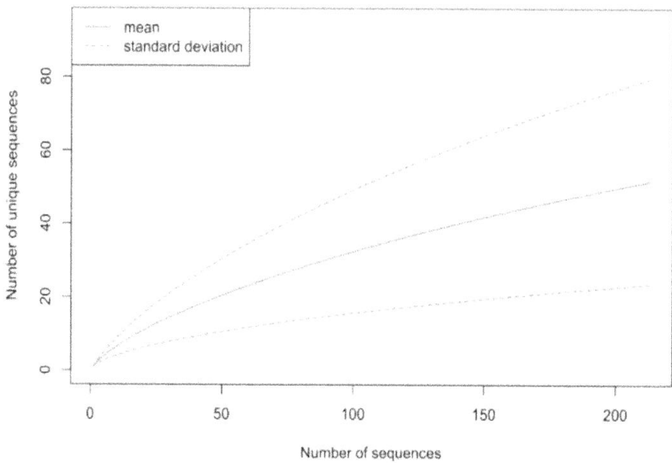

Figure S1. Rarefaction curve of the fungal community in the rhizosphere. Only the mean of all curves and the standard deviation are shown.

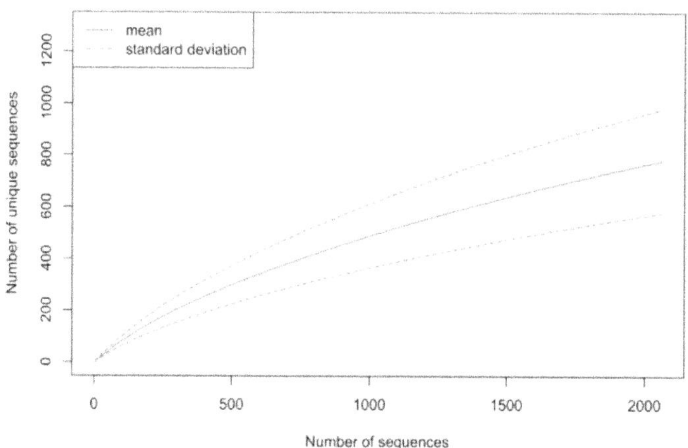

Figure S2. Rarefaction curve of the bacterial community in the rhizosphere. Only the mean of all curves and the standard deviation are shown.

Supplementary material for chapter 2.4

Table S1. Sampling dates for investigated edaphic properties and plant parameters.

Investigated Parameter/Harvest	H1	H2	Weekly
Water content	unplanted, rhizosphere soil	roots, aerial parts	-
RWC	-	leaves	-
N_{total}	unplanted, rhizosphere soil	unplanted, rhizosphere soil, roots, leaves	-
C_{total}	unplanted, rhizosphere soil	unplanted, rhizosphere soil, roots, leaves	-
pH	unplanted, rhizosphere soil	-	-
Plant height	-	-	yes
Thermal images	-	-	yes
Biomass	roots, aerial parts	roots, aerial parts	-

Abbreviations: RWC, relative water content; H1/H2, refers to harvest/sampling time.

Table S9. Effect of cropping regimes and compartment on bacterial and fungal community composition.

	Bacteria				**Fungi**			
	H1		**H2**		**H1**		**H2**	
	R^2	p	R^2	p	R^2	p	R^2	p
Unplanted soil								
C vs CF	0.061	0.834	NA	NA	0.108	0.358	NA	NA
C vs CS	0.285	0.093	NA	NA	0.219	0.293	NA	NA
CF vs CS	0.416	0.087	NA	NA	0.259	0.243	NA	NA
Rhizosphere soil								
MB vs MBF	0.243	0.263	0.270	0.587	0.098	0.695	0.237	0.125
MB vs XB	0.141	0.313	0.156	0.587	0.055	0.827	0.342	0.056
MB vs XBF	0.218	0.309	0.130	0.587	0.031	0.961	0.136	0.324
MBF vs XB	0.281	0.168	0.116	0.587	0.151	0.687	**0.296**	**0.036**
MBF vs XBF	0.031	0.741	0.104	0.587	0.085	0.764	0.148	0.250
XB vs XBF	0.214	0.498	0.031	0.868	0.061	0.764	0.228	0.125
MW vs MWF	0.305	0.168	0.099	0.587	0.191	0.687	**0.367**	**0.036**

Table continued on the following page

MW vs XW	0.324	0.168	0.033	0.800	0.175	0.687	0.169	0.176
MW vs XWF	0.232	0.263	0.096	0.587	0.063	0.764	0.093	0.554
MWF vs XW	0.235	0.309	0.047	0.587	0.044	0.827	**0.374**	**0.036**
MWF vs XWF	0.355	0.077	0.172	0.587	0.283	0.289	0.293	0.109
XW vs XWF	0.698	0.075	0.115	0.587	0.301	0.289	0.140	0.324
Roots								
MB vs MBF	0.080	0.889	**0.382**	**0.046**	0.069	0.836	0.410	0.054
MB vs XB	0.111	0.563	0.189	0.189	0.220	0.399	0.315	0.105
MB vs XBF	0.096	0.731	0.340	0.172	0.073	0.771	0.219	0.140
MBF vs XB	0.250	0.193	0.145	0.293	0.117	0.548	0.192	0.379
MBF vs XBF	0.167	0.319	0.235	0.187	0.050	0.857	0.340	0.057
XB vs XBF	0.046	0.731	0.123	0.345	0.214	0.441	0.284	0.067
MW vs MWF	0.178	0.497	0.156	0.189	0.017	0.857	0.157	0.267
MW vs XW	0.089	0.569	0.101	0.502	0.043	0.771	0.029	0.883
MW vs XWF	0.073	0.569	0.062	0.617	0.048	0.771	**0.507**	**0.024**
MWF vs XW	0.088	0.497	0.145	0.236	0.037	0.835	0.197	0.171
MWF vs XWF	0.162	0.497	0.245	0.086	0.056	0.771	**0.803**	**0.024**
XW vs XWF	0.058	0.753	0.201	0.131	0.025	0.916	**0.474**	**0.024**
Leaves								
MB vs MBF	0.846	0.175	NA	NA	0.216	0.289	0.076	0.646
MB vs XB	0.133	0.952	NA	NA	0.413	0.289	0.193	0.646
MB vs XBF	0.940	0.175	NA	NA	0.237	0.283	0.331	0.509
MBF vs XB	0.861	0.115	0.453	0.197	0.393	0.362	0.049	0.879
MBF vs XBF	0.159	0.977	0.253	0.197	0.122	0.581	0.156	0.554
XB vs XBF	0.945	0.133	0.880	0.350	0.208	0.597	0.277	1.000
MW vs MWF	0.510	0.133	0.947	0.053	0.135	0.454	0.323	0.068
MW vs XW	0.262	0.254	0.181	0.197	0.194	0.289	0.152	0.255
MW vs XWF	0.464	0.133	0.296	0.109	0.031	0.733	0.633	0.068
MWF vs XW	0.024	0.977	0.928	0.053	0.045	0.683	0.137	0.329
MWF vs XWF	0.026	0.977	0.875	0.053	0.111	0.487	0.264	0.281
XW vs XWF	0.003	0.994	0.141	0.327	0.160	0.364	0.364	0.068

Results of pairwise adonis testing for the different cropping regimes. Statistically significant differences ($p \leq 0.05$) between the treatments for each plant compartment are written in bold.

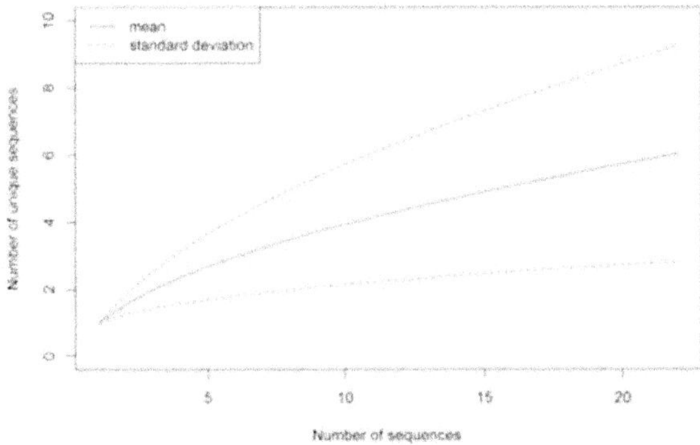

Figure S1. Rarefaction curve for bacterial leaf endophytes. Only the mean of all curves and the standard deviation are shown.

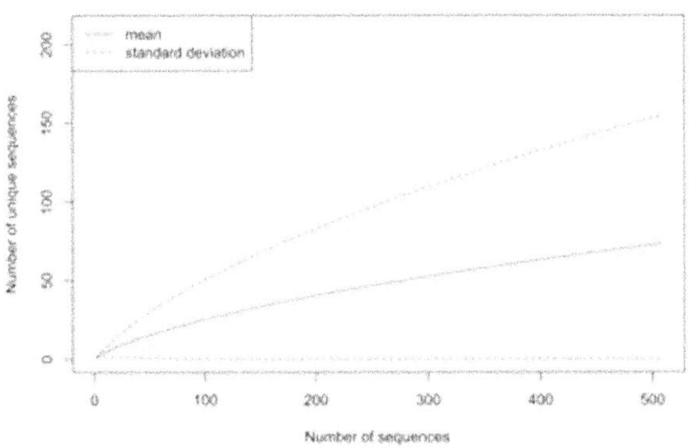

Figure S2. Rarefaction curve for bacterial root endophytes. Only the mean of all curves and the standard deviation are shown.

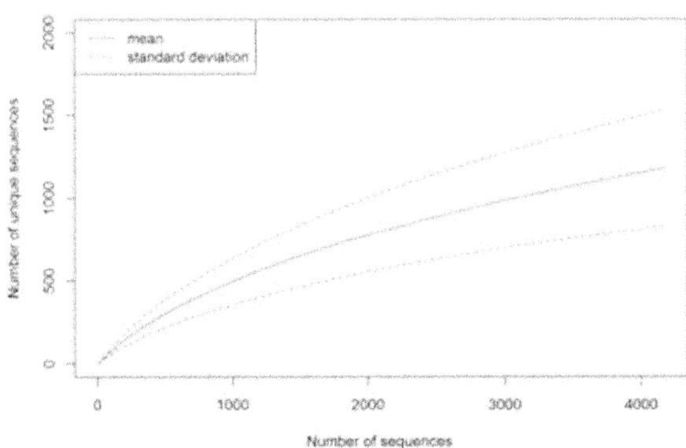

Figure S3. Rarefaction curve for bacteria in the rhizosphere soil. Only the mean of all curves and the standard deviation are shown.

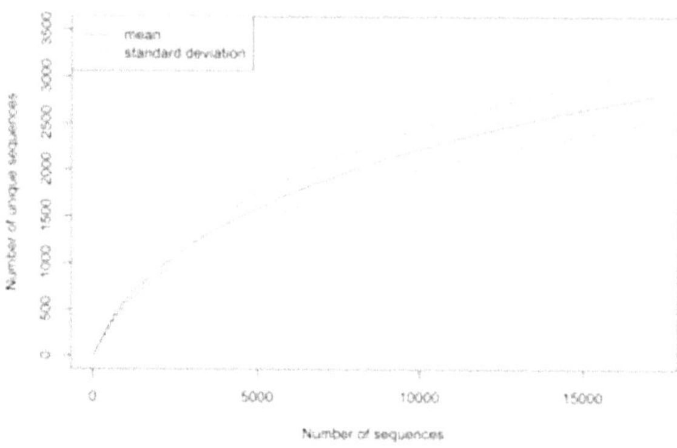

Figure S4. Rarefaction curve for bacteria in soil. Only the mean of all curves and the standard deviation are shown.

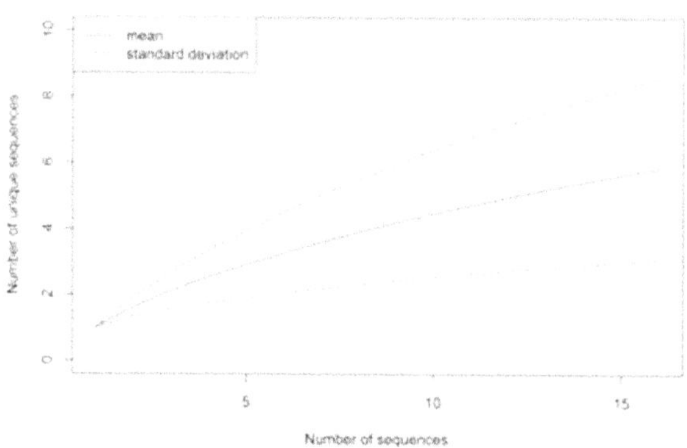

Figure S5. Rarefaction curve for fungal leaf endophytes. Only the mean of all curves and the standard deviation are shown.

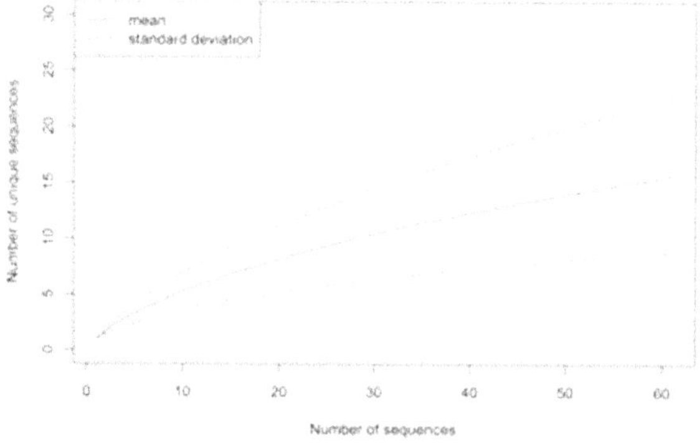

Figure S6. Rarefaction curve for fungal root endophytes. Only the mean of all curves and the standard deviation are shown.

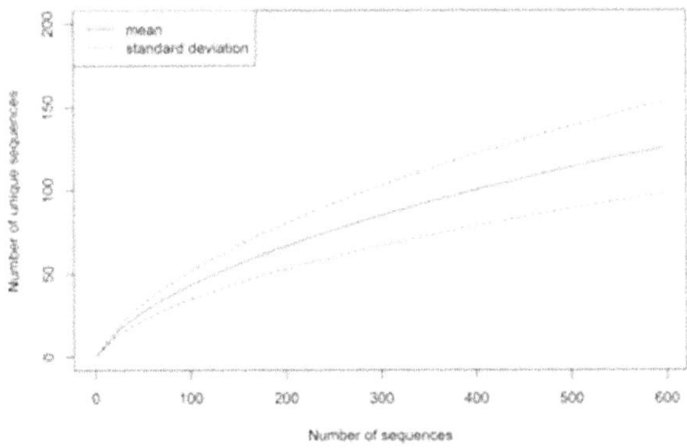

Figure S7. Rarefaction curve for fungi in the rhizosphere soil. Only the mean of all curves and the standard deviation are shown.

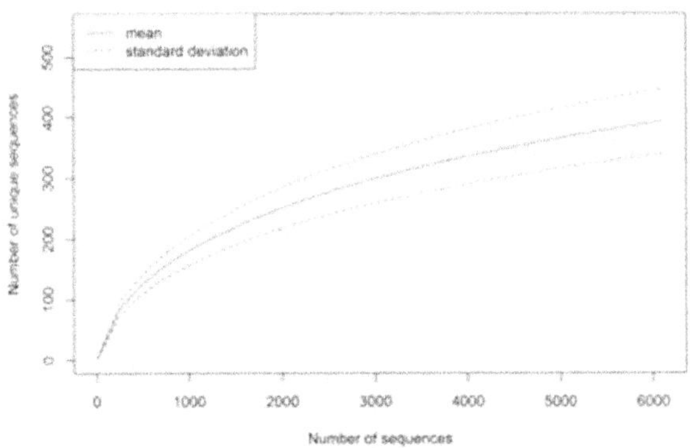

Figure S8. Rarefaction curve for fungi in soil. Only the mean of all curves and the standard deviation are shown.

Supplementary material for chapter 3

Table S.1: ANOVA table of carbon stable isotope discrimination ($\delta^{13}C$) of arable and grassland crops.

Domain	Species	Model	Factor	numDF	denDF	F-value	p-value	
Arable land	Winter faba bean	$\delta^{13}C$ ~	(Intercept)	1	367	25464.796	<.0001	***
		Site + Genotype + Cropping + Year	Site	1	367	107.911	<.0001	***
		+Site:Year	Genotype	7	367	12.500	<.0001	***
			Cropping	1	367	0.011	0.9163	
			Year	2	367	34.994	<.0001	***
			Site:Year	2	367	5.276	0.0055	**
	Winter wheat	$\delta^{13}C$ ~	(Intercept)	1	266	387196.700	<.0001	***
		Site + Stand + Year	Site	1	266	12.000	0.0006	***
		+ Site:Year	Stand	8	266	9.400	<.0001	***
			Year	2	266	24.000	<.0001	***
			Site:Year	2	266	9.000	0.0002	***
Grassland		$\delta^{13}C$ ~	(Intercept)	1	315	535398.000	<.0001	***
		Site + Stand + Year	Site	1	315	58.100	<.0001	***
		+ Site:Stand + Site:Year + Stand:Year	Stand	15	315	78.500	<.0001	***
			Year	2	315	374.600	<.0001	***
			Site:Stand	15	315	2.700	0.0007	***
			Site:Year	2	315	16.300	<.0001	***
			Stand:Year	30	315	6.200	<.0001	***

Signif. codes: 0.001 '***' 0.01 '**' 0.05 '*' 0.1 '.'

Table S.2: Net ecosystem exchange of CO_2 of the arable crops winter faba bean and winter wheat in pure and intercropped stands at the fertile and marginal site in the years 2015 and 2016. Additional wheat plots were fertilized with 240 kg N/ha. Means and standard errors were calculated from each measurement (n = 4).

	Fertile site					Marginal site			
	Net ecosystem exchange [µmol CO_2 m^{-2} s^{-1}]					Net ecosystem exchange [µmol CO_2 m^{-2} s^{-1}]			
	Faba bean	Intercropping	Wheat	Wheat + N		Faba bean	Intercropping	Wheat	Wheat + N
29.04.2015	7,37 ± 1,44	7,37 ± 2,96	6,12 ± 2,33						
07.05.2015	17,18 ± 2,43	8,69 ± 1,56	8,84 ± 2,61						
22.05.2015	12,64 ± 0,60	14,18 ± 4,07	16,17 ± 3,21						
30.05.2015	27,52 ± 5,22	13,71 ± 3,93	14,63 ± 2,00						
06.06.2015	17,61 ± 2,27	17,19 ± 1,98	24,55 ± 8,55						
15.06.2015	27,20 ± 2,87	32,39 ± 2,15	20,90 ± 5,81						
25.06.2015	18,87 ± 5,65	21,50 ± 1,33	14,23 ± 3,18						
03.07.2015	19,12 ± 3,26	23,74 ± 3,36	11,08 ± 2,33						
09.07.2015	30,76 ± 9,02	41,82 ± 5,08	12,56 ± 3,05						
18.07.2015	3,69 ± 2,31	5,84 ± 1,42	-3,28 ± 0,65						
08.04.2016	10,46 ± 1,29	9,98 ± 1,45	7,28 ± 2,89		**06.04.2016**	3,52 ± 1,11	6,45 ± 0,92	5,77 ± 1,58	
14.04.2016	13,33 ± 2,36	7,92 ± 0,66	5,46 ± 0,76		**18.04.2016**	8,07 ± 1,63	7,85 ± 0,81	5,32 ± 1,23	
20.04.2016	12,29 ± 1,38	13,72 ± 0,86	9,23 ± 1,58	3,48 ± 0,71	**22.04.2016**	8,72 ± 1,28	8,20 ± 1,32	6,05 ± 0,31	5,39 ± 0,20
04.05.2016	32,85 ± 6,93	35,57 ± 4,33	15,76 ± 2,28	12,77 ± 1,24	**02.05.2016**	8,73 ± 0,85	8,93 ± 0,55	4,65 ± 0,48	5,80 ± 0,58
09.05.2016	6,74 ± 3,16	6,09 ± 2,88	4,90 ± 1,92	18,99 ± 0,00	**10.05.2016**	31,26 ± 8,26	25,47 ± 3,85	15,69 ± 2,81	13,96 ± 7,05
					28.05.2016	21,59 ± 3,65	33,52 ± 2,90	32,00 ± 2,16	52,04 ± 6,78
08.06.2016	42,21 ± 11,16	38,45 ± 7,89	16,47 ± 3,10	26,79 ± 3,79	**07.06.2016**	19,51 ± 4,86	27,42 ± 1,67	12,01 ± 2,22	24,89 ± 0,63
24.06.2016	2,82 ± 9,79	30,76 ± 11,80	1,13 ± 4,00	3,78 ± 9,41	**20.06.2016**	12,44 ± 2,20	19,87 ± 6,04	9,45 ± 2,46	12,69 ± 0,00

Table S.3: Evapotranspiration of the arable crops winter faba bean and winter wheat in pure and intercropped stands at the fertile and marginal site in the years 2015 and 2016. Additional wheat plots were fertilized with 240 kg N/ha. Means and standard errors were calculated from each measurement (n = 4).

	Fertile site					Marginal site			
	Evapotranspiration [mmol H_2O m^{-2} s^{-1}]					Evapotranspiration [mmol H_2O m^{-2} s^{-1}]			
	Faba bean	Intercropping	Wheat	Wheat + N		Faba bean	Intercropping	Wheat	Wheat + N
29.04.2015	1.25 ± 0.19	1.49 ± 0.07	1.18 ± 0.17						
07.05.2015	0.89 ± 0.14	0.72 ± 0.12	0.57 ± 0.11						
22.05.2015	2.33 ± 0.10	2.68 ± 0.29	3.00 ± 0.40						
30.05.2015	2.44 ± 0.69	1.79 ± 0.56	1.82 ± 0.06						
06.06.2015	1.75 ± 0.31	1.83 ± 0.14	2.40 ± 0.83						
15.06.2015	2.39 ± 0.33	2.97 ± 0.36	2.38 ± 0.24						
25.06.2015	2.79 ± 0.40	3.08 ± 0.51	2.89 ± 0.37						
03.07.2015	3.37 ± 0.36	3.12 ± 0.87	2.59 ± 0.78						
09.07.2015	3.02 ± 0.92	3.57 ± 0.67	3.11 ± 0.43						
18.07.2015	2.97 ± 0.79	2.37 ± 0.21	1.34 ± 0.29						
08.04.2016	1.77 ± 0.29	1.96 ± 0.18	1.42 ± 0.23		**06.04.2016**	1.21 ± 0.33	1.31 ± 0.29	1.47 ± 0.53	
14.04.2016	1.33 ± 0.16	1.37 ± 0.28	0.98 ± 0.24		**18.04.2016**	1.38 ± 0.06	1.51 ± 0.12	1.41 ± 0.22	
20.04.2016	1.77 ± 0.32	1.93 ± 0.08	1.51 ± 0.15	1.65 ± 0.10	**22.04.2016**	1.49 ± 0.11	1.20 ± 0.09	1.21 ± 0.05	1.22 ± 0.03
04.05.2016	3.64 ± 0.27	3.58 ± 0.74	2.76 ± 0.65	2.58 ± 0.59	**02.05.2016**	2.19 ± 0.12	2.24 ± 0.09	1.87 ± 0.17	1.79 ± 0.28
09.05.2016	2.23 ± 0.22	2.07 ± 0.49	2.06 ± 0.49	2.99 ± 0.00	**10.05.2016**	4.64 ± 1.08	3.60 ± 0.60	3.05 ± 0.62	2.76 ± 1.02
					28.05.2016	3.64 ± 0.72	3.35 ± 0.97	4.81 ± 0.64	5.43 ± 0.77
08.06.2016	6.30 ± 0.90	6.47 ± 1.56	2.71 ± 0.48	3.97 ± 0.39	**07.06.2016**	3.31 ± 1.01	4.49 ± 0.29	3.46 ± 0.36	5.26 ± 0.83
24.06.2016	5.90 ± 1.49	5.69 ± 0.75	3.04 ± 0.52	4.07 ± 0.82	**20.06.2016**	2.54 ± 0.33	2.61 ± 0.70	2.47 ± 0.52	2.99 ± 0.00

Table S.4: Net ecosystem exchange of CO_2 of the grassland crops white clover and perennial ryegrass in pure and intercropped stands at the fertile and marginal site in the years 2015 and 2016. Additional ryegrass plots were fertilized with 240 kg N/ha. Means and standard errors were calculated from each measurement (n = 4).

	Fertile site					Marginal site			
	Net ecosystem exchange [µmol CO_2 m^{-2} s^{-1}]					**Net ecosystem exchange [µmol CO_2 m^{-2} s^{-1}]**			
	White clover	Intercropping	Ryegrass	Ryegrass + N		White clover	Intercropping	Ryegrass	Ryegrass + N
29.04.2015	15,29 ± 2,74	11,40 ± 1,09	10,09 ± 2,71						
07.05.2015	14,78 ± 2,44	13,92 ± 1,61	10,54 ± 1,88						
22.05.2015	15,59 ± 1,07	7,55 ± 3,99	4,75 ± 1,28						
30.05.2015	20,34 ± 6,07	9,66 ± 3,66	6,87 ± 1,44						
06.06.2015	12,00 ± 3,73	9,76 ± 2,43	8,41 ± 1,00						
15.06.2015	15,50 ± 4,05	10,50 ± 2,21	6,91 ± 3,03						
25.06.2015	10,88 ± 1,63	9,21 ± 1,80	5,73 ± 2,35						
03.07.2015	-9,46 ± 1,52	-6,63 ± 1,39	-2,07 ± 0,70						
09.07.2015	21,62 ± 6,85	5,16 ± 1,61	4,17 ± 1,17						
18.07.2015	16,48 ± 2,99	10,02 ± 2,51	4,09 ± 0,31						
08.04.2016	9,47 ± 2,03	5,24 ± 0,95	5,21 ± 1,34		**06.04.2016**	2,89 ± 0,57	6,09 ± 2,94	3,28 ± 0,51	
14.04.2016	11,07 ± 0,62	8,36 ± 1,00	6,45 ± 0,87		**18.04.2016**	10,78 ± 2,19	10,49 ± 1,66	3,72 ± 0,51	
20.04.2016	12,43 ± 2,66	11,00 ± 2,17	6,08 ± 0,63		**22.04.2016**	8,45 ± 1,35	3,46 ± 1,22	6,07 ± 2,48	
04.05.2016	12,22 ± 1,46	15,95 ± 1,74	8,87 ± 1,30	16,51 ± 2,76	**02.05.2016**	11,46 ± 0,55	10,76 ± 0,88	4,01 ± 0,96	11,92 ± 0,80
09.05.2016	4,59 ± 1,42	2,35 ± 0,55	1,10 ± 0,39	2,68 ± 0,89	**10.05.2016**	11,32 ± 1,32	9,14 ± 1,37	2,53 ± 0,56	8,69 ± 0,67
					28.05.2016	-2,58 ± 1,04	-4,65 ± 1,79	-3,61 ± 0,58	-1,01 ± 1,08
08.06.2016	13,81 ± 0,80	8,42 ± 1,94	-0,86 ± 2,23	7,16 ± 1,97	**07.06.2016**	8,19 ± 1,83	5,83 ± 0,76	1,21 ± 1,89	9,86 ± 1,77
24.06.2016	-4,44 ± 2,43	-10,64 ± 2,03	-4,82 ± 1,58	-8,62 ± 1,02	**20.06.2016**	13,19 ± 2,07	12,03 ± 2,52	3,88 ± 1,46	11,92 ± 3,23
07.07.2016	3,79 ± 5,99	6,48 ± 4,25	-2,24 ± 2,68	2,17 ± 2,70	**06.07.2016**	0,62 ± 0,67	1,07 ± 0,37	0,28 ± 0,26	0,85 ± 0,27
20.07.2016	11,21 ± 1,18	8,18 ± 3,23	-3,81 ± 3,50	-8,22 ± 5,95	**20.07.2016**	-4,36 ± 0,44	-14,26 ± 5,43	-8,87 ± 6,83	-6,93 ± 5,91

Table S.5: Evapotranspiration of the grassland crops white clover and perennial ryegrass in pure and intercropped stands at the fertile and marginal site in the years 2015 and 2016. Additional ryegrass plots were fertilized with 240 kg N/ha. Means and standard errors were calculated from each measurement (n = 4).

	Fertile site					Marginal site			
	Evapotranspiration [mmol H_2O m^{-2} s^{-1}]					**Evapotranspiration [mmol H_2O m^{-2} s^{-1}]**			
	White clover	Intercropping	Ryegrass	Ryegrass + N		White clover	Intercropping	Ryegrass	Ryegrass + N
29.04.2015	1,81 ± 0,47	1,92 ± 0,45	1,62 ± 0,51						
07.05.2015	0,96 ± 0,22	0,87 ± 0,20	1,06 ± 0,17						
22.05.2015	3,96 ± 0,42	3,44 ± 0,76	2,81 ± 0,37						
30.05.2015	3,40 ± 0,90	1,56 ± 0,81	2,67 ± 0,56						
06.06.2015	1,71 ± 0,09	1,89 ± 0,13	1,64 ± 0,22						
15.06.2015	2,50 ± 0,67	1,70 ± 0,29	2,15 ± 0,61						
25.06.2015	3,87 ± 0,34	3,72 ± 0,21	2,72 ± 0,14						
03.07.2015	1,43 ± 0,62	1,63 ± 0,70	1,41 ± 0,62						
09.07.2015	2,98 ± 0,49	1,85 ± 0,56	2,45 ± 0,73						
18.07.2015	2,55 ± 0,48	2,26 ± 0,38	2,23 ± 0,79						
08.04.2016	2,31 ± 0,64	1,39 ± 0,25	1,74 ± 0,17		**06.04.2016**	1,03 ± 0,21	1,03 ± 0,34	0,77 ± 0,18	
14.04.2016	1,38 ± 0,15	1,94 ± 0,24	1,62 ± 0,09		**18.04.2016**	1,99 ± 0,37	1,81 ± 0,11	1,26 ± 0,13	
20.04.2016	1,73 ± 0,17	2,40 ± 0,34	1,47 ± 0,17		**22.04.2016**	1,51 ± 0,11	0,79 ± 0,45	1,37 ± 0,21	
04.05.2016	2,75 ± 0,32	2,48 ± 0,22	1,80 ± 0,19	2,58 ± 0,61	**02.05.2016**	2,88 ± 0,06	2,78 ± 0,06	1,82 ± 0,25	2,74 ± 0,04
09.05.2016	1,08 ± 0,18	0,69 ± 0,04	0,42 ± 0,07	0,69 ± 0,11	**10.05.2016**	2,07 ± 0,46	1,81 ± 0,49	1,22 ± 0,34	1,70 ± 0,16
					28.05.2016	1,78 ± 0,01	1,71 ± 0,31	0,92 ± 0,27	1,29 ± 0,35
08.06.2016	3,67 ± 0,38	3,73 ± 0,52	3,39 ± 0,30	3,96 ± 0,85	**07.06.2016**	3,22 ± 0,30	2,89 ± 0,22	2,74 ± 0,18	3,29 ± 0,15
24.06.2016	2,60 ± 0,66	1,36 ± 0,16	2,51 ± 0,47	1,75 ± 0,14	**20.06.2016**	2,88 ± 0,17	3,00 ± 0,30	2,40 ± 0,72	2,74 ± 0,55
07.07.2016	2,98 ± 0,51	2,70 ± 0,25	1,84 ± 0,37	1,70 ± 0,44	**06.07.2016**	0,21 ± 0,05	0,68 ± 0,09	0,58 ± 0,13	0,57 ± 0,13
20.07.2016	2,30 ± 0,38	2,14 ± 0,51	1,61 ± 0,26	2,73 ± 0,18	**20.07.2016**	3,73 ± 0,19	2,80 ± 0,30	2,30 ± 0,38	2,22 ± 0,45

Table S.6: ANOVA table of the water use efficiency (WUE) of arable and grassland crops.

Domain	Site & Year	Model	Factor	numDF	denDF	F-value	p-value	
Arable land	Fertile site 2015	WUE ~ Stand * Date	(Intercept)	1	80	170.135	<.0001	***
			Stand	2	80	7.272	0.0013	**
			Date	9	80	20.354	<.0001	***
			Stand:Date	18	80	1.557	0.0924	.
	Fertile site 2016	WUE ~ Stand + Date	(Intercept)	1	54	136.549	<.0001	***
			Stand	3	54	2.878	0.0443	*
			Date	4	54	9.365	<.0001	***
	Marginal site 2016	WUE ~ Stand + Date	(Intercept)	1	66	179.183	<.0001	***
			Stand	3	66	17.066	<.0001	***
			Date	5	66	28.006	<.0001	***
Grassland	Fertile site 2015	WUE ~ Stand + Date	(Intercept)	1	97	40.897	<.0001	***
			Stand	2	97	19.884	<.0001	***
			Date	9	97	12.788	<.0001	***
	Fertile site 2016	WUE ~ Stand * Date	(Intercept)	1	60	13.967	0.0004	***
			Stand	3	60	5.835	0.0014	**
			Date	5	60	41.116	<.0001	***
			Stand:Date	15	60	2.828	0.0022	**
	Marginal site 2016	WUE ~ Stand + Date	(Intercept)	1	90	23.304	<.0001	***
			Stand	3	90	21.540	<.0001	***
			Date	6	90	22.703	<.0001	***

Signif. codes: 0.001 '***' 0.01 '**' 0.05 '*' 0.1 '.'

Table S.7: ANOVA table of the temperature difference between canopy and air (dT) of arable and grassland crops.

Domain	Site & Year	Model	Factor	numDF	denDF	F-value	p-value	
Arable land	Fertile site 2015	dT ~ Stand + Date	(Intercept)	1	79	44.745	<.0001	***
			Stand	2	79	4.19654	0.0185	*
			Date	7	79	17.62125	<.0001	***
	Fertile site 2016	dT ~ Stand + Date	(Intercept)	1	69	342.9017	<.0001	***
			Stand	3	69	7.3566	0.0002	***
			Date	5	69	43.4944	<.0001	***
	Marginal site 2016	dT ~ Stand + Date	(Intercept)	1	74	253.27253	<.0001	***
			Stand	3	74	21.54449	<.0001	***
			Date	6	74	11.62839	<.0001	***
Grassland	Fertile site 2015	dT ~ Stand + Date	(Intercept)	1	72	211.34446	<.0001	***
			Stand	2	72	4.42182	0.0154	*
			Date	7	72	13.24991	<.0001	***
	Fertile site 2016	dT ~ Stand + Date	(Intercept)	1	125	1363.0978	<.0001	***
			Stand	3	125	17.5125	<.0001	***
			Date	8	125	11.0543	<.0001	***
	Marginal site 2016	dT ~ Stand + Date	(Intercept)	1	147	570.4387	<.0001	***
			Stand	3	147	24.6234	<.0001	***
			Date	10	147	16.7866	<.0001	***

Signif. codes: 0.001 '***' 0.01 '**' 0.05 '*' 0.1 '.'

Table S.8: ANOVA table of grain and dry matter yield of arable and grassland crops.

Domain	Year	Model	Factor	numDF	denDF	F-value	p-value	
Arable land	2015	Grain yield ~ Site + Stand + Species + Site:Species + Stand:Species	(Intercept)	1	114	43.35407	<.0001	***
			Site	1	114	104.54092	<.0001	***
			Stand	1	114	9.78184	0.0022	**
			Species	8	114	5.56222	<.0001	***
			Site:Species	8	114	4.60014	0.0001	***
			Stand:Species	8	114	11.95356	<.0001	***
	2016	Grain yield ~ Site + Stand + Species + Site:Species + Stand:Species	(Intercept)	1	114	7138.428	<.0001	***
			Site	1	114	0.007	0.9333	
			Stand	1	114	82.079	<.0001	***
			Species	8	114	14.396	<.0001	***
			Site:Species	8	114	7.021	<.0001	***
			Stand:Species	8	114	5.196	<.0001	***
Grassland	2015	Dry matter yield ~ Site * Treatment	(Intercept)	1	109	546.5414	<.0001	***
			Site	1	109	0.9226	0.3389	
			Treatment	15	109	24.3634	<.0001	***
			Site:Treatment	15	109	2.2339	0.0091	**
	2016	Dry matter yield ~ Site * Treatment	(Intercept)	1	109	2534.4434	<.0001	***
			Site	1	109	44.8375	<.0001	***
			Treatment	15	109	22.5693	<.0001	***
			Site:Treatment	15	109	1.712	0.0587	.

Signif. codes: 0.001 '***' 0.01 '**' 0.05 '*' 0.1 '.'

CV

Annika Kathrin Meißner, geb. Lingner

Geboren am 06. Dezember 1988 in Göttingen

Nationalität: Deutsch

Beruflicher Werdegang

Seit 05/2019	Produktionsmanagerin Raps KWS SAAT SE & Co. KGaA, Einbeck Thema: Saatgutproduktion und Aufbereitungssteuerung von Linien- und Hybridrapssorten
09/2014 bis 04/2019	Wissenschaftliche Mitarbeiterin Georg-August-Universität Göttingen Institute of Applied Plant Nutrition (IAPN) und Abteilung Pflanzenernährung und Ertragsphysiologie Thema: Wassernutzungseffizienz verschiedener Genotypen von Acker- und Grünland-Pflanzen in Leguminosen-basierten Gemengen
11/2013 bis 08/2014	Wissenschaftliche Mitarbeiterin Georg-August-Universität Göttingen Abteilung Pflanzenernährung und Ertragsphysiologie Thema: Molekulare Mechanismen der Regulation der N_2-Fixierung von *Medicago truncatula*

Ausbildung

04/2014 bis 09/2018	Promotionsstudiengang für Agrarwissenschaften (PAG) Georg-August-Universität Göttingen Promotion am Institute of Applied Plant Nutrition (IAPN)
10/2011 bis 03/2014	Studium der Agrarwissenschaften, Masterabschluss (M.Sc.) Georg-August-Universität Göttingen Studienrichtung: Nutzpflanzenwissenschaften Masterarbeit: Ertrag und Stickstofffixierung von Ackerbohnen in Abhängigkeit von einer Schwefel- und Bordüngung

10/2008 bis 09/2011	Studium der Agrarbiologie, Bachelorabschluss (B.Sc.) Universität Hohenheim Profil: Pflanzenschutz Bachelorarbeit: Wirkung eines Acetolactat-Synthase-Hemmers auf verzweigte Aminosäuren in Spross und Wurzel der Ackerwinde

Praktika

02/2013 bis 03/2013	Wissenschaftliche Hilfskraft an der Georg-August-Universität Göttingen, Abteilung Pflanzenernährung und Ertragsphysiologie
08/2012 bis 05/2013	Bildagentur Landpixel.de in Rosdorf Thema: Redaktionelle Tätigkeiten im Bereich Fotografie; gefolgt von freier Mitarbeit bei der Bildagentur Landpixel.de
08/2010 bis 09/2010	Landwirtschaftlicher Betrieb Volker Sohnrey in Esebeck Thema: Haltung von Milchvieh und Ackerbau
07/2010 bis 08/2010	Landwirtschaftlicher Betrieb Rick und Werner Magerhans GbR in Göttingen Thema: Sauenhaltung, Ferkelproduktion und Ackerbau
08/2009 bis 09/2009	Institut für Zuckerrübenforschung in Göttingen Thema: Entwicklung von Konzepten zur Zuckerrübenproduktion

Lizenzen und Weiterbildungen

Drohnenschulung für Systeme bis 5 kg	erworben 05/2015
Ausbildung zum Ersthelfer im Betrieb	erworben 03/2014, 02/2016, 06/2018
Zertifikat Digitalfotografie und Bildbearbeitung	erworben 09/2012
Sachkundenachweis Pflanzenschutz	erworben 10/2011

Declarations

1. I, hereby, declare that this Ph.D. dissertation has not been presented to any other examining body either in its present or a similar form.

   Furthermore, I also affirm that I have not applied for a Ph.D. at any other higher school of education.

Göttingen, ...

Annika Meißner

2. I, hereby, solemnly declare that this dissertation was undertaken independently and without any unauthorised aid.

Göttingen, ...

Annika Meißner

www.ingramcontent.com/pod-product-compliance
Ingram Content Group UK Ltd.
Pitfield, Milton Keynes, MK11 3LW, UK
UKHW022001190726
13853UKWH00004B/1670